Bullying and Emotional Abuse in the Workplace

Over the last decade or so research into bullying, emotional abuse and harassment at work, as distinct from harassment based on sex or race and primarily of a non-physical nature, has emerged as a new field of study. Two main academic streams have emerged: a European tradition applying the concept of 'mobbing' or 'bullying' and the American tradition, primarily identified through concepts such as emotional abuse and mistreatment. One focuses on the perpetrator, the other on the victim. In addition, research in this field has also started in Australia and South Africa. All are brought together in this work, in a synthesis of research and analysis of practice in the field. This book also aims to document the existence and consequences of the problem of bullying, to explore its causes and to investigate the effectiveness of approaches aimed at mitigating and managing the problem, as well as to offer suggestions for further progress in this important new field.

Ståle Einarsen is Associate Professor in Work and Organizational Psychology in the Department of Psychosocial Science at the University of Bergen in Norway.

Helge Hoel is a lecturer in change management and cross-cultural management in Manchester School of Management at the University of Manchester Institute of Science and Technology (UMIST), UK.

Dieter Zapf (PhD) is Professor of Organizational Psychology in the Department of Psychology at the Johann Wolfgang Goethe-University, Frankfurt am Main, Germany.

Cary L. Cooper (PhD, CBE) is currently BUPA Professor of Organizational Psychology and Health in the Manchester School of Management, and Pro-Vice-Chancellor (External Activities) of the University of Manchester Institute of Science and Technology (UMIST), UK.

Bullying and Emotional Abuse in the Workplace

International perspectives in research and practice

Edited by Ståle Einarsen, Helge Hoel, Dieter Zapf and Cary L. Cooper

LONDON AND NEW YORK

First published 2003 by Taylor & Francis
11 New Fetter Lane, London EC4P 4EE

Simultaneously published in the USA and Canada
by Taylor & Francis Inc
29 West 35th Street, New York, NY 10001

Taylor & Francis is an imprint of the Taylor & Francis Group

Typeset in Sabon by Taylor & Francis Books Ltd
Printed and bound in Great Britain by St Edmundsbury Press,
Bury St Edmunds, Suffolk

British Library Cataloguing in Publication Data
A catalogue record for this book is available from the British Library

Library of Congress Cataloging in Publication Data
Bullying and emotional abuse in the workplace: international perspectives in research and practice ... [et al]
p.cm
Includes bibliographical references
1. Bullying in the workplace. I. Einarson, Stale.
HF5549.5.E43 B85 2002
158.7–dc21 2002074011

ISBN 0–415–25359–4

Contents

Illustrations

Tables

Figures

Contributors

Professor Robert A. Baron, Rensselaer Polytechnic Institute, USA

Professor Cary L. Cooper, University of Manchester Institute of Science and Technology, UK

Hope Daley, National Health and Safety Officer, UNISON, UK

Ståle Einarsen, University of Bergen, Norway

Sirkku Fallenius, Medivire Occupational Health Services Ltd, Finland

Professor Louise F. Fitzgerald, University of Illinois at Urbana-Champaign, USA

Helge Hoel, University of Manchester Institute of Science and Technology, UK

Adrienne B. Hubert, Hubert Consulting, The Netherlands

Mike Ironside, Keele University, UK

Karen Jagatic, Wayne State University, USA

Peter J. Jordan, Griffith University, Australia

Loraleigh Keashly, Wayne State University, USA

Lena Korppoo, Varma-Sampo Mutual Pension Insurance Company, Finland

Duncan Lewis, University of Glamorgan, UK

Andreas P. D. Liefooghe, Birkbeck College, University of London, UK

Kate Mackenzie Davey, Birkbeck College, University of London, UK

Susan Marais-Steinman, Foundation for the Study of Work Trauma, South Africa

Maj-Lis Mattila, Medivire Occupational Health Services Ltd, Finland

Paul McCarthy, Griffith University, Australia

Vicki Merchant, Merchant and Co., UK

Eva Gemzøe Mikkelsen, University of Aarhus, Denmark

Joel H. Neuman, State University of New York at New Paltz, USA

Branda L. Nowell, Michigan State University

Professor Dan Olweus, University of Bergen, Norway

Professor John B. Pryor, Illinois State University, USA

Charlotte Rayner, Staffordshire University Business School, UK

Jon Richards, National Officer, UNISON, UK

Denise Salin, Swedish School of Economic and Business Administration, Finland

Professor Roger Seifert, University of Keele, UK

Michael J. Sheehan, Griffith University, Australia

Anne Spurgeon, University of Birmingham, UK

Noreen Tehrani, Employee Support, Training and Development, UK

Maarit Vartia, Institute of Occupational Health, Finland

Professor David Yamada, Suffolk University Law School, USA

Professor Dieter Zapf, Johann Wolfgang Goethe-University Frankfurt, Germany

Preface

Over the last decade research into bullying, emotional abuse and harassment at work, primarily of a non-physical nature, as distinct from sexual and racial harassment, has emerged as a new field of study in Europe, Australia, South Africa and the USA. European research into this problem started in Scandinavia in the 1980s and spread during the late 1990s to many other European countries as well as to Australia. Using the concept of 'mobbing' in the context of work Heinz Leymann was inspired by progress in the study of aggressive behaviour among schoolchildren, aimed at singling out individual children for negative treatment, to pioneer systematic studies of similar types of behaviour in the world of work. The focus of this approach, which has gained considerable momentum in recent years, is the process whereby hostile and aggressive behaviour is directed systematically at one or more colleagues or subordinates, leading to a stigmatisation and victimisation of the target. Central to any definition of 'mobbing', referred to as 'bullying' in most English-speaking countries, is the enduring and repeated nature of the negative behaviour to which the target is being exposed.

Despite several commendable attempts, it was only with the work of Loraleigh Keashly in the 1990s that a coherent conceptual framework of workplace bullying emerged in the USA. The American psychiatrist Carroll Brodsky published an extensive report on the issue of bullying at work in 1976, entitled the *The harassed worker*, but it appears that this book only had an impact much later. Rather, the US research in the area of hostile behaviours that may be relevant to workplace bullying is found in a variety of literatures, using many different concepts such as workplace aggression, emotional abuse, generalised workplace abuse, mistreatment and workplace harassment. Applying the term 'emotional abuse', Keashly highlights the prolonged suffering identified with the concept of bullying at work. This approach places the conflict process right at the centre of attention by focusing on interpersonal conflict in the workplace, involving behaviour which is unwanted and unwelcome, and which is perceived as abusive, inappropriate and offensive by the recipient of such behaviour, as well as the standard of behaviour which may be seen as unreasonable, and which appears to violate an individual's rights.

Up to the present, a host of studies have documented the existence and negative consequences of this problem, both from the point of view of the individual victim and the organisation. From the data available, it is reasonable to believe that the majority of employees, at some time during their career, will be exposed to systematic bullying or abusive behaviour, either directly, or indirectly as observers or witnesses of such negative conduct. It has been suggested that a significant proportion of stress-related illness may be attributed to bullying. Whatever the real scale of the problem may be, bullying appears to affect individuals psychologically, physically and behaviourally. For industry it represents a serious cost problem. Increased sickness absenteeism and turnover rate are two likely outcomes. Poor performance and a decrease in productivity are other factors that are likely to emerge as a result of reduced commitment, creativity and general loss of morale.

The growing attention given to these issues may in part be explained by recent economic and social change. In order to survive, industry is faced with continuous pressure to downsize and restructure to sustain competitiveness in an increasingly global economy. Fewer people are left with more work in a climate of uncertainty. This uncertainty is further exacerbated by an increasing number of temporary contracts and voluntary redundancy schemes. Such factors bring about a climate in which insecurity becomes endemic, creating a work environment in which the potential for interpersonal conflict and bullying becomes overwhelming. This said, the harassment and emotional abuse of employees, subordinates and co-workers is hardly a new phenomenon. Mistreatment of workers in the workplace has always existed. Such mistreatment has even been built into the very nature of the employer–employee relationship, or has at least been a part of the process of managing and supervising labour, as illustrated in some of the chapters of this book.

The main aim of this book is to present the reader with a comprehensive review of the literature, the empirical findings, the theoretical developments and the experiences of leading international academics and practitioners in the field of bullying at work. In this volume the reader will find chapters examining the very concept of bullying at work, as well as related concepts such as school bullying and sexual harassment. The reader will also find chapters which document the existence and consequences of the problem. The book also explores a variety of approaches and explanatory models by researchers and practitioners from different countries and research traditions. There is discussion of the effectiveness of different theoretical approaches aimed at mitigating and managing the problem. Also, a wide range of practitioners from a number of countries present and discuss their work in relation to the prevention and management of bullying at work and the treatments of its targets.

The book has been structured in five parts, each with three to six chapters, as follows: (1) The problem; (2) The evidence; (3) Explaining the problem; (4) Managing the problem: 'best practice' and; (5) Remedial actions: a critical outlook. In part 1 the concept of bullying at work itself is introduced and discussed. First, the editors review and discuss the European perspective on bullying at work. Chapter 1 starts by looking at the development and rise of the concept in Europe. The various key defining characteristics of bullying are then discussed, such as frequency, duration, power balance, status of bullies and victims, objective vis-à-vis subjective bullying, and interpersonal vis-à-vis organisational bullying. This chapter also discusses those conceptual models of bullying that have so far dominated European research on bullying.

Second, Keashly and Jagatic review research findings, as well as key conceptual and methodological challenges, from the North American literature on hostile behaviours at work which are highly relevant to the issues of workplace bullying. An important aim of this chapter is to facilitate awareness of and communication between the various literatures on hostile workplace interactions. We hope that this chapter will serve as a catalyst to thinking and action with respect to hostile workplace interactions and bullying at work. We believe that the field of bullying at work may have much to learn from the North American literature on these topics.

The editors also believe that empirical findings, methodological issues and theoretical ideas, as well as knowledge on preventive measures from the related fields of research on school bullying and sexual harassment at work, may be highly valuable for those working in the field of bullying at work. Not only are the concepts and phenomena of bullying among children and the sexual harassment of women at work related, but research in these areas is much more extensive and well developed compared with the field of bullying at work. Therefore, we asked three prominent researchers in these fields to review their area of research for the benefit of the readers of this book. This has resulted in a chapter by Olweus on school bullying, and a chapter by Pryor and Fitzgerald reviewing the North-American research on sexual harassment at work. We believe that these chapters, in combination with those described above, present a comprehensive discussion on conceptual, theoretical and methodological, as well as preventive issues of high relevance to both practitioners and academics in the field of bullying at work.

Part 2, entitled 'The evidence', contains three chapters that review and summarise empirical evidence on bullying at work. Chapter 5, by Zapf, Einarsen, Hoel and Vartia, summarises descriptive empirical findings, looking at issues such as the frequency and the duration of bullying, gender and status of bullies and victims, how bullying is distributed across different sectors, and the use of different kinds of bullying strategies. Chapter 6, by Einarsen and Mikkelsen, reviews the literature on the potential negative effects of bullying on the health and well-being of the individual victim, as well as suggesting theoretical explanations for the relationships that have been found to exist between exposure to bullying and symptoms of ill health. Einarsen and Mikkelsen are particularly interested in the role of post-traumatic stress disorders in victims of bullying at work. In chapter 7, Hoel, Einarsen and Cooper review the emerging evidence of the relationship between bullying at work and negative organisational outcomes, such as absenteeism, turnover and productivity. The potential financial costs accrued in a typical bullying case are illustrated by means of a case study. The aim of all three chapters is to provide a state-of-the-art review, in terms of the empirical findings, by also reviewing empirical findings which so far have not been available to an English-speaking audience. Much research in this field is, and has been, published in languages other than English. This section of the book makes much of this valuable information accessible to an English-speaking readership.

So far, much work in the field of bullying at work has been of an empirical nature. A major aim of this book has also been further to develop our theoretical understanding of the problem. This is also particularly the aim of part 3, entitled 'Explaining the problem'. The first chapter in this section discusses the role of individual antecedents of bullying at work. The issue of individual causes of bullying at work has been a 'hot issue' of debate in both the popular press and in the scientific community. While some argue that individual antecedents, such as the personality of bullies and victims, may indeed be involved as causes of bullying, others have disregarded totally the role of individual characteristics as a cause of bullying. In chapter 8, Zapf and Einarsen provide a balanced discussion on the role of individual factors among both victims and bullies. Chapter 9 presents a social interactionist approach to bullying behaviours, looking at social and situational antecedents of aggressive behaviours at work. Here Neuman and Baron address a key question: why do societal norms against aggression fail to apply, or apply only weakly, where workplace bullying is concerned? The authors draw on a substantial literature devoted to interpersonal aggression, and examine what they believe to be important social antecedents of bullying. In chapter 10, Hoel and Salin look at potential organisational antecedents of bullying, focusing on quality of work environment, leadership, organisational culture and impact of change.

The last two chapters in this section present postmodern perspectives on bullying at work. First, Liefooghe and Mackenzie Davey argue for the importance of listening to a range of voices when explaining bullying at work. These authors argue particularly for the need to listen to how employees in general use the term bullying at work. The chapter describes a study from a UK bank on why so many employees reported being bullied. While employees recognised what they called interpersonal bullying, they also used the term to describe destructive organisational practices. The authors argue that while research hitherto has mainly regarded the organisation as a backdrop facilitating interpersonal bullying, these employee accounts point to the organisation playing a much more active role, something akin to 'institutionalised bullying'. Also using a postmodern perspective, McCarthy discusses how the concept of 'bullying at work' may be seen as a new way of giving meaning to experiences of distress at work. 'Bullying' is seen as a new signifier that emerges from a wider discourse which attributes distress in contemporary workplaces to unacceptable behaviours, and which gives expression to a variety of anxieties, fears and resentments at work. In mapping the rise of this new concept, McCarthy shows how understandings of the concept of bullying are a product of shifting alignments and tensions amongst diverse individual, organisational, professional and institutional interests.

Another major aim of this book is to provide some examples of how leading practitioners in different countries work in the field of bullying at work. Therefore, part 4, entitled 'Managing the problem', contains six chapters written by leading practitioners in the field. These chapters summarise the authors' experiences of working with individuals, organisations and in different countries with the aim of preventing and ameliorating bullying at work. In chapter 13, Richards and Daley discuss how anti-bullying policies may be developed, implemented and monitored. Chapter 14 describes the process of investigating complaints of bullying. Here, Merchant and Hoel advise on how to run an effective and ethical investigation in cases of alleged bullying at work. Chapter 15 examines how to manage the problem, and looks at the ways in which counsellors can be used to help address some of the problems facing targets of bullying. Through the use of illustrative cases, Tehrani looks at various counselling interventions and methods that may be used when working with individual targets. Tehrani also discusses the importance of developing appropriate rehabilitation programmes for targets of bullying, as well as the main issues facing independent counsellors working to reduce the impact of bullying in the workplace. The role of the occupational health services in cases of bullying is then discussed by Vartia, Korppoo, Fallenius and Mattila. Building on experiences in Finland, these authors provide an overview and discussion of the role of occupational health services in supporting individual victims of bullying, as well as in supporting groups and organisations in the prevention, handling and resolution of bullying situations in the workplace. Based on a Dutch study among professionals involved in tackling undesirable behaviour and bullying at work, Hubert presents a comprehensive model on how to prevent and manage bullying in larger organisations. This chapter describes five different phases involved in the tackling of bullying at work, from prevention to aftercare, as well as outlining specific responsibilities of the various professional disciplines involved in each of the five different phases In chapter 18, the final chapter in this part, Marais-Steinman discusses the challenges associated with the prevention and management of workplace bullying in South Africa. Marais-Steinman has years of practice working in this field in a developing country where violence and trauma are endemic, and she summarises her experience of dealing with workplace bullying, as well as discussing factors that make such work difficult in the particular cultural and socio-economic context of South Africa.

The final part of the book, entitled 'Remedial actions: a critical outlook', contains six theoretical chapters. They look at remedies for the prevention and management of bullying and emotional abuse at work from a theoretical perspective. Some chapters in this part also examine some inherent characteristics of organisations that may be related to the existence of bullying, such as the nature of employer–employee relationships, and the implications for the prevention and management of bullying following on from such organisational perspectives.

In chapter 19, Spurgeon discusses how and to what degree a risk management perspective may be used in the management of bullying at work, while, in chapter 20, Keashly and Nowell discusses the relationships between the concepts 'bullying' and 'conflict', and the value of a conflict perspective in the study and amelioration of workplace bullying. In chapter 21, Sheehan and Jordan examine how principles derived from theoretical frameworks related to the concepts of the learning organisation and bounded emotionality may be used in the prevention and management of bullying at work. Then, Lewis and Rayner investigate the possibilities and pitfalls of human resource management in the management of bullying at work. The central thesis of this chapter is that the very nature and rise of human resource management in the UK may in itself be related to the existence of bullying and may itself hinder a constructive management of the problem. Next, Ironside and Seifert provide an overview of the nature of the employment relationship, and in particular of its changing nature in the contemporary public sector. The chapter provides some examples of how trade unions are involved in handling the issue as part of mainstream union activity and representation, and concludes that where union representation is strong in the workplace, the best chance exists of reducing both the extent and the consequences of bullying. The penultimate contribution in this part identifies and discusses some of the central themes concerning bullying and the law. In this chapter, Yamada also investigates some national examples of how the law is used in order to prevent and respond to this problem. The book then ends with a chapter on future challenges in the study of workplace bullying which was written by the editors in collaboration with Loraleigh Keashly.

As editors we are delighted and pleased by the enthusiasm we encountered when approaching authors invited to participate. While some were invited to participate in the book due to earlier achievements in this area, others were asked to develop their thinking in this field by building on previous work in related fields or by building on their own experiences from practice. As leading scholars and practitioners, we believe that they all provide the reader with the best information currently available in their particular area. We also believe that this book provides a comprehensive and synoptic state-of-the-art overview of the field of bullying at work. Therefore, we hope that this book will be a useful tool for students and academics as well as practitioners in this new, intriguing and difficult problem area.

Part 1

The problem

1 The concept of bullying at work

The European tradition

Ståle Einarsen, Helge Hoel, Dieter Zapf and Cary L. Cooper

Introduction

Within a time-span of fewer than ten years, the concept of bullying at work has found a resonance within large sections of the European working population as well as in the academic community. A wide range of popular as well as academic books and articles has been published in many European languages (e.g. Ege, 1996; Einarsen *et al.*, 1994b; Field, 1996; Niedl, 1995; Leymann, 1993; Rayner *et al.*, 2002). Various conferences and symposia, academic as well as professional, have also been held each year to discuss and disseminate awareness of this problem (see for instance Rayner *et al.*, 1999; Zapf and Einarsen, 2001; Zapf and Leymann, 1996). From being a taboo in both organisational research and in organisational life, the issue of bullying and harassment at work became what has been called the 'research topic of the 1990s' (Hoel *et al.*, 1999).

Differing concepts have been in use in different European countries, such as 'mobbing' (Leymann, 1996; Zapf *et al.*, 1996), 'harassment' (Björkqvist *et al.*, 1994), 'bullying' (Einarsen and Skogstad, 1996; Rayner, 1997; Vartia, 1996), 'victimisation' (Einarsen and Raknes, 1997) and 'psychological terror' (Leymann, 1990a). However, they all seem to refer to the same phenomenon, namely the systematic mistreatment of a subordinate, a colleague, or a superior, which, if continued, may cause severe social, psychological and psychosomatic problems in the victim. Exposure to such treatment has been claimed to be a more crippling and devastating problem for employees than all other kinds of work-related stress put together, and is seen by many researchers and targets alike as an extreme type of social stress at work (Zapf *et al.*, 1996).

The purpose of this chapter is to discuss the European perspective on bullying or mobbing at work. We will start with some historical notes and we will then discuss various key characteristics of bullying such as the frequency, duration, power balance, quality and content of bullying behaviour, objective versus subjective bullying, intentionality of bullying, interpersonal versus organisational bullying, and bullying as a process. A definition of the concept will then be proposed and we will discuss and present various conceptual models of bullying at work.

The development of a new concept: some historical notes

A borrowed concept

The interest in the issue of workplace bullying originated in Scandinavia in the 1980s. The late Professor Heinz Leymann worked as a family therapist in the 1970s. Having experience with family conflicts, he decided to investigate direct and indirect forms of conflicts in the workplace (Leymann, 1995). Through empirical work in various organisations he encountered the phenomenon of mobbing, and wrote the first Swedish book on the subject in 1986, entitled *Mobbing – Psychological Violence at Work*. Through some articles and books on the negative health effects of destructive leadership by the late Norwegian professor of organisational psychology Svein Kile (e.g. 1990), the seemingly 'new' phenomenon of bullying, or 'mobbing' as it was referred to, attracted growing interest from the public, from those responsible for health and safety in the workplace, and from union representatives, as well as from researchers (Björkqvist, 1992; Björkqvist *et al.*, 1994; Einarsen and Raknes, 1991; Matthiesen *et al.*, 1989; Vartia, 1991).

However, until the early 1990s the interest in this subject was largely limited to the Nordic countries with only a few publications available in English (e.g. Leymann, 1990b). The American psychiatrist Carroll Brodsky had published an extensive report on the issue of bullying at work in 1976, called the *The Harassed Worker*, but this book did not have any impact until much later, at least not in Europe. However, on the arrival of the journalist Andrea Adams's (1992) book *Bullying at Work* and her documentary programmes on BBC radio, the issue was put firmly on the UK agenda. Many listeners apparently saw this as an opportunity to make sense of their own experience and many were willing to come forward and bring their experience into the public domain, as had happened in Scandinavia some years earlier. Heinz Leymann published a book aimed at a wider readership in German in 1993, again entitled *Mobbing – Psychological Terror at Work*, which disseminated awareness of the concept of mobbing through the media within a year. Likewise, the concern about the issue of adult bullying spread from Scandinavia to other countries such as Austria (Kirchler and Lang, 1998; Niedl, 1995), The Netherlands (Hubert and van Veldhoven, 2001;), Hungary (Kauscek and Simon, 1995), Italy (Ege, 1996) and as far afield as Australia (McCarthy *et al.*, 1996).

The term 'mobbing' was borrowed from the English word 'mob' and was originally used to describe animal aggression and herd behaviour, as noted in a later article by Munthe (1989). Heinemann (1972) adopted the term from the Swedish translation of Konrad Lorenz's (1968) book on aggression, to describe victimisation of individual children by a group of peers in a school setting. Despite a debate on the appropriateness of applying the term 'mobbing' to aggressive behaviour of individuals as well as of groups (e.g. Pikas, 1989) a majority of researchers stuck to the term 'mobbing' as a common descriptor for victimisation among school children (e.g. Munthe, 1989; Olweus 1987). In a similar way, Leymann (1986) borrowed the term 'mobbing' to describe bullying in the workplace. Internationally, the term 'mobbing' was later adopted in the German-speaking countries and The Netherlands as well as in some Mediterranean countries, whilst 'bullying' became the preferred term in English-speaking countries (Roland and Munthe, 1989).

A case of moral panic?

The introduction of and public interest in the concept of bullying at work seem to have emerged in a similar pattern across most countries (Rayner *et al.*, 2002). Such 'waves of interest' (Einarsen, 1998) are typically instigated by press reports on high-profile single court cases, or cases where one or more employee(s) publicly claim(s) to have been subjected to extreme mistreatment at work. Such public attention is then followed by a study of the magnitude of the problem or the consequences of such experiences. Such, and often preliminary, research findings generate further publicity and public attention, especially from union representatives and a large number of human resources specialists. The activity and determination of articulate and high-profile victims fuels the public debate. By exercising continuous pressure on the media in their broadest sense, and through numerous innovative initiatives utilising conference appearances, written publications and the world-wide web, these activists (e.g. Field, 1996) contribute to informing and educating the public, and effectively preventing the issue from disappearing from public view. Hence, in a very short time-span, anti-bullying policies and procedures have been produced and implemented in large numbers of organisations throughout many countries of Europe, in collaboration with, or under pressure from, local trade unions.

This swift development and these 'waves' of strong public attention may call into question whether the rise of the concept of bullying at work and the accompanying high figures on perceived exposure to bullying are fuelled by a case of 'moral panic'. The concept of moral panic refers to 'events the majority of members of society feel threatened by as a result of wrongdoers who threaten the moral order and that something must be done about such behaviour' (Lewis, 2001, p. 17). The fear or concern created by this 'public panic' is often significantly out of proportion compared to the actual threat of the particular event or phenomenon. The media focus and their use of inflated headlines and dramatic pictures and stories, politicians (including union representatives) calling for penalties, adopting a vindictive and self-righteous stance, and action groups (e.g. former victims) campaigning for actions against offenders create a demonising process where 'folk devils' and a 'disaster mentality' prevail (Lewis, 2001).

In the case of bullying at work the folk devils have been evident in the discourse of the 'psychopaths at work' in the UK (e.g. Field, 1996) and the discourse on the 'neurotic victim' in Scandinavian countries (see Einarsen *et al.*, 1994; Einarsen, 2000a; Leymann, 1996). Disaster mentality has been seen both in media headlines and in claims by union representatives that bullying is the most profound work environment problem of our time (Einarsen and Raknes, 1991).

Although the attention given to the phenomenon of bullying at work at times may have elements of such moral panic, Lewis (2001) has shown that perceptions of being exposed to bullying at work or perceptions of others being bullied at work do not appear to be a response to a general panic or fear produced by the popular media and the public debate. A study among 415 Welsh teachers in further and higher education showed that being a perceived target of bullying was not related to any particular knowledge of the public debate on bullying at work. Rather, it was based on their own personal experiences or on information provided by their colleagues. Hence, bullying at work seems to be a phenomenon that in fact prevails in many organisations. Although hardly a new phenomenon, many people seem to find this concept useful to describe and label situations at work where someone's behaviour is perceived to be systematically aimed at frustrating or tormenting an employee who is unable to defend him or herself, or escape from this situation (Einarsen *et al.*, 1994).

The concept of bullying at work

Bullying at work is about repeated actions and practices that are directed against one or more workers, that are unwanted by the victim, that may be carried out deliberately or unconsciously, but clearly cause humiliation, offence and distress, and that may interfere with job performance and/or cause an unpleasant working environment (Einarsen and Raknes, 1997). Hence, the concept of bullying at work relates to persistent exposure to negative and aggressive behaviours of a primarily psychological nature (Einarsen, 1996; Leymann, 1996). It describes situations where hostile behaviours that are directed systematically at one or more colleagues or subordinates lead to a stigmatisation and victimisation of the recipient(s) (Björkqvist *et al.*, 1994; Leymann, 1996).

Target orientation

From the beginning, the Scandinavian public debate had a target or victim perspective on bullying (Einarsen *et al.*, 1994b; Leymann, 1993, 1996). People were more alarmed by the tremendous health damage reported by the victims than by the often unethical behaviour of the perpetrators. Consequently, bullying or mobbing was seen from a stress perspective (Einarsen and Raknes, 1991; Leymann, 1996; Zapf *et al.*, 1996). Mobbing or bullying was understood as a subset of social stressors at work. The basic characteristic of social stressors is that they are related to the social relations of employees within an organisation. Applying concepts used in stress research, bullying can occur in the form of daily hassles (Kanner *et al.*, 1981). However, there are also many cases where single episodes of bullying are experienced as critical life events (Dohrenwend and Dohrenwend, 1974): for example, when a person is physically threatened, threatened with dismissal, or if someone's career prospects are destroyed.

If 'normal' social stressors occur in a department, it can be assumed that almost everybody will be negatively affected after some time. In a study by Frese and Zapf (1987) members of the same work group reported more similar levels of social stressors compared to members of different groups. Bullying, however, is targeted at particular individuals. These individuals will show severe health consequences after some time, whereas the perpetrators, observers or neutral bystanders may not be affected at all. Being singled out and stigmatised has been considered a key characteristic of bullying (Leymann, 1993, 1996; Zapf, 1999a). This is in contrast to approaches which focus on the aggressive behaviour of perpetrators. An aggressive perpetrator can frequently harass other persons; however, if this is, in the case of an abusive supervisor, distributed across several persons, this is clearly different from a victim or a few targeted victims being the focus of aggressive acts of one or more other persons. Although the concept of bullying is used both to describe the aggressive behaviour of certain perpetrators as well as the victimisation process of particular targets, the latter has been the main focus of the European perspective so far.

The frequency of negative behaviours

Definitions of bullying at work further emphasise two main features: repeated and enduring aggressive behaviours that are intended to be hostile and/or perceived as hostile by the recipient (Einarsen and Skogstad, 1996; Leymann, 1990b; Zapf *et al.*, 1996). In other words, bullying is normally not about single and isolated events, but rather about behaviours that are repeatedly and persistently directed towards one or more employees. Leymann (1990b, 1996) suggested that to be called 'mobbing' or bullying, such events should occur at least once a week, which characterises bullying as a severe form of social stress. In many cases this criterion is difficult to apply because not all bullying behaviours are strictly episodic in nature. For example, a rumour can circulate that may be harmful or even threaten to destroy the victim's career or reputation. However, it does not have to be repeated every week. In cases we have been made aware of, victims had to work in basement rooms without windows and telephone. Here, bullying consists of a permanent state rather than a series of events. Hence, the main criterion is that the behaviours or their consequences are repeated on a regular as opposed to an occasional basis.

The duration of bullying

A number of studies have shown the prolonged nature of the experience, with a majority of targets reporting an exposure time greater than twelve months (e.g. Leymann, 1992; UNISON, 1997; Zapf, 1999a; Zapf *et al.*, this volume). A mean duration time of eighteen months is reported in the case of Einarsen and Skogstad's (1996) Norwegian study and on average not less than 3.4 years in an Irish nation-wide study (O'Moore *et al.*, 2000).

The problem then arises of when to define operationally the duration of bullying behaviours. Leymann (1990b, 1996) suggested exposure for more than six months as an operational definition of bullying at work. Others have used repeated exposure to negative behaviours within a six-month period as the proposed timeframe (Einarsen and Skogstad, 1996). Leymann's strict criterion has been argued to be somewhat arbitrary, as bullying seems to exist on a continuum from occasional exposure to negative behaviours to severe victimisation resulting from frequent and long-lasting exposure to negative behaviours at work (Matthiesen *et al.*, 1989). Yet, the criterion of about six months has been used in many studies in order to differentiate between exposure to social stress at work and victimisation from bullying (e.g. Einarsen and Skogstad, 1996; Mikkelsen and Einarsen, 2001; Niedl, 1995; Vartia, 1996; Zapf *et al.*, 1996). The reason for this criterion for Leymann (1993, 1996) was to argue that mobbing leads to severe psychiatric and psychosomatic impairment, stress effects which would not be expected to occur as a result of normal occupational stressors such as time-pressure, role-conflicts or everyday social stressors. Hence, the period of 6 months was chosen by Leymann because it is frequently used in the assessment of various psychiatric disorders.

The duration of the bullying seems to be closely related to the frequency of bullying, with those bullied regularly reporting a longer duration of their experience than those bullied less frequently (Einarsen and Skogstad, 1996). This seems to be in line with a model of bullying highlighting the importance of conflict-escalation, with the conflict becoming more intense and more personalised over time (Zapf and Gross, 2001).

The nature of behaviours involved

The negative and unwanted nature of the behaviour involved is essential to the concept of bullying. Victims are exposed to persistent insults or offensive remarks, persistent criticism, personal or, even in some few cases, physical abuse (Einarsen, 2000b). Others experience social exclusion and isolation; that is they are given the 'silent treatment' or 'sent to Coventry' (Williams, 1997). These behaviours are 'used with the aim or at least the effect of persistently humiliating, intimidating, frightening or punishing the victim' (Einarsen, 2000b, p. 8).

Based on both empirical and theoretical evidence, Zapf (1999a) categorised five main types of bullying behaviour:

1. work-related bullying which may include changing the victim's work tasks in some negative way or making them difficult to perform;
2. social isolation by not communicating with somebody or excluding someone from social events;
3. personal attacks or attacks on someone's private life by ridicule or insulting remarks or the like;
4. verbal threats in which somebody is criticised, yelled at or humiliated in public; and
5. spreading rumours.

Moreover, physical violence, or threats of such violence, as well as attacks on people's political or religious beliefs were reported.

Although many researchers (e.g. Einarsen 1999; Leymann, 1996; Niedl, 1995; Vartia, 1991; Zapf *et al.*, 1996) include physical abuse in their categorisation of bullying, they all agree that the behaviours involved in workplace bullying are mainly of a psychological rather than a physical nature. In a study among male Norwegian shipyard workers, where 88 per cent had experienced some form of bullying behaviours during the last six months, only 2.4 per cent reported having been subjected to physical abuse or threats of such abuse (Einarsen and Raknes, 1997). In Zapf's (1999a) study, only about 10 per cent of the bullying victims reported physical violence or the threat of physical violence.

In line with research into school bullying (Olweus, 1991), a simple distinction has been made between direct actions, such as accusations, verbal abuse and public humiliation, on the one hand, and indirect acts of aggression, such as rumours, gossiping and social isolation, on the other (e.g. Einarsen *et al.*, 1994b; O'Moore *et al.*, 1998; see also Baron and Neuman, 1996). On the basis of empirical data from some 5,300 UK employees, Einarsen and Hoel (2001) differentiated between two main classes of bullying behaviours: work-related bullying and personal bullying. The former includes behaviours such as giving unreasonable deadlines or unmanageable workloads, excessive monitoring of work, or assigning meaningless tasks or even no tasks. Personal bullying consists of behaviours such as making insulting remarks, excessive teasing, spreading gossip or rumours, persistent criticism, playing practical jokes and intimidation.

Many of these acts may be relatively common in the workplace (Leymann, 1996). But when frequently and persistently directed towards the same individual they may be considered an extreme social source of stress (Zapf *et al.*, 1996), and become capable of causing severe harm and damage (Mikkelsen and Einarsen, 2002). The persistency of these behaviours also seems to drain the coping resources of the victim (Leymann, 1990b). The stigmatising effects of these activities, and their escalating frequency and intensity, make the victims constantly less able to cope with his or her daily tasks and co-operation requirements of the job, thus becoming continually more vulnerable and 'a deserving target' (Einarsen, 2000b, p.8). Hence, the frequency and duration of unwanted behaviours seem to be as important as the actual nature of the behaviours involved.

The imbalance of power between the parties

A central feature of many definitions of bullying is the imbalance of power between the parties (Einarsen *et al*, 1994b; Leymann, 1996; Niedl, 1995; Zapf *et al.*, 1996). Typically, a victim is constantly teased, badgered and insulted and perceives that he or she has little recourse to retaliate in kind (Einarsen, 1999). In many cases, it is a supervisor or manager who systematically, and over time, subjects subordinates to highly aggressive or demeaning behaviour (Rayner *et al.*, 2002). In other cases, a group of colleagues bully a single individual, who for obvious reasons finds it difficult to defend him or herself against this 'overwhelming' group of opponents.

The imbalance of power often mirrors the formal power structure of the organisational context in which the bullying scenario unfolds. This would be the case when someone is on the receiving end of negative acts from a person in a superior organisational position. Alternatively, the source of power may be informal, based on knowledge and experience as well as access to support from influential persons (Hoel and Cooper, 2000). The imbalance of power may also be reflected in the target's dependence on the perpetrator(s), which may be of a social, physical, economic or even psychological nature (Niedl 1995). An employee will in most cases be more dependent on his supervisor than vice versa. A single individual will be more dependent on the work group than the other way around. Thus, at times the perceptions of targets may be more dependent upon the actual instigator of a negative act than the act itself (Einarsen, 2000a). Einarsen (1999) argues that knowledge of someone's 'weak point' may become a source of power in a conflict situation. Bullies typically exploit the perceived inadequacies of the victim's personality or work performance, which in itself indicate powerlessness on the part of the victim.

However, one may argue that in a conflict situation, some individuals may initially feel that they are as strong as their opponent, but gradually come to realise that their first impression was wrong, or that their own or their opponent's moves have placed them in a weaker position. In other words, and at least from the position of targets, a power deficit has emerged. Equal balance of power in a harsh conflict may, therefore, be considered hypothetical, as the balance of power in such situations is unlikely to remain stable for any length of time. Hence, bullying may result from the exploitation of power by an individual or by a group, as well as from taking advantage of a power deficit on the part of the target.

Subjective versus objective bullying

According to Niedl (1995), the 'definitional core of bullying at work rests on the subjective perception made by the victim that these repeated acts are hostile, humiliating and intimidating and that they are directed at himself/herself' (p. 49). Yet, situations where one person offends, provokes or otherwise angers another person, often involve substantial discrepancies between the subjective perceptions and interpretations of the conflicting participants. Incidences that may be considered mildly offensive by one individual might be seen as serious enough to warrant a formal complaint by others (Tersptra and Baker, 1991). This reflects the discussion on the role of subjective appraisal in psychological stress theories (Lazarus and Folkman, 1984).

The distinction between subjective and objective experiences of bullying was first made by Brodsky (1976) and has been an important part of the discussion about the definition of bullying at work. Although most studies theoretically seem to regard bullying as an objective and observable phenomenon, which is not entirely in the 'eye of the beholder', with only a few exceptions the empirical data have so far been gathered by the use of self-report from victims (Einarsen *et al.*, 1994a; Vartia, 1996). So far, little is known about the 'interrater reliability' with regard to bullying, that is the agreement of the victim with some external observers.

Frese and Zapf (1988) define a subjective stressor as an event that is highly influenced by an individual's cognitive and emotional processing. An objective stressor is observed independently of an individual's cognitive and emotional processing. Of course, in many cases there will be a substantial overlap between the two, as has been found in studies among Scandinavian school children, where teachers tend to agree with subjective perceptions of victimisation made by children (Olweus, 1987).

Björkqvist *et al.* (1994) argue strongly against an approach where peer nominations are used as an objective measure of bullying. The economic importance of and dependence on a job would effectively prevent people from being honest in their assessment, in particular with regard to superiors in formal positions of power. It has also been argued that it is often difficult for the observer to stay neutral in cases of bullying (Einarsen, 1996; Neuberger 1999), a fact which is likely to influence the 'objectivity' of such ratings. In the course of time, social perceptions of the victim seem to change, turning the situation gradually into one where even third parties perceive it as a no more than fair treatment of a neurotic and difficult person (Einarsen *et al.*, 1994b; Leymann, 1992). As the behaviours involved in bullying are often of a subtle and discrete nature (e.g. not greeting, leaving the table when the victim arrives, not passing information, gossiping, etc.), sometimes even exhibited in private (Einarsen, 1999; Neuberger, 1999), they are not necessarily observable to others. Bullying is, therefore, often a subjective process of social reconstruction, and difficult to prove. Uninformed bystanders could interpret the respective behaviours completely differently (e.g. as forgetful, impolite, careless, making jokes). The significance of a particular behaviour may only be known by the perpetrator and the recipient. The fact that the parties have a past and a future together will have a bearing on the perceptions and interpretations of the exhibited behaviours as well as on the process development (Hoel *et al.*, 1999). Imbalances of power are also more evident from the perspective of those experiencing a lack of power. For instance, Hofstede (1980) measures the cultural dimension of power distance as the difference in power between two parties as perceived by the one in the least powerful position.

On the whole, we tend to agree with Lengnick-Hall (1995), who, in the case of sexual harassment, argues that an objective conceptualisation is, of course, necessary in connection with legal issues and cases of internal disciplinary hearings. However, subjective conceptualisations will be a better prediction of victims' responses and reactions, organisational outcomes such as turnover and absenteeism, as well as organisational responses. Also, subjective conceptualisations must suffice to evoke organisational interventions and attempts at problem-solving and mediation.

Intentionality of bullying

Considerable disagreement exists with regard to the issue of intent (Hoel *et al.*, 1999). The role of intent in bullying is linked both to whether the perceived negative action was intended in the first place and to the likely harmful outcome of the behaviour. Some contributors, notably those whose approach is anchored in aggression theory, consider intent to cause harm on the part of the perpetrator as a key feature of bullying (Björkqvist *et al.*, 1994). In other words, where there is no intention to cause harm, there is no bullying. The problem is that it is normally impossible to verify the presence of intent (Hoel *et al.*, 1999). For the same reason intent is excluded from most definitions of sexual harassment (e.g. Fitzgerald and Shullman, 1993).

A further issue to consider in connection with intent is the distinction between instrumental aggression and affective aggression (Hoel and Cooper, 1999). Whilst the bullying behaviour may be conscious and intended, there may be no intent to cause harm on the part of the perpetrator. As such the bullying behaviour may be considered instrumental to achieving a certain goal or objective. However, whereas intent may be a controversial feature of bullying definitions, there is little doubt that perception of intent is important as to whether an individual decides to label their experience as bullying or not.

Interpersonal versus organisational bullying

Following on from the above presentation, bullying is an interpersonal phenomenon that evolves from a dynamic interaction between at least two parties. Bullying is exhibited by one or more persons, directed towards another individual, and perceived and reacted to by this individual. However, from the work of Liefooghe and Mackenzie Davey (2001) one may argue that the concept may also refer to incidences of what we may call 'organisational bullying' or 'structural mobbing' (Neuberger, 1999). Organisational bullying refers to situations in which organisational practices and procedures perceived to be oppressive, demeaning and humiliating are employed so frequently and persistently that many employees feel victimised by them. Again we are talking about persistent negative events and behaviours that wear down, frustrate, frighten or intimidate employees. However, in these situations managers individually or collectively enforce organisational structures and procedures that may torment, abuse or even exploit employees. Hence, bullying in these cases does not strictly refer to interpersonal interactions, but rather to indirect interactions between the individual and management.

One may of course question the fruitfulness of such a use of the bullying concept, especially since most authors see bullying as an interpersonal phenomenon. Furthermore, we have elsewhere argued that the term bullying is easily misused (Einarsen, 1998; Zapf, 1999a). However, in a study in a UK telecommunications company, using focus groups, employees did use the term bullying to account for grievances and discontentment with the organisation and its procedures (Liefooghe and Mackenzie Davey, 2001). Examples of such kinds of procedures were the excessive use of statistics (such as performance targets), rules regarding call-handling times, penalties for not hitting targets (such as withdrawing possibilities for overtime) and use of sickness policy. These employees acknowledged the existence of bullying as an interpersonal phenomenon, but also used the term as an emotive and highly charged term, which helped them to highlight their discontent at what they perceived to be increasingly difficult work situations. Brodsky (1976) also argued that excessive work pressure was to be seen as one particular kind of bullying.

Bullying as a process

Empirical studies indicate that bullying is not an 'either or' phenomenon, but rather a gradually evolving process (Björkqvist, 1992; Einarsen, 2000b; Leymann, 1990b; Zapf and Gross, 2001). During the early phases of the bullying process, victims are typically subjected to aggressive behaviour that is difficult to pin down because of its indirect and discrete nature. Later on more direct aggressive acts appear (Björqkvist, 1992). The victims are clearly isolated and avoided, humiliated in public by excessive criticism or by being made a laughing-stock. In the end both physical and psychological means of violence may be used.

In line with Leymann (1990b), Einarsen (1999) identified four stages of process development and referred to them as *aggressive behaviours*, *bullying*, *stigmatisation* and *severe trauma*. In many cases the negative behaviours in the first phase may be characterised as indirect aggression. They may be 'subtle, devious and immensely difficult to confront' (Adams, 1992, p. 17) and sometimes difficult to recognise for the persons being targeted (Leymann, 1996). Where bullying evolves out of a dispute or a conflict, it may even at times be difficult to tell who may turn out to be the victim (Leymann, 1990b). The initial phase, which in some cases can be very brief, tends to be followed by a stage of more direct negative behaviours, often leaving the target humiliated, ridiculed and increasingly isolated (Leymann, 1990b, 1996). As a result, the targets become stigmatised and find it more and more difficult to defend themselves (Einarsen, 1999). At this point in the process the victims may suffer from a wide range of stress symptoms. According to Field, himself a victim of bullying,

> the person becomes withdrawn, reluctant to communicate for fear of further criticism. This results in accusations of 'withdrawal', 'sullenness', 'not co-operating or communicating', 'lack of team spirit', etc. Dependence on alcohol, or other substances can then lead to impoverished performance, poor concentration and failing memory, which brings accusations of 'poor performance.
>
> (Field, 1996, p. 128)

It has also been noted that the erratic and obsessive behaviour of many victims in this phase may frequently cut them off from support within their own working environment, exacerbating their isolation and the victimisation process (Leymann, 1986). The situation is frequently marked by helplessness, and, for many, lengthy sickness absences may be necessary to cope with the situation (Einarsen *et al.*, 1994b; Zapf and Gross, 2001). When the case reaches this stage, victims are also often left with no role in the workplace, or with little or no meaningful work provided.

Leymann (1990b) refers to this last stage as 'expulsion', where victims are either forced out of the workplace directly, by means of dismissal or redundancy, or indirectly, when the victims consider their work situation so impossible that they decide to leave 'voluntarily' (constructive dismissal). In a study among German victims of severe bullying, Zapf and Gross (2001) revealed that victims of such bullying strongly advise other victims to leave the organisation and seek support elsewhere.

Despite the severity of the situation, neither management nor work colleagues are likely to interfere or take action in support of the victim. On the contrary, if victims complain, they frequently experience disbelief and questioning of their own role. Such prejudices against victims often extend beyond management to include work colleagues, trade union representatives and the medical profession (Leymann, 1996; O'Moore *et al.*, 1998).

Such rather depressing outcomes have been observed in many countries (Björqkvist 1992; Einarsen *et al.*, 1994b; Leymann, 1990b;1996; Zapf and Gross, 2001). The prejudices against the victim produced by the bullying process seem to cause the organisation to treat the victim as the source of the problem (Einarsen, 1999). When the case comes to their attention, senior management, union representatives or personnel administrations tend to accept the prejudices produced by the offenders, thus blaming victims for their own misfortune. Third parties and managers may see the situation as no more than fair treatment of a difficult and neurotic person (Leymann, 1990b).

A definition of bullying at work

Building on this line of argument we suggest the following definition of bullying (cf. Einarsen and Skogstad, 1996; Leymann, 1996; Zapf, 1999b):

> Bullying at work means harassing, offending, socially excluding someone or negatively affecting someone's work tasks. In order for the label bullying (or mobbing) to be applied to a particular activity, interaction or process it has to occur repeatedly and regularly (e.g. weekly) and over a period of time (e.g. about six months). Bullying is an escalating process in the course of which the person confronted ends up in an inferior position and becomes the target of systematic negative social acts. A conflict cannot be called bullying if the incident is an isolated event or if two parties of approximately equal 'strength' are in conflict.

The measurement of bullying at work

It follows from this discussion of the concept of bullying that its measurement is not an easy task and may be done according to a host of different criteria (see also Cowie *et al.*, 2001). Most studies measure subjective bullying by either one of two methods: perceived exposure to bullying behaviours, or perceived victimisation from bullying at work. Studies measuring exposure to bullying behaviours follow an approach developed by Leymann (1992), in which respondents are presented with an inventory of negative behaviours identified with bullying. Bullying is then operationalised by defining the criteria for when a person is being bullied. The two most used instruments in this tradition are the Leymann Inventory of

Psychological Terror (LIPT) (Leymann, 1990a) and the Negative Acts Questionnaire (Einarsen and Raknes, 1997; Hoel and Cooper, 2000; Mikkelsen and Einarsen, 2001). The second method is derived from research on bullying among children (Olweus, 1994), presenting respondents with a definition of bullying, followed up by questions regarding frequency and duration of exposure for those who identified or labelled their experience as bullying according to the definition given. This approach was introduced by Einarsen and Raknes (1991) and was later used in several studies (Einarsen and Skogstad, 1996; Hoel and Cooper, 2000; Mikkelsen and Einarsen, 2001; O'Moore, 2000).

Linking back to the previous discussion about subjective versus objective bullying (cf. Frese and Zapf, 1988), measuring bullying by means of an inventory of behaviours connected with bullying may be considered a somewhat more 'objective' method than one which requires individuals to 'label' their own experience as bullying. Yet, as the subjective experience of being bullied seems to be at the heart of the problem (cf. Niedl, 1995), the labelling approach may be equally valid. Hence, Einarsen (1996) argues that the optimal measurement of bullying at work includes both methods, as this will bring information on both the nature and the intensity of the perceived behaviours, as well as on the subjective perception of being victimised by these behaviours (see also Mikkelsen and Einarsen, 2001; Salin, 2001 for a discussion).

Conceptual models of bullying at work

The Leymann model

Heinz Leymann (1990b; 1993; 1996), who has been influential in many European countries, argued strongly against individual factors as antecedents of bullying, especially when related to issues of victim personality. Instead he advocated a situational outlook, where organisational factors relating to leadership, work design and the morale of management and workforce are seen as the main factors. He asserted that four factors are prominent in eliciting bullying behaviours at work (Leymann, 1993): (1) deficiencies in work-design, (2) deficiencies in leadership behaviour, (3) the victim's socially exposed position, and (4) low departmental morale. Leymann (1996) also acknowledged that poor conflict management might be a source of bullying, but in combination with inadequate organisation of work. However, he again strongly advocated that conflict management is an organisational issue and not an individual one (Leymann, 1996). Conflicts only escalate into bullying when the managers or supervisors either neglect or deny the issue, or if they are involved in the group dynamics just like everyone else, thereby fuelling conflict. Since bullying takes place within a situation regulated by formal behavioural rules and responsibilities, it is always and by definition the responsibility of the organisation and its management.

Some research has been conducted on the work-environment hypothesis of Leymann. Some thirty Irish victims of bullying described their workplace as a highly stressful and competitive environment, plagued with interpersonal conflicts and a lack of a friendly and supportive atmosphere, undergoing organisational changes and managed by means of an authoritarian leadership style (O'Moore *et al.*, 1998). In a Norwegian survey of 2,200 members of seven different trade unions, both victims and observers of bullying at work reported a lack of constructive leadership, lack of possibilities to monitor and control their own work tasks and, in particular, a high level of role-conflict (Einarsen *et al.*, 1994a). In a Finnish survey, victims and observers of bullying described their work unit as having the following features: a poor information flow, an authoritative way of settling differences of opinion, lack of discussions about goals and tasks, and lack of opportunity to influence matters affecting oneself (Vartia, 1996). A few studies have also shown a link between organisational changes and bullying at work (e.g. McCarthy, 1996).

It should be noted, however, that these results are based on cross-sectional studies that do not allow us to interpret relations as cause and effect. Although we believe that in many cases organisational deficiencies contribute substantially to the development of bullying, it is equally plausible that severe social conflicts at work may be the cause rather than the effect of organisational problems (Zapf, 1999b). Conflicts may, for example, negatively affect the information flow and thus impair 'leader–member' relationships. Moreover, relations between bullying and low levels of control can also be explained by the fact that restricting someone's opportunities to affect decision making has been described as a bullying strategy (Leymann, 1990a).

Moreover, Einarsen *et al.* (1994a) found that work-environment factors only explained 10 per cent of the variance in the prevalence of workplace bullying within seven different organisational settings, and in no subsetting was this greater than 24 per cent. Thus, there is certainly room for alternative explanations. Leymann himself never presented any empirical evidence for his strong focus on organisational factors and his disregard for the role of personality. Hence, bullying is neither the product of chance nor of destiny (Einarsen *et al.*, 1994a). Instead, bullying should be understood primarily as a dyadic interplay between people, where neither situational nor personal factors are entirely sufficient to explain why bullying develops. Although one may agree that the organisation and its management are responsible for intervening in cases of interpersonal conflict and bullying, this may still be caused by a wide range of factors, both on an individual level, and on dyadic, group, organisational and societal levels (Hoel and Cooper, 2001; Zapf, 1999b). Zapf provided some preliminary evidence for these various potential causes of bullying by identifying subgroups of bullying victims for which certain causal factors such as the organisation or characteristics of the victim were likely to dominate.

Assuming that the concept of bullying at work refers to a range of situations and contexts where repeated aggressive behaviour may occur and where the targets are unable to defend themselves, Einarsen (1999) introduced the concepts of *dispute-related* and *predatory* bullying to broaden the perspectives and to account for the two main classes of situations where bullying may seem to originate.

Predatory bullying

In cases of 'predatory bullying', the victim has personally done nothing provocative that may reasonably justify the behaviour of the bully. In such cases the victim is accidentally in a situation where a predator is demonstrating power or is exploiting the weakness of an accidental victim. The concept of petty tyranny proposed by Ashforth (1994) seems to refer to such kinds of bullying. Petty tyranny refers to leaders who lord their power over others through arbitrariness and self-aggrandisement, the belittling of subordinates, lack of consideration and the use of an authoritarian style of conflict management. In some organisations, bullying is more or less institutionalised as part of the leadership and managerial practice, sometimes in the guise of 'firm and fair' management (Brodsky, 1976). However, 'firm and fair' may easily turn into 'harsh and unfair' management, which again may turn into bullying and the victimisation of subordinates.

A person may also be singled out and bullied due to the fact that he or she belongs to a certain outsider group, for instance by being the first woman in a local fire brigade. If perceived as a representative of a group or a category of people who are not approved by the dominant organisational culture, such employees may indeed be bullied without doing anything other than merely showing up at work (Archer, 1999). A study in UK fire brigades revealed an environment where bullying of females, non-whites and non-conforming whites prevailed as a mechanism to ensure the preservation and dominance of the white male culture (Archer, 1999). As such, the individual victim of bullying was in fact a coincidental target.

An employee may even be bullied by being an easy target of frustration and stress caused by other factors. In situations where stress and frustration are caused by a source that is difficult to define, inaccessible, or too powerful or respected to be attacked, the group may turn its hostility towards a suitable scapegoat (Thylefors, 1987). Björkqvist (1992) argues that such displaced aggressiveness may act as a collective defence mechanism in groups where much unstructured aggression and hostility prevails.

Allport's (1954) description of the process involved in acting out prejudices seems to describe very well how such bullying evolves. In his first phase, called 'antilocation', prejudicial talk starts, but is restricted to small circles of the 'in-group' and taking place 'behind the back of the victim'. This is followed by a second phase in which one moves beyond talking and starts to avoid the victim. In the third phase the victim is openly harassed and discriminated against by being alienated and excluded or subjected to offensive remarks and jokes. In the fourth phase physical attacks occur, which may lead to the final stage called 'extermination'. Although victims of bullying are not literally killed, some do commit suicide (Leymann, 1990b); others are permanently expelled from working life (Leymann, 1996) or at least driven out of their organisation (Zapf and Gross, 2001). Hence, these examples of predatory bullying suggest the following subcategories: exposure to a destructive and aggressive leadership style, being singled out as a scapegoat, and the acting out of prejudice.

Dispute-related bullying

'Dispute-related bullying', on the other hand, occurs as a result of highly escalated interpersonal conflicts (Einarsen, 1999; Zapf and Gross, 2001). Although interpersonal struggles and conflicts are a natural part of all human interaction and must not be considered bullying, there may be a thin line between the battles between two parties in an interpersonal conflict and the aggressive behaviour used in bullying (see also Zapf and Gross, 2001). The difference between interpersonal conflicts and bullying is not to be found in what is done or how it is done, but rather in the frequency and duration of what is done (Leymann, 1996), as well as the ability of both parties to defend themselves in this situation (Zapf, 1999a). In some instances, the social climate at work turns sour and creates differences that may escalate into harsh personalised conflicts and even 'office wars', where the total destruction of the opponent is seen as the ultimate goal to be gained by the parties (van de Vliert, 1998).

In highly escalated conflicts both parties may deny the opponent any human value, thus clearing the way for highly aggressive behaviours. If one of the parties acquires a disadvantaged position in this struggle, he or she may of course become the victim of bullying (Zapf, 1999b). It may also be the case that one of the parties exploits his or her own power or a potential power imbalance, leading to a situation where one of the parties is unable to defend itself or retaliate against increasingly aggressive behaviours. The defenceless position will then lead to a victimisation of one of the parties.

Interpersonal conflicts where the identity of the protagonists is at stake – for instance when one of the parties attacks the self-esteem or self-image of the other – are often characterised by intense emotional involvement (Glasl, 1994). The latter includes feelings of being insulted, fear, suspicion, resentment, contempt, anger and so forth (van de Vliert, 1998). In such cases people may subject each other to bullying behaviour or resent the behaviour of their opponent to a degree where they feel harassed and victimised even though there are few observable signs of bullying behaviour by the alleged offender. It may also be true that claiming to be a victim of bullying may be a very effective strategy in interpersonal conflicts, in some cases even used by both parties. The conflict escalation model of Glasl (1982, 1994) has been proposed as a model suitable to explain how conflicts may escalate into bullying (Einarsen *et al.*, 1994b; Neuberger, 1999; Zapf and Gross, 2001). The model differentiates between three phases and nine stages (Figure 1.1). According to this model, conflicts are inevitable in organisations, and under certain circumstances even fruitful, contributing to innovation, performance and learning (de Dreu, 1997). However, if allowed to escalate, conflicts may turn into 'office wars' and become extremely harmful and destructive both on an individual as well as an organisational level (see Figure 1.1).

In the first stages of a conflict, the parties are still interested in a reasonable resolution of conflict about tasks or issues. Although they may experience and acknowledge interpersonal tension, they mainly focus on co-operation to solve the problems in a controlled and rational manner. However, this becomes increasingly more difficult as the interpersonal tensions escalate (Zapf and Gross, 2001).

The second phase is characterised by a situation where the original issue of the conflict has more or less vanished, while the interpersonal tension between the parties and their increasingly difficult relationship becomes the heart of the problem. Now the issue of the conflict is more to do with 'who is the problem' than 'what is the problem'. The parties cease to communicate and start to seek allies and support from others. They become increasingly more concerned about their own reputation and about 'losing face' and they experience moral outrage against their oppo-

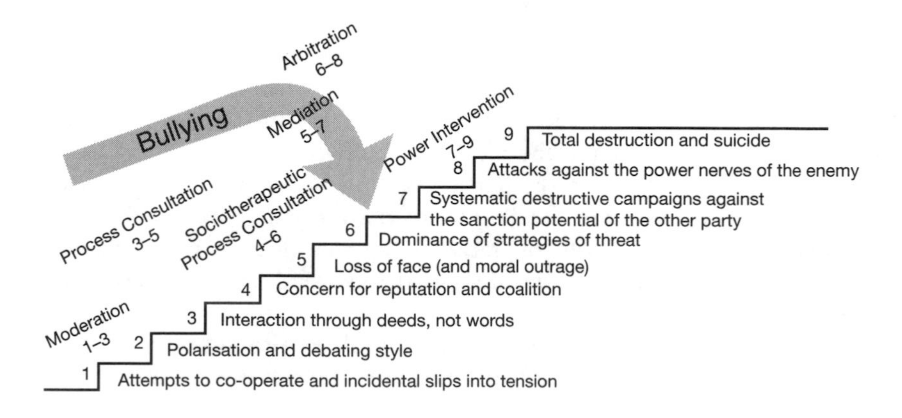

Figure 1.1 The conflict escalation model of Glasl (1994)

Source: After Zapf and Gross (2001, p. 501).

nent(s), perceiving them as immoral, as having a personality deficit or as being plain stupid. At this point, disrespect, lack of trust and finally overt hostility evolve. Ultimately, the interaction is dominated by threat as well as openly hostile and aggressive behaviour. In the following phase the confrontations become increasingly more destructive, until the total destruction of the opponent is the sole aim of the parties. In this struggle both the parties are willing to risk their own welfare, even their own existence, in order to destroy the opponent.

Zapf and Gross (2001) argue that bullying may be seen as a kind of conflict at the boundary between phases two and three. In their interview study of nineteen German victims of bullying, fourteen victims reported a continuously escalating situation, in which the situation became worse over the course of time. Almost 50 per cent of the victims described a sequence of escalation resembling Glasl's model.

Although Glasl argues that the latter stages of the model may not be reached in organisations, we will argue that they are in fact reached in the more extreme cases of bullying. Some victims commit suicide (Leymann, 1990b), and many consider it (Einarsen *et al.*, 1994b). Some victims go to court even though they may be unable to afford a solicitor and are likely to lose. Others refuse a reasonable settlement out of court because they do want to take their employer to court at any cost (cf. the case described in Diergarten, 1994). Some work groups even take pleasure in the suffering of the victim.

A theoretical framework

In the following section, we will argue that a complex social phenomenon such as bullying is characterised by multi-causality, involving a range of factors found at many explanatory levels, depending on whether we focus on the behaviour of the actor or on the perceptions, reactions and responses of the target (see also Einarsen, 1999, 2000b; Hoel and Cooper, 2001; Zapf, 1999b). On an individual level both the personality of the perpetrator as well as the victim may be involved as causes of both bullying behaviour and perceptions of being bullied. Individual factors may also contribute to a potential lack of coping on behalf of the victim as well as other emotional and behavioural reactions to the perceived treatment. On a dyadic level the focus is on the relationship and the interaction between the alleged perpetrator and the alleged victim. Since a power differential between the parties is central to the definition of bullying, a dyadic perspective is vital to the understanding of the concept of bullying at work. According to Brodsky (1976) many cases of bullying involve an artless teaser who meets a humourless target. Also, to focus on a potential clash or mismatch in terms of personalities and power may be as relevant as to focus on the pathological and deviant personality of the perpetrator or the victim. On a dyadic level we may also focus on the dynamics of conflict escalation and the dynamic transaction between the perpetrator and the victim in the course of the conflict (Glasl, 1994; Zapf and Gross, 2001). In most cases, and especially those of a dispute-related kind, the victim is not an entirely passive recipient of negative acts and behaviours (Hoel and Cooper, 2001). The responses by the victims are likely to impinge upon the further responses of the perpetrator. As shown by Zapf and Gross (2001), those victims who successfully coped with bullying fought back with similar means less often and avoided further escalation of the conflict. Less successful victims in terms of coping often contributed to the escalation of the bullying by their aggressive counterattacks and 'fights for justice' (p. 497).

On a social group level, bullying may be explained in terms of 'scapegoating' processes in groups and organisations. Such 'witch-hunting' processes arise when groups displace their frustration and aggression on to a suitable and less powerful group member. Being seen as an outsider or part of a minority may be one criterion for this choice (Schuster, 1996). Another may be outdated behaviours that do not keep pace with the developments of the group. Also, being too honest or unwilling to compromise may also contribute to attaining the role of a scapegoat (Thylefors, 1987). On the organisational level, many factors may contribute to explain cases of bullying at work (Hoel and Cooper, 2001). Archer (1999) has shown how bullying may become an integrated part of an organisational culture, while Zapf *et al.* (1996) have shown that requirements for a high degree of co-operation combined with restricted control over one's own time, may contribute to someone becoming a victim of bullying. This situation may lead to many minor interpersonal conflicts and may simultaneously undermine the possibilities for conflict resolution. Similarly, Vartia (1996) has shown that the work environment in organisations with bullying is characterised by a general atmosphere experienced by employees as strained and competitive, where everyone pursues their own interests.

In Figure 1.2 we present a theoretical framework that identifies the main classes of variables to be included in future research efforts and future theoretical developments in this field. The model may also be utilised in order to guide and structure future organisational action programmes. This model pinpoints another level of explanation, the societal level, consisting of national culture, and historical, legal and socio-economic factors (see also Hoel and Cooper, 2001). Although it has not been much studied yet, the occurrence of bullying must always be seen against such a background (see also McCarthy, and Ironside and Seifert, this volume). The high pace of change, the intensifying workloads, increasing work hours and uncertainty with regard to future employment that characterise contemporary working life in many countries influence the level of stress of both perpetrator and victim. Hence, both the level of aggression and coping resources may be influenced by such factors. In addition, the tolerance of organisations and their management of cases of

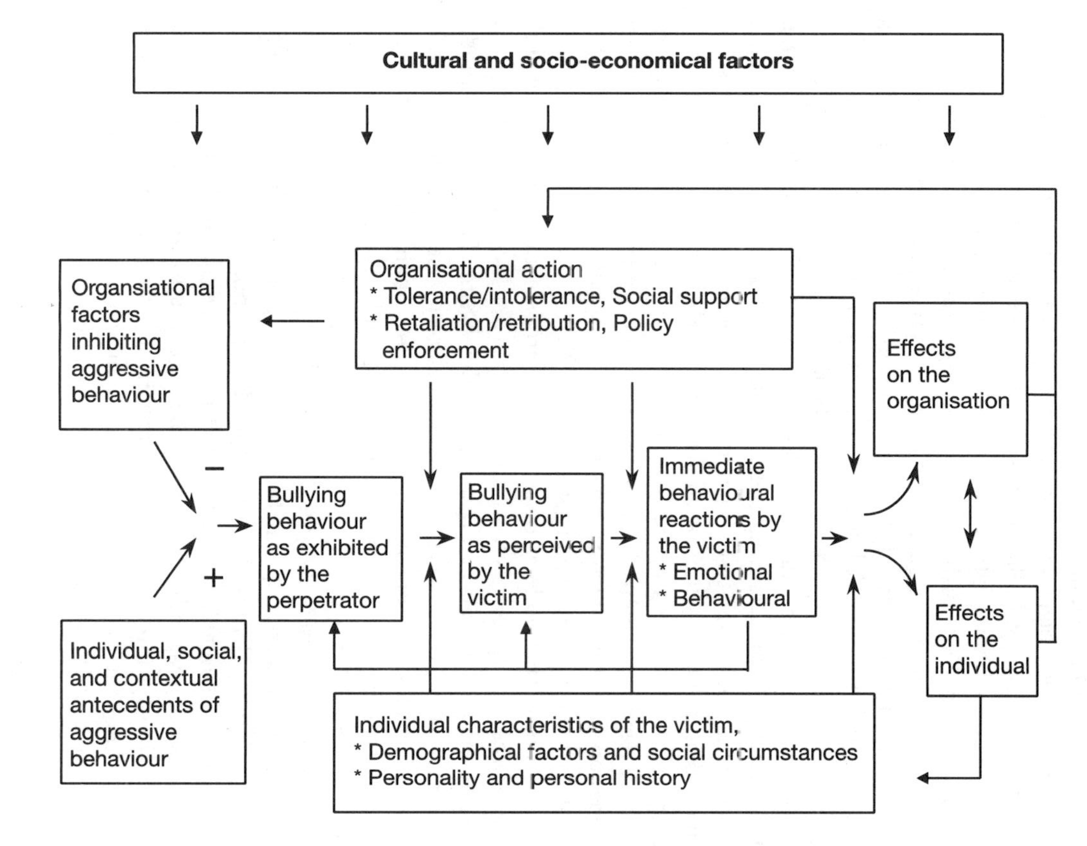

Figure 1.2 A theoretical framework for the study and management of bullying at work

bullying must also to some extent be seen in light of prevailing societal factors.

Following on from the debate on objective and subjective bullying, this model distinguishes between the nature and causes of bullying behaviour as exhibited by the alleged offender, and the nature and causes of the perceptions of the target of these behaviours. Furthermore, it distinguishes between the perceived exposure to bullying behaviours, and the reactions to these kinds of behaviour.

Looking at the behaviour of the perpetrator first, Brodsky (1976) claimed that although bullies may suffer from personality disorders, they will only act as bullies when the organisational culture permits or even rewards this kind of misbehaviour. Although there may be situational, contextual as well as personal factors that may cause a manager or an employee to act aggressively towards subordinates or colleagues, such behaviour will not be exhibited systematically if there are factors in the organisation that hinder or inhibit such behaviours. On the basis of survey data on the experiences and attitudes of British union members, Rayner (1998) concluded that bullying prevails due to an organisational tolerance of such behaviour. Ninety-five per cent of the respondents in her study claimed that bullying was caused by the fact that 'bullies can get away with it' and 'victims are too scared to report it'. Hence, bullying behaviour may be a result of the combination of a propensity to bully due to either personal or situational factors, and the lack of organisational inhibitors of bullying behaviour (see also Pryor *et al.*, 1993).

Furthermore, the model shows that such organisational factors, as well as an effective support system for victims, are key factors that may moderate the perceptions and reactions of the victim. The presented model argues that attention to such organisational response patterns and other contextual issues within the organisation are highly important when understanding the many different aspects of bullying at work. The latter part of the model has clearly an individual, subjective and most of all a reactive focus (Einarsen, 2000b). Although bullying at work may to some degree be a subjectively experienced situation in which the meaning assigned to an incident will differ, depending on both the persons and the circumstances involved, this part of the model highlights the necessity for any strategy against bullying to take the perceptions and reactions of the victims seriously and as a real description of how they experience their work environment. Second, this part of the model argues for inclusion of a rehabilitation programme in an effective organisational strategy against bullying.

This theoretical framework also gives some credit to the dynamic process involved in the interaction between perpetrator, victim and organisation (see above). Even Leymann (1986, 1992) argued that the stress reaction of the victim to the perceived bullying, and the consequential effects on the victim, may backfire and justify the treatment of the victim.

The process of stigmatisation may also alter the perception of the victim, which again may change how an organisation tolerates, reacts to and manages a particular case of bullying. Hence, the behaviour of the perpetrator, the personal characteristics of the victim as well as responses of the organisation to bullying may be altered in the course of the process. We believe that knowledge of the escalation and the dynamics of interaction involved in the victimisation process is essential to the understanding of this phenomenon.

Conclusions

Some authors have argued that there is in fact a difference between the UK concept of bullying and the Scandinavian and German concept of mobbing at work (Leymann, 1996). According to Leymann (1996) the choice of the term 'mobbing at work' in preference to 'bullying' was a conscious decision. It reflected the fact that the phenomenon in question very often refers to subtle, less direct aggression as opposed to the more physical aggression commonly identified with the term 'bullying', but with the same debilitating and stigmatising effects. The term 'bullying' is also particularly concerned with aggression from someone in a managerial or supervisory position (Zapf, 1999b).

Although, the concept of 'bullying' as used in English-speaking countries and the term 'mobbing' as used in many other European countries may have some semantic differences and connotations, to all intents and purposes they refer to the same phenomenon. Any differences in the use of the terms may be related as much to cultural differences in the phenomenon in the different countries than to real differences in the concepts. One may argue that while the term 'bullying' may better fit predatory kinds of situations, the term 'mobbing' may be more attuned to dispute-related cases. Then again, Scandinavian and German cases of bullying seem mainly to be of a dispute-related kind (Einarsen, *et al.*, 1994b; Leymann, 1996; Zapf and Gross, 2001). The strong focus on bullying by managers in the UK may indicate that predatory bullying is the most prevalent kind of bullying taking place in that particular culture. While managers are involved as perpetrators in 50 per cent of the cases in Norway, 80 per cent of the cases in the UK involve a manager or supervisor bullying a subordinate (Rayner *et al.*, 2002). In any case, the nature of the bullying situation seems to be the same – the persistent and systematic victimisation of a colleague or a subordinate through repeated use of various kinds of aggressive behaviours over a long period of time and in a situation where the victim has difficulty in defending him/herself.

Since much of the data on the process of bullying stems from the victims of long-lasting bullying, the processes described in this chapter may seem more deterministic than they are in real life. Most conflicts in organisations do not escalate into bullying. Not all cases of bullying will prevail for years, leading to severe trauma in the victim. However, so far we know little of the processes and factors involved when cases of bullying and potential bullying take alternative routes. New research efforts must therefore be directed at such issues.

Bibliography

Adams, A. (1992) *Bullying at work. How to confront and overcome it.* London: Virago Press.

Allport, G. (1954) *The nature of prejudice.* Reading, MA: Addison Wesley.

Archer, D. (1999) Exploring 'bullying' culture in the para-military organisation. *International Journal of Manpower, 20,* 94–105.

Ashforth, B. E. (1994) Petty tyranny in organizations. *Human Relations, 47,* 755–778.

Baron, R. A. and Neuman, J. H. (1996) Workplace violence and workplace aggression: Evidence on their relative frequency and potential causes. *Aggressive Behavior, 2,* 161–173.

Björkqvist, K. (1992) Trakassering förekommer bland anställda vid AA (Harassment exists among employees at Abo Academy). *Meddelande från Åbo Akademi, 9,* 14–17.

Björkqvist, K., Österman, K. and Hjelt-Bäck, M. (1994) Aggression among university employees. *Aggressive Behavior, 20,* 173–184.

Brodsky, C. M. (1976) *The harassed worker.* Lexington, MA: D. C. Heath and Company.

Cowie, H., Naylo, P., Rivers, I., Smith, P. K. and Pereira, B. (2002) Measuring workplace bullying. *Aggression and Violent Behavior, 7,* 33–51.

de Dreu, C. K. W. (1997) Productive conflict: The importance of conflict management and conflict issue. In C. K. W. de Dreu and E. van de Vliert (eds), *Using conflict in organizations* (pp. 9–22). London: Sage Publications.

Diergarten, E. (1994) *Mobbing. Wenn der Alltag zum Alptraum wird* (Mobbing: When everyday life becomes a nightmare). Cologne: Bund Verlag.

Dohrenwend, B. S. and Dohrenwend, B. P. (eds) (1974) *Stressful life events: Their nature and effects.* New York: Wiley.

Ege, H. (1996) *Mobbing. Che cosé il terrore psicologico sull posto di lavoro.* Bologna: Pitagora Editrice.

Einarsen, S. (1996) *Bullying and harassment at work: epidemiological and psychosocial aspects.* PhD thesis, University of Bergen: Department of Psychosocial Science.

—— (1998) Bullying at work: The Norwegian lesson. In C. Rayner, M. Sheehan and M. Barker (eds), *Bullying at work 1998 research update conference: Proceedings.* Stafford: Staffordshire University.

—— (1999) The nature and causes of bullying. *International Journal of Manpower, 20,* 16–27.

—— (2000a) Bullying and harassment at work: unveiling an organizational taboo. In M. Sheehan, S. Ramsey and J. Partick (eds), *Transcending boundaries. Integrating people, processes and systems.* Brisbane: The School of Management, Griffith University.

—— (2000b) Harassment and bullying at work: A review of the Scandinavian approach. *Aggression and Violent Behavior, 4,* 371–401.

Einarsen, S. and Hoel, H. (2001) The validity and development of the Revised Negative Acts Questionnaire. Paper presented at the 10th European Congress of Work and Organizational Psychology, Prague, 16–19 May.

Einarsen, S. and Raknes, B. I. (1991) *Mobbing i arbeidslivet* (Bullying at work). Research Centre for Occupational Health and Safety: University of Bergen, Norway.

—— (1997) Harassment at work and the victimization of men. *Violence and Victims*, *12*, 247–263.

Einarsen, S., Raknes, B. I. and Matthiesen, S. B. (1994a) Bullying and harassment at work and their relationships to work environment quality. An exploratory study. *The European Work and Organizational Psychologist*, *4*, 381–401.

Einarsen, S., Raknes, B. I., Matthiesen, S. B. and Hellesøy, O. H. (1994b) *Mobbing og harde personkonflikter. Helsefarlig samspill pa arbeidsplassen* (Bullying and severe interpersonal conflicts. Unhealthy interaction at work). Soreidgrend: Sigma Forlag.

Einarsen, S. and Skogstad, A. (1996) Prevalence and risk groups of bullying and harassment at work. *European Journal of Work and Organizational Psychology*, *5*, 185–202.

Field, T. (1996) *Bullying in sight*. Wantage: Success Unlimited.

Fitzgerald, L. F. and Shullman, S. (1993) Sexual harassment: a research analysis and agenda for the 1990's. *Journal of Vocational Behaviour*, *42*, 5–27.

Frese, M. and Zapf, D. (1987) Eine Skala zur Erfassung von Sozialen Stressoren am Arbeitsplatz (A scale for the assessment of social stressors at work). *Zeitschrift für Arbeitswissenschaft*, *41*, 134–141.

—— (1988) Methodological issues in the study of work stress: Objective vs. subjective measurement of work stress and the question of longitudinal studies. In C. L. Cooper and R. Payne (eds), *Causes, coping, and consequences of stress at work* (pp. 375–411). Chichester: Wiley.

Glasl, F. (1982) The process of conflict escalation and roles of third parties. In G. B. J. Bomers and R. Peterson (eds), *Conflict management and industrial relations* (pp. 119–140). Boston: Kluwer-Nijhoff.

—— (1994) *Konfliktmanagement. Ein Handbuch für Führungskräfte und Berater* (Conflict management. A handbook for managers and consultants), 4th edn. Bern: Haupt.

Heinemann, P. (1972) *Mobbning – Gruppvåld bland barn och vuxna* (Mobbing – group violence by children and adults). Stockholm: Natur and Kultur.

Hoel, H. and Cooper, C. L. (1999) *The role of 'intent' in perceptions of workplace bullying*. Presented at the 9th European Congress on Work and Organizational Psychology: Innovations for Work, Organization and Well-being. 12–15 May, Espoo-Helsinki, Finland.

—— (2000) *Destructive conflict and bullying at work*. Unpublished report, University of Manchester, Institute of Science and Technology.

—— (2001) Origins of bullying: Theoretical frameworks for explaining bullying. In N. Tehrani (ed.), *Building a culture of respect: Managing bullying at work* (pp. 3–19). London: Taylor and Francis.

Hoel, H., Rayner, C. and Cooper, C. L. (1999) Workplace bullying. In C. L. Cooper and I. T. Robertson (eds), *International review of industrial and organizational psychology*, vol. 14 (pp. 195–230). Chichester: Wiley.

Hofstede, G. (1980) *Culture's consequences*. New York: Sage Publications.

Hubert, A. B., and van Veldhoven, M. (2001) Risk sectors for undesired behaviour and mobbing. *European Journal of Work and Organizational Psychology*, *10*, 415–424.

Kanner, A. D., Coyne, J. C., Schaefer, C. and Lazarus, R. S. (1981) Comparison of two modes of stress measurement: Daily hassles and uplifts versus major life events. *Journal of Behavioral Medicine*, *4*, 1–39.

Kaucsek, G. and Simon, P. (1995) Psychoterror and risk-management in Hungary. Paper presented as a poster at the 7th European Congress of Work and Organizational Psychology, 19–22nd April, Györ, Hungary.

Kile, S. M. (1990) *Helsefarleg leiarskap. Ein eksplorerande studie* (Health endangering leadership. An exploratory study). University of Bergen: Department of Psychosocial Science.

Kirchler, E. and Lang, M. (1998) Mobbingerfahrungen: Subjektive Beschreibung und Bewertung der Arbeitssituation (Bullying experiences: Subjective description and evaluation of the work situation). *Zeitschrift für Personalforschung*, *12*, 352–262.

Lazarus, R. S. and Folkman, S. (1984) *Stress, appraisal and coping*. New York: Springer.

Lengnick-Hall, M. L. (1995) Sexual harassment research: a methodological critique. *Personnel Psychology*, *48*, 841–864.

Lewis, D. (2000) Bullying at work: A case of moral panic? In M. Sheehan, S. Ramsey and J. Partick (eds), *Transcending boundaries. Integrating people, processes and systems*. Brisbane: The School of Management, Griffith University.

—— (2001) Perceptions of bullying in organizations. *International Journal of Management and Decision Making*, 2, 1, 48–63.

Leymann, H. (1986) *Vuxenmobbning – psykiskt våld i arbetslivet* (Bullying – psychological violence in working life). Lund: Studentlitterature.

—— (1990a) *Handbok för användning av LIPT-formuläret för kartläggning av risker för psykiskt vald* (Manual of the LIPT questionnaire for assessing the risk of psychological violence at work). Stockholm: Violen.

—— (1990b) Mobbing and psychological terror at workplaces. *Violence and Victims*, *5*, 119–126.

—— (1992) *Från mobbning til utslagning i arbetslivet* (From bullying to exclusion fro working life). Stockholm: Publica.

—— (1993) *Mobbing – Psychoterror am Arbeitsplatz und wie man sich dagegen wehren kann* (Mobbing – psychoterror in the workplace and how one can defend oneself). Reinbeck bei Hamburg: Rowohlt.

—— (1995) Einführung: Mobbing. Das Konzept und seine Resonanz in Deutschland (Introduction: Mobbing. The concept and its resonance in Germany). In H. Leymann (ed.), *Der neue Mobbingbericht. Erfahrungen und Initiativen, Auswege und Hilfsangebote* (pp. 13–26). Reinbeck bei Hamburg: Rowohlt.

—— (1996) The content and development of mobbing at work. *European Journal of Work and Organizational Psychology*, *5*, 165–184.

Liefooghe, A. P. D. and Mackenzie Davey, K. (2001) Accounts of workplace bullying: The role of the organization. *European Journal of Work and Organizational Psychology*, *10*, 375–392.

Lorenz, K. (1968) Aggression. Dess bakgrund och natur (On aggression). Stockholm: Natur & Kultur.

Matthiesen, S. B., Raknes, B. I. and Røkkum, O. (1989) Mobbing på arbeidsplassen (Bullying in the workplace). *Tidsskrift for Norsk Psykologforening*, *26*, 761–774.

McCarthy, P., Sheehan, M. J. and Wilkie, W. (eds) (1996) *Bullying: From backyard to boardroom*. Alexandria, Australia: Millenium Books.

Mikkelsen, G. E. and Einarsen, S. (2001) Bullying in Danish work-life: Prevalence and health correlates. *European Journal of Work and Organizational Psychology*, *10*, 393–413.

—— (2002) Basic assumptions and symptoms of post-traumatic stress among victims of bullying at work. *European Journal of Work and Organizational Psychology*, *11*, 87–111.

Munthe, E. (1989) Bullying in Scandinavia. In E. Roland and E. Munthe (eds), *Bullying: An international perspective* (pp. 66–78). London: David Fulton Publishers.

Neuberger, O. (1999) *Mobbing. Übel mitspielen in Organisationen* (Mobbing. Playing bad games in organisations), 3rd edn. Munich and Mering: Hampp.

Niedl, K. (1995) *Mobbing/Bullying am Arbeitsplatz. Eine empirische Analyse zum Phänomen sowie zu personalwirtschaftlich relevanten Effekten von systematischen Feindseligkeiten* (Mobbing/bullying at work. An empirical analysis of the phenomenon and of the effects of systematic harassment on human resource management). Munich: Hampp.

Olweus, D. (1987) Bullying/victim problems among school children in Scandinavia. In J. P. Myklebust and R. Ommundsen (eds), *Psykologprofesjonen mot ar 2000* (pp. 395–413). Olso: Universitetsforlaget.

—— (1991) Bullying/victim problems among school children. In I. Rubin and D. Pepler (eds), *The development and treatment of childhood aggression* (pp. 411–447). Hillsdale, NJ: Erlbaum.

—— (1994) Annotation: Bullying at school – Basic facts and effects of a school based intervention program. *Journal of Child Psychology and Psychiatry*, *35*, 1171–1190.

O'Moore, M. (2000) *National Survey on bullying in the workplace*. Dublin: The Anti-Bullying Research Centre, Trinity College.

O'Moore, M., Seigne, E., McGuire, L. and Smith, M. (1998) Victims of bullying at work in Ireland. In C. Rayner, M. Sheehan and M. Barker (eds), *Bullying at work 1998 research update conference: Proceedings*. Stafford: Staffordshire University.

Pikas. (1989) The common concern method for the treatment of mobbing. In E. Roland and E. Munthe (eds), *Bullying: An international perspective* (pp. 91–104). London: David Fulton Publishers.

Pryor, J. B., LaVite, C. and Stoller, L. (1993) A social psychological analysis of sexual harassment: The person/situation interaction. *Journal of Vocational Behavior* (Special Issue), *42*, 68–83.

Rayner, C. (1997) The incidence of workplace bullying. *Journal of Community and Applied Social Psychology*, *7*, 199–208.

—— (1998) Workplace bullying: Do something! *The Journal of Occupational Health and Safety – Australia and New Zealand*, *14*, 581–585.

Rayner, C., Hoel, H. and Cooper, C. L. (2002) *Workplace bullying. What we know, who is to blame, and what can we do?*. London: Taylor & Francis.

Rayner, C., Sheehan, M. and Barker, M. (1999) *Bullying at work 1998 research update conference: Proceedings*. Stafford: Staffordshire University.

Roland, E. and Munthe, E. (eds) (1989) *Bullying. An international perspective*. London: D. Fulton.

Salin, D. (2001) Prevalence and forms of bullying among business professionals: A comparison of two different strategies for measuring bullying. *European Journal of Work and Organizational Psychology*, *10*, 425–441.

Schuster, B. (1996) Rejection, exclusion, and harassment at work and in schools. *European Psychologist*, *1*, 293–317.

Terpstra, D. E. and Baker, D. D. (1991) Sexual harassment at work: the psychosocial issues. In M. J. Davidson and J. Earnshaw (eds), *Vulnerable workers: Psychosocial and legal issues* (pp. 179–201). Chichester: Wiley.

Thylefors, I. (1987) *Syndabockar. Om utstötning och mobbning i arbetslivet* (Scapegoats. On expulsion and bullying in working life). Stockholm: Natur och Kulture.

UNISON (1997) *UNISON members' experience of bullying at work*. London: UNISON.

van de Vliert, E. (1998) Conflict and conflict management. In P. J. D. Drenth, H. Thierry and C. J. J. Wolff (eds), *Handbook of work and organizational psychology*, vol. 3: *Personnel Psychology*, 2nd edn (pp. 351–376). Hove: Psychology Press.

Vartia, M. (1991) Bullying at workplaces. In S. Lehtinen, J. Rantanen, P. Juuti. A. Koskela, K. Lindström, P. Rehnström and J. Saari (eds), *Towards the 21st century. Proceedings from the International Symposium on Future Trends in the Changing Working Life* (pp. 131–135). Helsinki: Institute of Occupational Health.

—— (1996) The sources of bullying – psychological work environment and organizational climate. *European Journal of Work and Organizational Psychology*, *5*, 203–214.

Williams, K. (1997) Social ostracism. In R. M. Kowalski (ed.), *Aversive interpersonal behaviors* (pp. 133–170). New York: Plenum Press.

Zapf, D. (1999a) Mobbing in Organisationen. Ein Überblick zum Stand der Forschung (Mobbing in organisations. A state of the art review). *Zeitschrift für Arbeits- and Organisationspsychologie*, *43*, 1–25.

—— (1999b) Organizational, work group related and personal causes of mobbing/bullying at work. *International Journal of Manpower*, *20*, 70–85.

Zapf, D. and Einarsen, S. (eds) (2001) Bullying in the workplace. Recent trends in Research and practice. *Special Issue of the European Journal of Work and Organizational Psychology*, vol. 10 (pp. 369–525). Hove: Psychology Press.

Zapf, D. and Gross, C. (2001) Conflict escalation and coping with workplace bullying: A replication and extension. *European Journal of Work and Organizational Psychology*, *10*, 497–522.

Zapf, D. and Leymann, H. (eds.) (1996) Mobbing and victimization at work. *Special Issue of the European Journal of Work and Organizational Psychology*, vol. 5 (pp. 161–307). Hove: Psychology Press.

Zapf, D., Knorz, C. and Kulla, M. (1996) On the relationship between mobbing factors, and job content, the social work environment and health outcomes. *European Journal of Work and Organizational Psychology*, *5*, 215–237.

2 By any other name

American perspectives on workplace bullying

Loraleigh Keashly and Karen Jagatic

Introduction

The intent of this chapter is to share research findings, as well as key conceptual and methodological challenges, from the North American literature on hostile behaviours at work that we believe are relevant to workplace bullying. One of the challenges we face in undertaking this task is that research on hostile workplace behaviours can be found in a variety of literatures (organisational/management, public health, substance abuse, sociological, medical, nursing, legal, and educational, to name a few), and under a variety of names, such as 'workplace aggression' (Neuman and Baron, 1997), 'workplace incivility' (Andersson and Pearson, 1999), 'emotional abuse' (Keashly, 1998), 'generalised workplace abuse' (Richman *et al.*, 1999) and 'workplace harassment' (Brodsky, 1976). Our survey is limited to research in which the behaviours and the relationships under investigation:

1 include non-physical forms of hostility and aggression (i.e. this review is not about physical violence at work, but about aggression in various forms);
2 do not include the sexual harassment literature as that is comprehensively discussed elsewhere (e.g. Fitzgerald and Shullman, 1993);
3 focus on actions that occur between organisational insiders, such as co-workers, supervisors and subordinates;
4 cause harm, broadly defined (i.e. physical, psychological, individual, organisational).

Even within these parameters, we are aware that our survey is not exhaustive of all relevant literature on the American side of the pond; rather, it is exemplary of it.

A secondary intent of this chapter is to facilitate awareness of and communication among the various literatures that speak to hostile workplace interactions. A serious difficulty is that a number of literatures focused on hostile workplace behaviours appear to have developed in parallel to one another with little interconnection. For example, little interchange has occurred between the workplace aggression literature as presented in organisational psychology and management research domains (e.g. Andersson and Pearson, 1999; Robinson and Bennett, 2000; Giacalone and Greenberg, 1997; Glomb, 2001, in press; Neuman and Baron, 1997; O'Leary-Kelly *et al.*, 1996), and the workplace abuse and harassment research from areas such as substance abuse, organisational development, medicine, nursing, conflict, psychology and sociology (e.g. Graydon *et al.*, 1994; Keashly, 1998; Price Spratlen, 1995; Richman *et al.*, 1999; Rospenda *et al.*, 2000; Schneider *et al.*, 2000a). Further, related literatures such as on workplace conflict and social influence are not often referenced in either of these bodies of research. However, their concepts, which include coercive influence, escalatory processes, intervention and management approaches (Tedeschi and Felson, 1994), are highly relevant. In addition, the gap between the research literature and the more practice-oriented literature (as reflected primarily in popular writing) is also wide, although, fortunately, it is being bridged by efforts on both sides (e.g. Bassman, 1992; Davenport *et al.*, 1999; Lewis and Zare, 1999; Namie and Namie, 2000; Wright and Smye, 1996). This concern of parallel literatures is also true for the limited North American familiarity with the more global workplace bullying literature emerging from Europe, the UK and Australia. As a result, we have found some duplication of effort both conceptually and empirically has occurred that may have been alleviated by familiarity with each other's work and thinking. Our hope is that by discussing some of these various works, this chapter will serve a catalytic function of moving thinking and action with respect to hostile workplace interactions forward in a more co-ordinated fashion.

The labelling and definitional dilemma

The variety of constructs falling under the rubric of hostile workplace behaviours has expanded significantly in the past six years due to a welcome increase in research attention (Keashly, 1998). In Box 2.1, we have provided a sampling of definitions from the most recent work. While workplace deviance encompasses both hostile behaviours and other forms of deviant workplace behaviours (e.g. theft, sabotage), the other definitions focus on types and combinations of hostile behaviours that occur at work to varying degrees. Beyond the variability in breadth and type of behaviours, these definitions differ most fundamentally in the acknowledgement of contextual features related to the 'experience' of these behaviours. More specifically, these definitions vary in the degree to which elements of time, intention, power differences, and norm violation are incorporated as central features of the construct they reflect (Keashly, 1998). These features are key elements of the definition of workplace bullying used in this book. Thus, the direct applicability of North

Box 2.1: Select definitions of hostile workplace behaviours

Harassment (Brodsky, 1976)

'repeated and persistent attempts by one person to torment, wear down, frustrate or get a reaction from another. It is treatment which persistently provokes, pressures, frightens, intimidates, or otherwise discomforts another person' (p. 2).

Workplace deviance (Robinson and Bennett, 1995)

'voluntary behavior that violates significant organizational norms and, in so doing, threatens the well-being of the organization or its members, or both' (p. 555).

Workplace aggression (Baron and Neuman, 1996)

'efforts by individuals to harm others with whom they work, or have worked, or the organizations in which they are currently, or were previously, employed. This harm-doing is intentional and includes psychological as well as physical injury' (p. 38).

Generalised workplace abuse (Richman *et al.*, 1997)

'violations of workers' physical, psychological and/or professional integrities ... nonsexual yet psychologically demeaning or discriminatory relationships' (p. 392).

Workplace incivility (Andersson and Pearson, 1999)

'low-intensity deviant behavior with ambiguous intent to harm the target, in violation of workplace norms for mutual respect. Uncivil behaviors are characteristically rude and discourteous, displaying a lack of regard for others' (p. 457).

Abusive supervision (Tepper, 2000)

'subordinates' perceptions of the extent to which supervisors engage in the sustained display of hostile verbal and nonverbal behaviors, excluding physical contact' (p. 178).

Ethnic harassment (Schneider *et al.*, 2000a)

'threatening verbal conduct or exclusionary behavior that has an ethnic component and is directed at a target because of his or her ethnicity ... behaviors that may be encountered on a daily basis and may contribute to a hostile environment, particularly for ethnic minorities' (p. 3).

Workplace bullying (Namie and Namie, 2000)

'is the deliberate, hurtful and repeated mistreatment of a Target (the recipient) by a bully (the perpetrator) that is driven by the bully's desire to control the Target ... encompasses all types of mistreatment at work ... as long as the actions have the effect, intended or not, of hurting the Target, if felt by the Target' (p. 17).

Emotional abuse at work (Keashly, 2001)

'interactions between organizational members that are characterized by repeated hostile verbal and nonverbal, often nonphysical behaviors directed at a person(s) such that the target's sense of him/herself as a competent worker and person is negatively affected' (p. 212).

American research findings on hostile workplace behaviours to the discussion of workplace bullying will be dependent upon the consideration and assessment of these elements in this research. Thus, we have structured our review in terms of these fundamental elements.

We will begin with a brief description of the range and types of behaviours referred to in various literatures to give a sense of what has been considered as hostile or aggressive conduct. From there, we will focus on the notion of hostile relationships as revealed by attention to single versus patterned and persistent hostile behaviours. This distinction has important implications for the degree to which research on understanding hostile incidents can be extended to understanding hostile relationships such as workplace bullying. We will then discuss research evidence regarding the three other contextual elements of intent, power and norm violation. Intent has been critical in many definitions yet is rarely ever assessed empirically. Any discussion of relationship requires an exploration of the role of power in the experience. We will focus on moving from the simplistic assessment of relative power to include the dynamics of power. The element of norm violation will be examined by reviewing evidence on organisational culture and hostile work climate that suggest these behaviours and relationships may actually be consistent with norms rather than deviant from them. We will then turn our attention to the effects of these hostile behaviours and relationships by summarising the findings on the range and types of harm. Throughout these sections we will identify methodological and measurement challenges in the research on hostile workplace interactions.

The behavioural domain: actual conduct

For quite some time, research attention was focused almost exclusively on workplace violence (e.g. homicide, assault; Baron and Neuman, 1996; Glomb, 2001). This attention has been fuelled by several high-profile workplace shootings and schoolyard killings that have occurred in the USA, particularly in recent years (Denenberg and Braverman, 1999). It has been with a sense of shock that Americans have come to realise that workplaces and schools are not the safe places they were assumed to be. Journalistic analyses of these incidents have found evidence in some cases that the actors had been marginalised, harassed or bullied by other workers and/or managers over an extended period of time prior to their deadly acts. For instance, in early April 1999, Pierre LeBrun shot and killed four co-workers and then shot himself at a public transportation facility in Ottawa, Ontario, Canada (McLaughlin, 2000). The coroner's inquest revealed that LeBrun had been mocked and teased over the years for stuttering, and that the company had done little in response to his complaints. This identification of hostile mistreatment by co-workers as contributing to violent episodes has resulted in increased media attention to and public awareness of bullying as a workplace phenomenon.

In recent years, there has been a growing recognition that physical violence was merely the 'tip of the iceberg' concerning hostile behaviour at work (Neuman and Baron, 1997). This expansion of the range and type of hostile behaviours studied was spurred by Baron and Neuman's (1996) application of Buss's (1961) conceptualisation of human aggression to hostile workplace behaviours. Buss (1961) argued that aggressive behaviour could be conceptualised along three dimensions: physical–verbal, active–passive and direct–indirect. When fully crossed, there are eight categories of behaviour. We have used this framework to categorise the types of behaviour that have been investigated in the literature that we reviewed (see Table 2.1). Several studies utilising this expanded domain (e.g. Greenberg and Barling, 1998; Neuman and Baron, 1997; Keashly and Rogers, 2001) have documented what has been long known by workers: most hostile behaviour in the workplace is verbal, indirect and passive. These types of behaviour have been variously labelled as 'psychological aggression' (Barling, 1996), 'emotional abuse' (Keashly, 1998) or 'generalised workplace abuse' (Richman *et al.*, 1999). As a result of the accumulating evidence that workplace hostility has many behavioural faces, most research on hostile behaviour at work in the USA and Canada now examines a variety of behaviours reflecting these dimensions concurrently (e.g. Glomb, 2001; Richman *et al.*, 1999; Keashly and Rogers, 2001; Schneider *et al.*, 2000b). This broadened focus enhances the applicability of North American research on hostile workplace behaviours to workplace bullying.

Incidence of hostile workplace behaviours

Despite the diversity of definitions, we believe there is sufficient overlap in the behaviours studied to hazard sharing some incidence data regarding hostile behaviour in American workplaces. The Northwest National Life Insurance survey (1993) is frequently cited as an illustration of the pervasiveness of aggression in American workplaces. Based on a sample of 600 full-time workers, the researchers concluded that one in four workers reported being harassed, threatened or physically attacked on the job in the previous twelve months. More recently, in a state-wide survey of Michigan residents, Keashly and Jagatic (2000) reported that about 59 per cent of the representative working sample indicated they had experienced at least one type of emotionally abusive behaviour at the hands of fellow workers. Almost 20 per cent of the sample reported they had been exposed to five or more different emotionally abusive behaviours. Regarding a more general experience of 'mistreatment', 27 per cent reported being mistreated by a fellow worker in the previous twelve months (62 per cent were notably bothered by the treatment) and 42 per cent (64 per cent

Table 2.1 Examples of workplace abuse and aggression behaviours

Behavioural category	*Behavioural examples*	*Source*
Verbal /active /direct	Name calling, use of derogatory terms	Baron and Neuman (1996); Robinson and Bennett (2000); Cortina *et al.* (2001); Glomb (2001); Price Spratlen (1995); Keashly and Rogers (2001); Tepper (2000)
	Subject to insulting jokes	Baron and Neuman (1996); Glomb (2001); Keashly and Jagatic (2000)
	Belittled intellectually, talked down to	Cortina *et al.* (2001); Keashly and Jagatic (2000); Price Spratlen (1995); Richman *et al.* (1997); Keashly and Rogers (2001); Tepper (2000)
	Criticised harshly, attacked verbally in private or public; put down in front of others	Aquino *et al.* (1999); Cortina *et al.* (2001); Keashly and Jagatic (2000); Price Spratlen (1995); Richman *et al.* (1997); Robinson and Bennett (1995); Keashly and Rogers (2001); Tepper (2000)
	Sworn at	Aquino *et al.* (1999); Glomb (2001); Keashly and Jagatic (2000); Price Spratlen (1995); Richman *et al.* (1997); Keashly and Rogers (2001)
	Lied to, deceived	Aquino *et al.* (1999); Robinson and Bennett (1995); Tepper (2000)
	Yelled at, shouted at	Baron and Neuman (1996); Keashly and Jagatic (2000); Price Spratlen (1995); Richman *et al.* (1997); Keashly and Rogers (2001)
	Interrupted when speaking, working	Glomb (2001); Keashly and Jagatic (2000); Richman *et al.* (1997)
	Pressured to change personal life, beliefs, opinions	Cortina *et al.* (2001); Glomb (2001)
	Flaunting status	Baron and Neuman (1996); Keashly and Jagatic (2000)
Verbal /active /indirect	Treated unfairly	Robinson and Bennett (1995)
	Subject to false accusations, rumours	Aquino *et al.* (1999); Baron and Neuman (1996); Keashly and Jagatic (2000); Richman *et al.* (1997); Robinson and Bennett (1995); Keashly and Rogers (2001); Tepper (2000)
	Attempts made to turn others against the target	Glomb (2001); Keashly and Rogers (2001); Tepper (2000)
Verbal /passive /direct	You or your contributions ignored; silent treatment	Aquino *et al.* (1999); Baron and Neuman (1996); Cortina *et al.* (2001); Keashly and Jagatic (2000); Price Spratlen (1995); Richman *et al.* (1997); Keashly and Rogers (2001); Tepper (2000)

Verbal /passive /indirect	Had memos, phone calls ignored	Baron and Neuman (1996); Keashly and Jagatic (2000); Keashly and Rogers (2001)
	Been given little or no feedback, guidance	Keashly and Jagatic (2000)
	Deliberately excluded	Cortina *et al.* (2001); Keashly and Jagatic (2000); Price Spratlen 1995); Richman *et al.* (1997); Keashly and Rogers (2001); Schneider *et al.* (2000a)
	Failing to pass on information needed by the target	Baron and Neuman (1996)
Physical /active /direct	Glared at	Glomb (2001); Keashly and Jagatic (2000); Robinson and Bennett (1995); Keashly and Rogers (2001)
	Physically assaulted (e.g. kicked, bitten, hit)	Aquino *et al.* (1999); Baron and Neuman (1996); Glomb (2001); Keashly and Jagatic (2000); Robinson and Bennett (1995); Keashly and Rogers (2001)
	Subject to sexual harassment	Bennett and Robinson (2000); Keashly and Jagatic (2000); Robinson and Bennett (1995); Keashly and Rogers (2001)
	Subject to racial harassment	Aquino *et al.* (1999); Bennett and Robinson (2000); Keashly and Jagatic (2000); Schneider *et al.* (2000a, 2000b).
Physical /active /indirect	Theft or destruction of property	Aquino *et al.* (1999); Baron and Neuman (1996); Glomb (2001); Keashly and Jagatic (2000); Robinson and Bennett (1995); Keashly and Rogers (2001)
	Deliberately assigned work overload	Robinson and Bennett (1995)
	Deliberately consuming resources needed by target	Baron and Neuman (1996)
Physical /passive / indirect	Expected to work with unreasonable deadlines, lack of resources	Robinson and Bennett (1995); Keashly and Rogers (2001)
	Causing others to delay action on matters of importance to target	Baron and Neuman (1996)

notably bothered) reported being mistreated at some point in their working career. In their study of workplace incivility, Cortina *et al.* (2001) found that 71 per cent of employees in a federal court system reported some experience with these behaviours in the previous five years.

These data are gathered from individuals who have experienced these behaviours. Of particular interest is incidence from the perspective of engaging in hostility. Greenberg and Barling (1998) found that over 75 per cent of male workers sampled reported engaging in some form of psychologically aggressive behaviours against their fellow employees. Similar rates were found in Glomb's (2002) study of specific angry incidents in which participants reported on their own behaviour. For incidence information at a more behavioural level, we have provided rates of specific behaviours from several studies in Table 2.2. It is clear from these different operationalisations of incidence that hostile workplace behaviours of a variety of forms, as well as behaviours identified as bullying, are a notable part of many people's working experiences in North America.

Discrete event or pattern: hostile relationships

The critical dimension in extending North American research findings to workplace bullying involves the issue of hostility as a discrete event versus a pattern of events. It is with this dimension that we found the greatest difference among concepts utilised in the literature. For example, workplace aggression theorising has tended to focus implicitly, if not explicitly, on understanding single incidents or aggregate levels of aggression (Glomb, 2002; Neuman and Baron, 1997; O'Leary-Kelly *et al.*, 1996). Generally, what has been largely absent from this literature has been the notion of hostile acts occurring over time (i.e. persistent hostility; see Tepper, 2000 and Glomb, 2002, for exceptions). Conversely, the workplace harassment and abuse literatures (e.g. Keashly, 2001; Namie and Namie, 2000; Richman *et al.*, 1999, 2001) have focused exclusively on persistent hostility.

Persistent hostility can be conceived as having four facets: repetition (frequency), duration (over some extended time period), pattern (variety of behaviours co-occurring) and escalation (moving from less to more severe behaviours).

Repetition and duration

Hostile workplace behaviours have been operationalised in both literatures by their frequency over some specified time period. Thus, even though these two literatures differ conceptually in how central the facet of time is to their particular constructs, they both measure *repetitive* hostile behaviours.

Although the facet of *duration* is included in definitions of workplace abuse (e.g. Keashly, 2001; Namie and Namie, 2000; Tepper, 2000), it has been essentially ignored from a measurement perspective in both literatures. Duration appears primarily as a timeframe over which respondents are asked to assess the frequency of their experience (e.g. in the past twelve months; two years; five years). In fact, with the noted exception of Robinson and Bennett's (2000) measure of workplace deviance, the anchors on the frequency scales are not tied to specific time referents such as monthly, weekly or daily. Such specificity would be necessary for determining duration of exposure.

The importance of duration of exposure to hostile behaviour is highlighted by three recent studies. First, in a web-based survey of people who self-identified as having been bullied at work, Namie (2000) reported the average exposure to bullying behaviours as 16.5 months. This is a much longer time period than the one-year timeframe most of the workplace aggression and abuse research typically uses. Second, in a longitudinal study of university employees, Rospenda *et al.* (2000) assessed exposure to generalised workplace abuse and sexual harassment over a two-year period by surveying respondents at two points, one year apart. Participants who reported exposure to generalised workplace abuse at both time periods (i.e. two years' duration) were found to be more likely to report one or more indicators of problem drinking at time 2, compared to participants who experienced abuse only at one time or not at all. Third, in their study of women exposed to multiple forms of harassment, Schneider *et al.* (2000b) found that the longer the period over which they experienced the behaviours, the more upset they became. Based on this limited evidence, it is clear that duration of exposure is a key aspect of the experience of hostile behaviour. This relative lack of attention to duration clearly limits the applicability of findings to workplace bullying as defined and measured in the European literature (Leymann, 1990).

Patterning and escalation

The frequency indicator as currently utilised also does not capture the element of a *pattern* of hostility. This would require a recognition that behaviours may co-occur and may be sequenced, as opposed to occurring independently or in isolation (Glomb and Miner, in press; Keashly and Rogers, 2001; Schneider *et al.*, 2000b). For example, Barling (1996) suggests that just as with family violence, psychological aggression may well precede, and therefore be predictive of, physical violence in the workplace. Research assessing a variety of forms of harassment concurrently (e.g. Richman *et al.*, 1999; Keashly and Rogers, 2001; Rospenda *et al.*, 2000; Schneider *et al.*, 2000b) recognises this idea of variety and the potential for a pattern. For example, Schneider *et al.* (2000b) found that employees who experienced one form of harassment were also likely to experience other forms of harassment. In fact, abusive *relationships* may

Table 2.2 Incidence of workplace aggression items

Study	*Behaviour*	*Incidence rates (per cent)*	
Robinson and Bennett (2000)	1 Made fun of someone at work	57.3	
(self-report)	2 Said something hurtful to someone at work	38.0	
(n 352 full-time employees, and n 133; per cent includes once to daily; 1 yr timeframe)	3 Made an ethnic, religious or racial remark or joke at work	43.0	
	4 Cursed at someone at work	29.8	
	5 Lost your temper while at work	69.2	
	6 Played a mean prank on someone at work		7.3
	7 Acted rudely toward someone at work	39.6	
	8 Publicly embarrassed someone at work	10.8	
Glomb (2002)		*Actor*	*Target*
(self-report)	1 Using an angry tone of voice	74.5	83.2
(n 74 from 3 organisations; per cent of respondents who were target and actor, who displayed behaviour during single incident)	2 Dirty looks /angry facial expressions	64.5	82.5
	3 Sabotaging someone's work	5.6	25.6
	4 Making angry gestures	44.6	65.5
	5 Avoiding another person	67.2	65.9
	6 Making another person look bad	16.7	56.1
	7 Yelling or raising your voice	56.7	71.3
	8 Withholding information from another person	21.9	62.7
	9 Swearing at another person	28.3	44.9
	10 Talking behind someone's back	58.2	77.0
	11 Withholding resources	9.9	38.5
	12 Physically assaulting another	3.4	6.0
	13 Using hostile body language	22.1	41.1
	14 Insulting, criticising another (including sarcasm)	49.5	62.0
	15 Failing to correct false information about another	16.1	46.6
	16 Interrupting /'cutting off' a person while speaking	51.9	70.5
	17 Getting 'in someone's face'	22.5	43.3

Study	Behaviour			
	18	Spreading rumours	24.1	50.0
	19	Throwing something	16.1	18.3
	20	Making threats	9.7	23.9
	21	Damaging another person's property	3.1	11.6
	22	Whistle-blowing or telling supervisors about the negative behaviour of others	33.4	40.7
	23	Belittling another's opinions in front of other people	22.0	43.5
	24	Flaunting status or power over another	13.8	48.9
Keashly and Jagatic (2000) *(n 689; general working population; per cent includes sometimes to very often; target perspective; 1 yr timeframe)*	1	Glared at you in a hostile manner	21.5	
	2	Flaunted their status	24.3	
	3	Ignored you or your contributions	22.7	
	4	Talked down to you	20.5	
	5	Interrupted /prevented you from expressing yourself	19.6	
	6	Failed to return phone calls /respond to memos	18.1	
	7	Gave you the silent treatment	17.1	
	8	Blamed you for others' errors	21.2	
	9	Verbally attacked you	13.0	
	10	Spread gossip /rumours about you	11.6	
	11	Failed to deny false rumours about you	10.6	
	12	Put you down in front of others	9.0	
	13	Yelled /shouted at you in a hostile manner	9.3	
	14	Excluded you from important activities /meetings	8.4	
	15	Swore at you in a hostile manner	7.1	
	16	Gestured at you obscenely /offensively	6.0	
	17	Played mean pranks on you	6.8	

only become clear when whole patterns of behaviour are examined (Bassman, 1992). In the related literature on perceptions of justice, Lind (1997) argues that people's judgements of fair or unfair treatment are based on the patterning of everyday social interaction. As an illustration of the 'whole being more than a sum of its parts' notion of hostile relationships, Keashly (2001) found that while the respondents could articulate specific and discrete examples of behaviour, they struggled with communicating that their experience was more than this specific focus suggested. It was necessary for participants to describe the entire set of behaviours and their interrelationships.

A relatively simple indicator of patterning arising from frequency data involves counting the number of different events that have occurred to each respondent. A few authors have used this indicator to illustrate patterning (Glomb, 2002; Keashly and Jagatic, 2000; Keashly *et al.*, 1994), although they have interpreted its meaning differently. For example, while we have defined it as one of a pattern or a variety of behaviours, Glomb (2002) views it as a measure of severity, and ties it to escalatory sequencing by examining both the number and the forms of behaviour involved. She then classifies study respondents as to whether they report psychologically aggressive behaviours, physically aggressive behaviours, or a combination of both, varying in severity. This measure of patterning is limited, however, as it does not allow the differentiation between situations of multiple behaviours occurring within a single incident versus multiple incidents. This distinction between single-incident and enduring hostile interactions is important as these are qualitatively different phenomena that may have different antecedents and outcomes (Glomb, 2002; Keashly, 1998).

Glomb's (2002) work forms the bridge to the fourth facet of aggression over time – *escalation*. While escalation is a key aspect of the conflict literature (e.g. Rubin *et al.*, 1994), it has only recently made its way into the research literature on workplace aggression (Andersson and Pearson, 1999; Baron and Neuman, 1996; Bassman and London, 1993; Glomb, 2001; Glomb and Miner, in press). Discussions of escalation have implicit in them assumptions of dynamic interaction between an actor(s) and a target, mutuality of these actions and increasing severity of behaviour. Andersson and Pearson (1999) describe an incivility spiral in which parties start out with a retaliatory exchange of uncivil behaviours ('tit for tat') until one party perceives that the other's behaviour directly threatens his or her identity (i.e. the tipping point). At this point, the parties engage in increasingly more coercive and severe behaviours with presumably greater risk of injury.

Glomb and Miner (in press) make distinctions between independent aggressive acts and escalatory or sequenced aggression, with the latter focused on the movement from low-level, relatively minor behaviours to more severe and damaging behaviours such as physical violence. This recent innovation in the workplace aggression and abuse literatures links to some extent with European research on the evolving process of workplace bullying. This evolving process involves an escalation of the bully's behaviours from indirect and subtle behaviours, to more direct psychologically aggressive acts, to ultimately severe psychological and physical violence (e.g. Einarsen, 1999).

While the discussion of dynamic escalatory processes in the workplace aggression literature focuses attention on the missing element of the relationship between the parties, it appears to imply the ability of both parties *mutually* to engage in negative behaviours. Glomb (2001) argues that actors become targets and targets become actors such that there are no pure actors or pure targets; rather, there are actor-targets. This interpretation is also reflected in research examining the elements of victims' personality (e.g. negative affectivity, conflict management style) that may relate to their experiences of being treated aggressively, thus highlighting the victim's role as a precipitator of and active contributor to events (Aquino, 2000; Aquino *et al.*, 1999). This assumption of mutuality in the escalatory process is consistent with the literature on escalated conflict where at some point, regardless of who initiated the actions, the actions become increasingly reciprocated and occur at more severe levels (e.g. Rubin *et al.*, 1994). However, this interpretation stands in sharp contrast to the construct of workplace bullying which involves a clearly identified actor (bully) and an identifiable target (victim). Indeed, it could be argued that what is being described in the escalation of aggression is the escalation of a conflict, rather than a situation of the escalatory processes involved in workplace bullying. In this latter situation, the actor is primarily the provocateur, while the target is attempting to defend or deter further behaviour (coping and management behaviours), rather than to initiate or retaliate (see Einarsen, 1999). This focus on mutuality may well be an artifact of having focused on anger-related or reactive aggression that is reciprocal in nature. This is in contrast to more proactive and purposeful instigating of aggression, which tends to be more unidirectional and which is more consistent with descriptions of workplace bullying (Opotow, 2000).

Measurement and methods of studying hostile relationships

In the American literature, the reliance on aggregate measures of hostility does not allow for the differentiation between hostile behaviours on the part of different actors or multiple behaviours on the part of a single actor. Indeed, much of workplace aggression and abuse research utilising behavioural checklists does not specifically focus on aggressive behaviours from particular actors (e.g. Ashforth, 1997; Cortina *et al.*, 2001; Keashly *et al.*, 1994; Keashly *et al.*, 1997; Tepper, 2000). This makes it difficult to operationalise facets of pattern and escalation which require the establishment and assessment of a specific relationship between actors and targets. In addition, it is not unreasonable to suppose that a variety of behaviours coming from one actor may be experienced differently than behaviours coming from a variety of actors. Indeed, simply measuring frequency without relating it to (a) specific actor(s) is measuring something much different, i.e. hostile workplace climate (Schneider *et al.*, 2000b) or ambient hostility (Robinson and O'Leary-Kelly, 1998).

To achieve a better understanding of these more specific hostile relationships, and indeed the 'experience' of hostile behaviours generally, more qualitative methods, such as interviews focused on the details of specific incidents and relationships, are more appropriate than the broader survey methodology (Glomb, 2002; Keashly, 1998). For example, Keashly (2001) used semi-structured interviews to probe targets about their experiences with abusive colleagues. She focused on identifying those features of the hostile interaction that lead targets to label their experiences as abusive. Among other findings, this study revealed that prior history with the actor (i.e. an enduring relationship) was key in targets' labelling of their experience. Interestingly, respondents considered the actor's history with other targets in the workplace in their assessment of their experience as well. This finding is consistent with Namie's (2000) survey of targets that found that 77 per cent of actors were identified as having also harassed others. Further, Schneider *et al.* (2000b) found that having been a bystander to others' harassment experience was related to a degree of upset at one's own experience of harassment. This aspect of actor history with other targets highlights the importance of exploring more fully the persons involved in these behaviours.

With regard to specific incident descriptions, Keashly and Rogers (2001) had respondents complete checklists on each of five working days. Respondents were then interviewed in detail about one or two particular incidents. Interviewees were asked about the status and gender of the actor, the history with the actor, the importance and salience of the situation, as well as how stressful the situation was. This information permitted them to explore the role that features of the hostile relationship play in respondents' appraisal of these incidents as threatening and stressful. Using a similar method, Glomb (2002) had participants discuss actual 'angry incidents' that they were *involved* in, as either the target or the actor. They were questioned about personal characteristics, job attitudes and experience, and interpersonal conflict, anger and aggression. Individuals were then asked to select a particular incident that affected them the most ('specific aggressive incident'), and to speak in detail about the people involved, their actions and the outcomes of these actions. The information garnered suggests a progression of hostility such that severe behaviours are preceded by 'less severe' behaviours. These results would not have been revealed by sole reliance on the more aggregate measures of hostile workplace behaviours.

A very recent methodological innovation of computer modelling of complex non-linear systems has been utilised specifically to address questions about the patterns of relationships among different types of hostile behaviours and how these patterns unfold over time (i.e. escalation; Glomb and Miner, in press). In brief, data from real organisations are compared with computer-model virtual organisations. The virtual organisation data are based on different theoretical models of hostile workplace behaviours that hypothesise varying interconnections between key attitudinal antecedents such as negative affect and the hostile behaviours. The results of these different models are then compared to the interconnections within the actual data to attempt to discern which theoretical model best fits and, hence, best describes the empirical data. The models examined varied in the degree to which and the manner in which various hostile behaviours were linked (e.g. independent to sequenced escalation).

Glomb and Miner gathered empirical data via questionnaires and included a behavioural checklist of twenty-four hostile behaviours. A valuable innovation was the inclusion of a scale developed to assess, rather than assume, the extent to which each of the behaviours would be considered aggressive. While noting that the various hostile behaviours were clearly intercorrelated, the results of this study did not find strong support for any of the models. The authors suggest this may reflect difficulties in setting the original parameters regarding timeframe and sources of variability in the empirical data not present in the simulated data. The important point in sharing Glomb and Miner's (in press) approach is to illustrate how facets of patterning and escalation are finally being conceptually acknowledged and empirically addressed in the workplace aggression and abuse literatures. What is clear in examining the various methods is that a comprehensive picture of hostile workplace relationships and the experience of hostile interactions requires the use of multiple methodologies (Glomb, 2002; Keashly, 1998).

It is also important to address the notion of severity or intensity of the hostile behaviours. A consensus seems to have developed that these various forms of hostile behaviours arrange themselves along a continuum of increasing severity. For example, behaviours captured under the rubrics of 'emotional abuse', 'psychological aggression' and 'incivility' are often characterised as low intensity or low level (see Andersson and Pearson, 1999; Glomb, 2002) relative to forms of physical violence such as assault, rape and homicide, which are considered 'extreme' forms. While we do not necessarily disagree with this depiction, this assignment of varying degrees of severity is rarely based on empirical evidence. Two recent exceptions are the work of Glomb and Miner (in press) and Rogers (1998).

Glomb and Miner included ratings of severity of each of eighteen behaviours when assessing patterning of hostile behaviours within computer models. The ranking of these behaviours by severity ratings was consistent with what has been assumed in the literature. More specifically, talking behind someone's back was rated as less severe, while angry gestures, insulting/criticising and yelling were rated as moderately severe. Making threats and physical assaults were rated as most severe. In another study, Rogers (1998) had a group of people with work experience rate forty-nine physically, emotionally and sexually abusive behaviours on a variety of dimensions, including severity. She found that while all emotionally abusive and sexually harassing behaviours tended to be rated as less severe than the physical behaviours of rape, homicide and assault, some emotionally abusive items had similar ratings to some physical items (e.g. being publicly belittled, verbally insulted versus bumped into, grabbed and subjected to threatening gestures). Thus, severity did vary across types, though it also varied within each type of hostile behaviour. Further, these types of behaviour overlapped in terms of severity ratings. This is a much more complex understanding of severity than has been assumed thus far in discussions of workplace aggression and abuse.

Intent (harm by design)

A cornerstone of the definition of workplace aggression is that the behaviour(s) must be intended to cause harm, thus distinguishing it from behaviours that may cause harm but were not intended to do so (accidental; Neuman and Baron, 1997). In the current definition of workplace bullying, intent to harm is revealed in the reference to deliberate or premeditated action. However, other authors suggest that intent is not a defining element. To the extent there are systemic or structural aspects to these behaviours or relationships, as with racism and sexism, the actors may be conforming to broader norms without intending harm (Bassman, 1992; Keashly, 1998; 2001; Richman *et al.*, 2001; Wright and Smye, 1996). Tepper (2000) suggests that some elements of abusive supervision do not involve hostility and desire to harm, but are more a result of indifference. Similarly, Jagatic and Keashly (1998), in a preliminary study of graduate experiences of faculty hostility, found that students tended to report more behaviours involving neglect, rather than actively hostile behaviours. This is not to suggest these behaviours do not cause harm; rather, the intent to harm is not clear or is non-existent. Indeed, Andersson and Pearson (1999) distinguish workplace incivility from workplace aggression by arguing they are referring to behaviour for which intent is ambiguous. From the perspective of the targets, Keashly (2001) found that intent did not figure prominently in their experience of feeling abused.

While this conceptual debate continues, the issue is moot at the measurement level. Rarely do any of the workplace aggression and abuse studies measure either actual or perceived intent. It appears as though intent is considered implicit in the particular behaviours studied. However, it could be argued that many of the behaviours encompassed under workplace aggression and abuse are ambiguous for intent. Rather, it is the context in which the behaviours occur (e.g. patterning, duration, relationship to actor, organisation norms) that affect attribution of intent (Keashly 2001; Keashly *et al.*, 1994). This attribution in turn affects both researchers' and workers' assessments as to whether the behaviour is experienced as hostile. The concept of threat appraisal of stressors from the organisational stress literature (e.g. Barling, 1996; Lazarus and Folkman, 1984) could be extremely useful in helping to discern which factors relate to perception of intent to harm and how this attribution subsequently relates to the subjective experience of the hostile event and the outcomes of that experience (Barling, 1996). Regarding the latter connection, Keashly and Rogers (2001) found that those incidents in which the actor was perceived as intending harm were evaluated as more threatening, and therefore more hostile, than those where no intention was perceived. This appraisal was subsequently related to greater perceived stress.

Another important aspect to consider when discussing intent is the distinction between intent and motive. While many behaviours may be intended to harm someone, the motive for that harm may be different. For example, Neuman and Baron (1997) highlight Buss's (1961) distinction between reactive (annoyance-motivated, anger-related) and instrumental (proactive or incentive-motivated) aggression. In the former, the goal is to harm the person. In the latter, harming the person is a means of obtaining something of value, such as promotions, or valued self-presentational goals such as heightened self-image. The focus on the psychopathology and personality (e.g. power-addicted; controlling) of bullies in popular writing (e.g. Hornstein, 1995; Namie and Namie, 2000) suggests that such hostile treatment is viewed as instrumental in nature and, in some cases, fun for the actor. Further, references to these hostile behaviours as exercises of power (e.g. Namie and Namie, 2000; Cortina *et al.*, 2001; Lewis and Zare, 1999), and as efforts to control and create target dependency (e.g. Bassman, 1992; Hornstein, 1995) are consistent with an instrumental aggression perspective. It may be that workplace bullying is more illustrative of instrumental and proactive aggression and would be suggestive of different antecedents than might be true for reactive aggression.

Power

In the North American literature, power has not explicitly been a key aspect of definitions of hostile workplace behaviour, with the exception of those concerned with supervisor behaviour (e.g. Ashforth, 1997; Hornstein, 1995; Tepper, 2000). However, power is embedded in the manner in which authors discuss actor motive as well as the vulnerability or protective factors for targets. As suggested in the earlier discussion of motive and instrumental aggression, one conceptualisation of power is the unwanted exertion of influence over another (i.e. coercive action, Tedeschi and Felson, 1994). Thus, hostile behaviours are used to demonstrate the actor's ability to control others. Supportive of the notion of power as control, almost 60 per cent of the targets responding to a web-based survey of workplace bullying indicated that they were being bullied because they refused to be subservient (Namie, 2000). This conceptualisation of power rarely appears in definitions of the constructs related to hostile workplace behaviours, but rather in the theoretical discussions of why some people behave in these ways (see Neuman and Baron, this volume, for a discussion of antecedents).

Another manifestation of power is much more concrete and is focused on identifying the relative power differential between the actor and the target within the organisational and social contexts as a potential vulnerability or protective factor. In the workplace aggression and abuse literatures, power has typically been operationalised as organisational position and occasionally as gender (e.g. Aquino, 2000; Cortina *et al.*, 2001; Keashly, 1998; Schneider *et al.*, 2000a, 2000b). The proposition is that those in low-power positions (subordinates, entry-level employees; women) are more vulnerable to being the target of hostile behaviours than those in higher power positions (supervisors, bosses; men). Conversely, those in high-power positions are hypothesised as more likely to be the instigators of hostile workplace behaviours. Thus, by virtue of position and the access to resources and influence that entails, the potential exists for the abuse of power (Bassman, 1992; Cortina *et al.*, 2001). Support for this proposition is mixed, suggesting that operationalising power in terms of organisational position may be too limiting. Regarding gender, Namie (2000) reports that the majority of bullies in his web survey of targets were women, yet victims were equally likely to be men or women. Cortina *et al.* (2001) report that women were more likely to be targets of workplace hostility. Conversely, Richman *et al.* (1999) and Keashly *et al.* (1994, 1997) note that men and women report similar rates of exposure to these behaviours. Regarding organisational position, some studies found that bosses are identified as actors more often than not (e.g. Keashly *et al.*, 1994; Namie, 2000). However, several studies have found that co-workers are the most frequent source of hostile workplace behaviours (e.g. Cortina *et al.*, 2001; Keashly and Rogers, 2001; Neuman and Baron, 1997; Richman *et al.*, 1999). Aquino (2000), in a study of co-worker victimisation, reports that higher status workers were just as likely to be victimised by their co-workers as lower status workers. Some studies have even identified subordinates as active perpetrators (Keashly *et al.*, 1997; Namie, 2000).

A third manifestation of power within the context of hostile workplace behaviours focuses on the underlying feature of these previous operationalisations by defining power as a process of dependency, creating a dominant–subordinate structure in the relationship. Bassman (1992) suggests 'one common thread in all abusive relationships is the element of dependency. The abuser controls some important resources in the victim's life, and the victim is therefore dependent on the abuser' (p. 2). Tepper (2000) echoes this theme in his work on abusive supervisors when he theorises that targets remain in these abusive relationships due to economic dependence, learned helplessness and fear of the unknown. Reduced job mobility has been linked to intensification of the experience of abusive supervision (Tepper, 2000). In essence, worker dependency on others creates a situation for power (and the misuse of that power in the form of hostile treatment) to become an issue (Keashly, 2001).

The notion of the central role of power as dependency in the experience of workplace abuse is illustrated in analogies used to portray the experience. In Keashly's (2001) interview study with targets, respondents used analogies to feeling treated as a child or feeling diminished or discounted to describe their experiences in these hostile interactions. Hornstein (1995) speaks of how 'brutal bosses' infantilise their subordinates by controlling their behaviour off the job. These are characteristic feelings and features of dominant subordinate relationships.

While we have some inkling from the research examining organisational position and gender as target and instigator characteristics that power is an important feature of hostile workplace experiences, much of the work has been conceptual in nature. Though the nature and types of effects on targets reveal a diminishment and disempowerment pattern (see discussion on effects), there is little systematic empirical documentation of the process by which target dependencies or vulnerabilities are formed and utilised by the actors. Some preliminary suggestion of how this process may work comes from interviews with targets. Keashly (2001) found that behaviours from more powerful others affected targets by reducing their ability to respond. For example, if the actor had important information the target required to complete a task, the target felt less willing to confront the actor for fear that critical information would be withheld. While preliminary, evidence of the process of dependency and its relationship to effects is critical for making the distinction between aggression or conflict between equals and workplace bullying, as noted in the European literature (Einarsen, 1999). Further, documentation of the process of abuse of power and disempowerment will be critical in identifying prevention and intervention efforts directed at reducing the presence of these types of behaviours and relationships.

Violation of norms

The theme that abusive behaviours or patterns of behaviour violate some set of understood norms is pervasive in much of the workplace aggression literature. For example, Andersson and Pearson (1999) define uncivil behaviours as those that are in violation of workplace norms for mutual respect. Workplace deviance, which includes interpersonal aggression, refers to behaviours that violate norms (Robinson and Bennett, 2000).

Hornstein's (1995) belief is that 'brutal' bosses violate norms of decency and civility. However, other authors explicitly note that some of these behaviours may not be deviant from organisational norms at all (Keashly, 2001; Tepper, 2000).

As noted earlier in the discussion of intent, to the extent that these behaviours are systemic in nature, by definition, they are conforming rather than deviating from organisational and social norms. Neuman and Baron (this volume) note that norms that typically counteract aggressive behaviour either do not apply or apply weakly in the case of workplace bullying. The organisational culture evidence (e.g. Andersson and Pearson, 1999; Bassman, 1992; Brodsky, 1976; Keashly and Jagatic, 2000) and the work on organisational conditions as antecedents to aggression appear to be supportive of this claim.

Organisational culture

Many researchers have suggested that an organisation's culture plays a large and important role in the manifestation of hostile behaviours at work (e.g. Brodsky, 1976; Keashly and Jagatic, 2000; O'Leary-Kelly *et al.*, 1996). However, few empirical studies have been conducted in which organisational culture is measured and the relationship tested among its dimensions and workplace hostility. Keashly and Jagatic (2000), studying a representative sample of workers from the state of Michigan, found that higher rates of emotionally abusive behaviours were reported in organisations in which respondents perceived that employee involvement was not facilitated, morale was low, teamwork was not encouraged and supervision was problematic. Taking a somewhat different tack, Schneider *et al.* (2000b) conceived of a hostile workplace climate as one characterised by multiple forms of harassment as assessed from both target and bystander perspectives (see Robinson and O'Leary-Kelly, 1998, for a similar approach). They found that female employees exposed to multiple types of harassment reported more negative outcomes than those reporting fewer types of harassment. Further, Cortina *et al.* (2001) found that an organisational climate characterised by fair interpersonal treatment was negatively related to personal experiences of rude, uncivil behaviour.

While this limited research suggests that a relationship between overall organisational culture and hostile workplace behaviours may indeed exist, the direction of the relationship is unclear. Does something about the organisation's culture lead to greater incidence of hostility, or does a prevalence of abuse poison an organisation's culture? It is also possible that this relationship is bi-directional (Glomb, 2001). The American literature offers several theoretical explanations for organisational culture as both a cause and an effect of hostile workplace behaviour and relationships.

One argument in favour of organisational culture as supportive of hostile work behaviour comes from Brodsky (1976), who suggests that hostile behaviours may be a result of the belief in industrial society that 'workers are most productive when subjected to the goad or fear of harassment' (p. 145). This statement implies that harassment is viewed as functional by management, and perhaps necessary, to achieve productivity and acceptable performance from employees. As a result, actors of abuse know their behaviour is justifiable within the cultural norms, and make no effort to temper it. Further, targets of abuse may tolerate mistreatment as they assume it is necessary in the name of productivity (i.e. part of the job).

The second influence of organisational culture on hostile work behaviour involves what Brodsky refers to as a 'sense of permission to harass' (1976, p. 84). Brodsky suggests that harassment at work cannot occur without the direct or indirect agreement of management. This argument thus includes the involvement of individuals within the broader organisation, apart from the actor and target involved in the abusive situations. Andersson and Pearson's (1999) notion of an incivility spiral provides an explanation for the 'permission to harass' implied by some organisational cultures. According to these researchers, a 'climate of informality' pervades some organisations and eventually encourages employees to behave toward one another in a disrespectful manner. Conversely, employees in organisations with a more formal organisational culture tend to behave more respectfully towards each other, as behavioural norms are clearly understood and therefore less likely to be misinterpreted or pushed aside. Andersson and Pearson (1999) propose that in a climate of informality, incivility spirals are formed in which one employee treats another employee disrespectfully in response to the disrespectful behaviour to which he or she had been subjected. These incivility spirals, in turn, lead to secondary incivility spirals that affect other employees who observe the initial disrespectful behaviour. Ultimately, an organisation becomes uncivil at a point in which a number of these incivility spirals occur simultaneously and perpetuate each other, causing employees to believe that the organisation itself disrespects its employees.

Literature on moral exclusion may help clarify this dilemma as to whether hostile work behaviours are deviant or normative. Opotow (2000) argues that 'aggression and violence result not only from underdeveloped norms, or from sidestepping and gradual disengagement of norms, but also from failing to view others as included within our scope of justice' (p. 417). She notes that a process of moral exclusion views those excluded as outside the group to whom the norms apply and therefore as 'expendable, undeserving, and eligible targets of exploitation, aggression, and violence' (p. 417). Extending this concept to workplace hostility, it may well be that 'generally' there are strong social and organisational norms against aggressive behaviours (Neuman and Baron, 1997). However, there appear to be circumstances under which certain targets are excluded from the social group and thus subjected to hostile behaviour. From the sexual harassment (e.g. Fitzgerald and Shullman, 1993; Richman *et al.*, 1999; Keashly and Rogers, 2001) and racial harassment (e.g. Schneider *et al.*, 2000a) literatures, women and people of colour report being treated in both exclusionary and hostile ways. Power as dependency, and the motive to control and exercise power, as discussed previously, would be features of a process of moral exclusion. In addition, references to these behaviours as a negation of social worth, dignity and equal human status fit with this notion of exclusion (Hornstein, 1995; Keashly, 2001; Namie and Namie, 2000). What this suggests is that defining or assessing behaviours as deviant or normative requires that a specific referent group is identified, and that the norms of behaviour are articulated, rather than assumed. We have argued that in some organisations as indexed by ambient levels of hostility and uninformative policies with negligent enforcement, the behaviours discussed in this chapter would be normative even though they would be considered deviant from the broader societal perspective of civility (Hornstein, 1995).

What is clear from this discussion of the relationship between organisational culture and hostile workplace behaviour is that these behaviours may not be deviant from workplace norms. Rather, they may be consistent with them. This conclusion challenges definitions of workplace hostility that label it as deviant behaviour to provide evidence of the existence of the norms that are being contravened (Wright and Smye, 1996).

An interesting question is 'what are the implications of defining these behaviours as deviant from, rather than consistent with, organisational norms?' To the extent that these behaviours are demonstrated to be organisationally deviant (e.g. organisational tolerance is low, Glomb, 2001), prevention and intervention focus primarily on the actor of the behaviour (e.g. pre-hiring screening; anger management training). If these behaviours are seen as organisationally consistent (e.g. organisational tolerance is high; a permissive climate for hostility), then prevention and intervention work requires wholesale organisational culture and overall structural change manifested through formal policies and procedures that are swiftly and effectively implemented.

Effects of exposure to hostile workplace behaviours

A consistent theme in the reviewed literatures is that these behaviours/relationships cause harm both individually and organisationally, as well as in both the short and long term. Much of the workplace aggression literature has tended to focus on antecedents, rather than effects, of these behaviours (Glomb, 2001). However, a limited but growing number of studies of workplace aggression, along with the workplace abuse literature whose focus has been primarily on documenting effects, have generated lengthy and well-established lists of effects (see Table 2.3).

There are some interesting observations to make regarding these behaviours. First, the effects of these seemingly minor behaviours are extensive, cutting across all manner of functioning from personal (psychological, cognitive, physical) to interpersonal (aggressive behaviours, marital and family conflict) to professional/job (satisfaction, turnover, withdrawal) to organisational (productivity, commitment). Second, targets may in turn become hostile and aggressive, speaking to the potential of hostile behaviour to deteriorate into an escalatory spiral of increasingly severe behaviours in which the target becomes the actor (Andersson and Pearson, 1999; Glomb, 2002). Glomb (2001) reports that being a target of hostile behaviour is a predictor of engaging in hostile behaviour. Anecdotal accounts of workplace homicide suggest that early mistreatment by fellow workers contributes to these deadly responses (e.g. McLaughlin, 2000). Third, the effect of alcohol use and abuse is also worthy of comment because of its potentially unique relationship to prolonged exposure to workplace hostility. Many of the effects noted are fairly characteristic of exposure to any workplace stressor (Barling, 1996). However, Richman and her colleagues (Richman *et al.*, 1997, 1999; Rospenda *et al.*, 2000) argue that chronic stressors such as workplace bullying are more predictive of disorders or diseases that develop slowly over time, such as alcohol abuse. In their two-year longitudinal study of generalised workplace abuse, they found that individuals exposed to chronic abuse were more likely to manifest drinking problems than those who had been exposed to it on a more time-limited basis. This distinction is important in both identifying and assessing a range of effects of enduring hostile relationships.

Fourth, the overall nature of the effects indicates a deterioration or disabling of the target, the people around him or her, and the organisation. In fact, several of these effects (hypervigilance, intrusive imagery, avoidance behaviours) are considered symptomatic of post-traumatic stress disorder (PTSD). While PTSD has been defined as a response to a single, overwhelming event (e.g. workplace shooting, natural disaster), it has been argued to apply to targets who experience prolonged hostile interactions (e.g. Namie and Namie, 2000). For example, Namie (2000) reports that 31 per cent of women and 21 per cent of men responding to his survey on workplace bullying reported exhibiting all three trauma symptoms. Similar arguments are made in the European literature on workplace bullying (Hoel *et al.*, 1999). What is clear from these studies is that the more frequent the exposure to hostile workplace behaviour, the greater the negative effects. Even seemingly minor behaviours can have significant negative effects when they occur frequently and over extended time periods (Glomb *et al.*, in press; Keashly, 1998; Richman *et al.*, 2001; Schneider *et al.*, 2000b). Thus, exposure to these behaviours undermines the personal and organisational resources to do well, rendering people ineffective in many spheres. With rare exception (Rospenda *et al.*, 2000; Tepper, 2000), the empirical evidence of these connections is correlational in nature. Clearly, longitudinal research is needed to establish the nature of these links.

Table 2.3 Some effects of workplace abuse and harassment on targets

Category	*Effect*	*Source*
Direct		
Negative mood	Anger, resentment	Barling (1996); Bassman (1992); Brodsky (1976); Northwest National Life Insurance Company (1993)
	Anxiety	Barling, 1996; Brodsky (1976); Keashly *et al.* (1994); Northwest National Life Insurance Company (1993); Tepper (2000)
	Depressed mood	Barling (1996); Brodsky (1976); NNL (1993)
Cognitive distraction	Concentration	Barling (1996); Brodsky (1976); Namie (2000)
Fear of violence	Fear	Barling (1996); Rogers and Kelloway (1997); Schat and Kelloway (2000)
Indirect		
Decreased psychological well-being	Lowered self-esteem	Ashforth (1997); Brodsky (1976); Cortina *et al.* (2001); Hornstein (1995); Price Spratlen (1995)
	Problem drinking	Namie (2000); Richman *et al.* (1999, 2001); Rospenda *et al.* (2000)
	Depression	Namie (2000); Tepper (2000)
	Overall emotional health	Keashly and Jagatic (2000); Namie (2000); Rogers and Kelloway (1997); Schat and Kelloway (2000); Schneider *et al.* (2000a)
	Life satisfaction	Glomb (2001); Schneider *et al.* (2000a)
	Post-traumatic stress disorder	Namie (2000); Schneider *et al.* (2000a)
Poor psychosomatic function	Physical ill health (general)	Barling (1996); Brodsky (1976); Hornstein (1996); Namie (2000); Northwest National Life Insurance Company (1993); Price Spratlen (1995); Richman *et al.* (1999); Rogers and Kelloway (1997); Schat and Kelloway (2000);

Reduced organisational functioning	Decreased job satisfaction	Barling (1996); Brodsky (1976); Cortina *et al.* (2001); Glomb (2001); Keashly and Jagatic (2000); Keashly *et al.* (1994); Northwest National Life Insurance Company (1993); Price Spratlen (1995); Schneider *et al.* (2000a); Tepper (2000)
	Greater turnover	Bassman (1992); Northwest National Life Insurance Company (1993); Pearson *et al.* (2000); Tepper (2000)
	Work withdrawal behaviours	Glomb (2001); Schat and Kelloway (2000); Schneider *et al.* (2000a)
	Greater Intention to Leave	Ashforth (1997); Cortina *et al.* (2001); Keashly and Jagatic (2000); Keashly *et al.* (1994); Schneider *et al.* (2000a); Pearson *et al.* (2000); Rogers and Kelloway (1997); Tepper (2000)
	Increased absenteeism	Bassman (1992); Pearson *et al.* (2000); Price Spratlen (1995); Schneider *et al.* (2000a)
	Decreased productivity	Ashforth (1997); Bassman (1992); Northwest National Life Insurance Company (1993); Pearson *et al.* (2000); Price Spratlen (1995); Robinson and Bennett (1995); Schat and Kelloway (2000)
	Decreased organisational commitment	Pearson *et al.* (2000); Tepper (2000)

The research drawn on here has tended to focus on relating exposure directly to these measured effects and has not addressed the distinction between immediate reactions and longer term effects (Barling, 1996; Rospenda *et al.*, 2000; Tepper, 2000). Barling (1996) argues that research needs to distinguish between and measure direct and indirect effects. Specifically, he argues that the direct effects of the psychological experience of hostile behaviours are negative mood (anger, anxiety, depressive symptoms), cognitive distraction (attention, concentration, intrusive imagery) and fear of violence. To the extent that these immediate effects are not alleviated, they will result in the more long-term and extensive effects, such as decreased psychological well-being (depression, poor family function), poor psychosomatic functioning (sleep problems, headaches, gastrointestinal disorders), reduced organisational functioning such as psychological attachment (organisational commitment, absenteeism, turnover), emotional exhaustion, poor job performance and accidents. In addition to explicating the evolution of these effects, this approach legitimates these more 'minor' effects as worthy of serious attention, and identifies places for action in helping targets deal with the experience of hostile workplace behaviours (e.g. counselling, leaves of absence) before the damage becomes more extensive.

While the research on effects has tended to focus on the immediate target and the organisation, the victim net is now being cast more widely. Barling (1996) distinguishes between primary (targets) and secondary victims (witnesses). Glomb (2001) also includes family and friends of the target in her notion of co-victimisation. Recent research (e.g. Neuman and Baron, 1997; Rogers and Kelloway, 1997; Schneider, 1996) has found that witnesses of hostile behaviour directed at others exhibit similar negative effects, such as fear that they will be next, frustration over not being able to intervene, and anger at the organisation for not controlling the actors. Andersson and Pearson's (1999) theorising on workplace incivility spirals suggests that hostility between workers can spill over on to the observers, fuelling secondary spirals (cascade effect). Recognising these 'other victims', Barling (1996), drawing on the work–family conflict literature, suggests including indicators of family functioning and marital relationships as possible outcomes of exposure to hostile workplace behaviours. Glomb (2001) throws the impact net even wider by suggesting that similar negative effects can be expected for perpetrators of these behaviours.

The recognition that workplace hostility can seriously and extensively affect both those within and outside the organisation immediately, as well as in the long-term, requires that research expand the range and type of effects and the time period over which they are investigated to capture truly the multivariate nature of workplace hostility's outcomes. Critical next steps for the North American literature are the identification of: (1) the mechanisms through which exposure to hostile behaviours causes these harmful effects, for example mediators such as fear of recurrence and organisational justice (e.g. Barling, 1996; Tepper, 2000); and (2) the factors that enhance or mitigate these effects such as targets' coping responses and organisational responses to complaints (Keashly, 2001; O'Leary-Kelly *et al.*, 1996; Richman *et al.*, 2001).

Conclusion

In this chapter, we have sought to identify and share relevant concepts and research from the North American literatures on workplace aggression and abuse. As an overall assessment, there are several findings that research on this side of the pond has to offer discussions on workplace bullying. First, consideration of a variety of hostile behaviours and a range of effects has now become the mainstay in the literature, providing more points of behavioural comparison with the other global literatures and creating opportunities to explore the inherently multivariate nature of this phenomenon. Second, the conceptualisation of workplace hostility as a chronic workplace stressor has permitted connections into the organisational stress literature, which is rich with models of how exposure to negative social stressors such as workplace hostility affects individuals and organisations. This enriches and makes possible informed discussion of what would be appropriate avenues for prevention and intervention work. Third, the utilisation of diverse measures and methodologies such as surveys, interviews, longitudinal studies and computer modelling bodes well for more accurate and precise data, as well as the development and testing of more complex models of these hostile behaviours.

The most serious limitation of the research reviewed is the almost exclusive focus on specific or frequent incidents of hostile behaviour from undefined actors, with little attention paid to persistent or patterned hostile behaviours from an identifiable actor which are experienced differently by the targets, and are suggestive of different antecedents and outcomes. This seeming lack of attention may be a reflection of the dramatic pull extraordinary single events of physical violence have had on the American public. Many of the behaviours examined here pale in comparison.

The growing evidence is that these hostile behaviours are more frequent, and, if persistent, can have profound negative implications for individual and organisational well-being. This may be one of those rare situations in which the research can help drive public opinion, rather than the other way around. The efforts of several activists such as Gary and Ruth Namie (2000; www.bullybusters.org), through the popular literature, the media, unions and community trainings, are raising public consciousness of these behaviours and communicating the message of their potential for great damage. Thus, we are eager to learn more about workplace bullying from our colleagues around the world.

Bibliography

Andersson, L. M. and Pearson, C. M. (1999) Tit for tat? The spiraling effect of incivility in the workplace. *Academy of Management Review*, *24*, 452–471.

Aquino, K. (2000) Structural and individual determinants of workplace victimization: The effects of hierarchical status and conflict management style. *Journal of Management*, *26*, 171–193.

Aquino, K., Grover, S., Bradfield, M. and Allen, D.G. (1999) The effects of negative affectivity, hierarchical status, and self-determination on workplace victimization. *Academy of Management Journal*, *42*, 260–272.

Ashforth, B. E. (1997) Petty tyranny in organizations: A preliminary examination of antecedents and consequences. *Canadian Journal of Administrative Sciences*, *14*, 126–140.

Barling, J. (1996) The prediction, psychological experience, and consequences of workplace violence. In G. VandenBos and E. Q. Bulatao (eds), *Violence on the job: Identifying risks and developing solutions* (pp. 29–50). Washington, DC: American Psychological Association.

Baron, R. A. and Neuman, J. H. (1996) Workplace violence and workplace aggression: Evidence on their relative frequency and potential causes. *Aggressive Behavior*, *22*, 161–173.

—— (1998) Workplace aggression – the iceberg beneath the tip of workplace violence: Evidence on its forms, frequency and targets. *Public Administration Quarterly*, *21*, 446–464.

Bassman, E. (1992) *Abuse in the workplace*. Westport, CT: Quorum Books.

Bassman, E. and London, M. (1993) Abusive managerial behavior. *Leadership and Organization Development Journal*, *14*, 2, 18–24.

Brodsky, C. M. (1976) *The harassed worker.* Toronto: Lexington Books, D. C. Heath and Company.

Buss, A. H. (1961) *The psychology of aggression*. New York: Wiley.

Cortina, L. M., Magley, V. J., Williams, J. H. and Langhout, R. D. (2001) Incivility in the workplace: Incidence and impact. *Journal of Occupational Health Psychology*, *6*, 64–80.

Davenport, N., Schwartz, R. D. and Elliott, G. P. (1999) *Mobbing: Emotional abuse in the American workplace*. Ames, IA: Civil Society.

Denenberg, R. V. and Braverman, M. (1999) *The violence-prone workplace: A new approach to dealing with hostile, threatening, and uncivil behavior*. Ithaca, NY: Cornell University Press.

Einarsen, S. (1999) The nature and causes of bullying at work. *International Journal of Manpower*, *20*, 1/2, 16–27.

Fitzgerald, L. F. and Shullman, S. L. (1993) Sexual harassment: A research analysis and agenda for the 90's. *Journal of Vocational Behavior*, *42*, 5–27.

Giacalone, R. A. and Greenberg, J. (1997) *Antisocial behavior in organizations*. Thousand Oaks, CA: Sage.

Glomb, T. M. (2001) Workplace aggression: Antecedents, behavioral components and consequences. Manuscript under review, University of Minnesota, Minneapolis.

—— (2002) Workplace aggression: Informing conceptual models with data from specific encounters. *Journal of Occupational Health Psychology*, *7*, 1, 20–36.

Glomb, T. M. and Miner, A. G. (in press) Exploring patterns of aggressive behaviors in organizations: Assessing model-data fit. In J. M. Brett and F. Drasgow

(eds), *The psychology of work: Theoretically based empirical research*. Mahwah, NJ: Lawrence Erlbaum.
Glomb, T. M., Steel, P. D. G. and Arvey, R. D. (in press). Office sneers, snipes and stab wounds: Antecedents, consequences, and implications of workplace violence and aggression. In R. G. Lord, R. Klimoski and R. Kanfer (eds), *Emotions at work*. San Francisco: Jossey-Bass.
Graydon, J., Kasta, W. and Khan, P. (1994) Verbal and physical abuse of nurses. *Canadian Journal of Nursing Administration*, November–December, 70–89.
Greenberg, L. and Barling, J. (1998) Predicting employee aggression against coworkers, subordinates, and supervisors: The roles of person behaviors and perceived workplace factors. *Journal of Organizational Behavior*, *20*, 897–913.
Hoel, H., Rayner, C. and Cooper, C. L. (1999) Workplace bullying. In C. L. Cooper and I. T. Robertson (eds), *International Review of Industrial and Organizational Psychology* (pp. 195–229). Chichester: John Wiley.
Hornstein, H. (1995) *Brutal bosses and their prey*. New York: Riverhead Books.
Jagatic, K. and Keashly, L. (1998) The nature and effects of negative incidents by faculty members toward graduate students. Unpublished manuscript, Wayne State University, Detroit.
Keashly, L. (1998) Emotional abuse in the workplace: Conceptual and empirical issues. *Journal of Emotional Abuse*, *1*, 85–117.
—— (2001) Interpersonal and systemic aspects of emotional abuse at work: The target's perspective. *Violence and Victims*, *16*, 3, 233–268.
Keashly, L. and Jagatic, K. (2000) The nature, extent, and impact of emotional abuse in the workplace: Results of a statewide survey. Paper presented at the Academy of Management Conference, Toronto, Canada.
Keashly, L. and Rogers, K. A. (2001) Aggressive behaviors at work: The role of context in appraisals of threat. Manuscript under review, Wayne State University, Detroit.
Keashly, L., Harvey, S. and Hunter, S. (1997) Emotional abuse and role state stressors: Relative impact on residence assistants' stress. *Work and Stress*, 11, 35–45.
Keashly, L., Trott, V. and MacLean, L. M. (1994) Abusive behavior in the workplace: A preliminary investigation. *Violence and Victims*, *9*, 125–141.
Lazarus, R. S. and Folkman, S. (1984) *Stress, appraisal, and coping*. New York: Springer.
Lewis, G. and Zare, N. (1999) *Workplace hostility: Myth and reality*. Muncie, IN: Accelerated Development Inc.
Leymann, H. (1990) Mobbing and psychological terror at workplaces. *Violence and Victims*, *5*, 119–126.
Lind, E. A. (1997) Litigation and claiming in organizations: Antisocial behavior or quest for justice? In R. A. Giacalone and J. Greenberg (eds), *Antisocial behavior in organizations* (pp. 150–171). Thousand Oaks, CA: Sage.
McLaughlin, J. (2000) Anger within. *O.H.S. Canada*, *16*, 8, 30–36.
Namie, G. (2000) U.S. Hostile Workplace Survey 2000. Paper presented at the New England Conference on Workplace Bullying, Suffolk University Law School, Boston, MA.
Namie, G. and Namie, R. (2000) *The bully at work: What you can do to stop the hurt and reclaim your dignity on the job*. Naperville, IL: Sourcebooks.

Neuman, J. H. and Baron, R. A. (1997) Aggression in the workplace. In R. A. Giacalone and J. Greenberg (eds), *Antisocial behavior in organizations* (pp. 37–67). Thousand Oaks, CA: Sage Publications, Inc.

Northwest National Life Insurance Company (1993) *Fear and violence in the workplace*. Minneapolis, MN: Northwest National Life Insurance Company.

O'Leary-Kelly, A. M., Griffin, R. W. and Glew, D. J. (1996) Organization-motivated aggression: A research framework. *Academy of Management Review*, *21*, 225–253.

Opotow, S. (2000) Aggression and violence. In M. Deutsch and P. T. Coleman (eds), *The handbook of conflict resolution* (pp. 403–427). San Francisco: Jossey-Bass.

Pearson, C, Andersson, L. and Porath, C. (2000) Assessing and attacking workplace incivility. *Organizational Dynamics*, Fall, 129–137.

Price Spratlen, L. (1995) Interpersonal conflict which includes mistreatment in a university workplace. *Violence and Victims*, *10*, 285–297.

Richman, J. A., Rospenda, K. M., Flaherty, J. A. and Freels, S. (2001) Workplace harassment, active coping, and alcohol-related outcomes. *Journal of Substance Abuse*, *13*, 3, 347–366.

Richman, J. A., Rospenda, K. M., Nawyn, S. J. and Flaherty, J. A. (1997) Workplace harassment and the self-medication of distress: A conceptual model and case illustrations. *Contemporary Drug Problems*, *24*, 179–199.

Richman, J. A., Rospenda, K. M., Nawyn, S. J., Flaherty, J. A., Fendrich, M., Drum, M. L. and Johnson, T. P. (1999) Sexual harassment and generalized workplace abuse among university employees: Prevalence and mental health correlates. *American Journal of Public Health*, *89*, 3, 358–363.

Robinson, S. L. and Bennett, R. J. (1995) A typology of deviant workplace behaviors: A multidimensional scaling study. *Academy of Management Journal*, *38*, 555–572.

—— (2000) Development of a measure of workplace deviance. *Journal of Applied Psychology*, *85*, 3, 349–360.

Robinson, S. L. and O'Leary-Kelly, A. (1998) Monkey see, monkey do: The influence of work groups on the antisocial behavior of employees. *Academy of Management Journal*, *41*, 658–672.

Rogers, K.A. (1998) Toward an integrative understanding of workplace mistreatment. Unpublished dissertation, University of Guelph, Ontario, Canada.

Rogers, K. A. and Kelloway, E. K. (1997) Violence at work: Personal and organizational outcomes. *Journal of Occupational Health Psychology*, *12*, 63–71.

Rospenda, K. M., Richman, J. A., Wislar, J. S and Flaherty, J. A. (2000) Chronicity of sexual harassment and generalized workplace abuse: Effects on drinking outcomes. *Addiction*, *95*, 12, 1805–1820.

Rubin, J. Z., Pruitt, D. G. and Kim, S. H. (1994) *Social conflict: Escalation, stalemate, and settlement*, 2nd edn. New York: McGraw-Hill.

Schat, A. C. H. and Kelloway, E. K. (2000) Effects of perceived control on the outcomes of workplace aggression and violence. *Journal of Occupational Health Psychology*, *5*, 3, 386–402.

Schneider, K. T. (1996) Bystander stress: Effects of sexual harassment on victims' co-workers. Paper presented at a symposium on Responses to Sexual Harassment, American Psychological Association annual convention, Toronto, Ontario, Canada.

Schneider, K. T., Hitlan, R. T. and Radhakrishnan, P. (2000a). An examination of the nature and correlates of ethnic harassment experiences in multiple contexts. *Journal of Applied Psychology*, *85*, 1, 3–12.

Schneider, K. T., Hitlan, R. T., Delgado, M., Anaya, D. and Estrada, A. X. (2000b). Hostile climates: The impact of multiple types of harassment on targets. Paper presented at the Society for Industrial and Organizational Psychology, April 2000, New Orleans, LA.

Tedeschi, J. T. and Felson, R. B. (1994) *Violence, aggression, and coercive actions*. Washington, DC: American Psychological Association.

Tepper, B.J. (2000) Consequences of abusive supervision. *Academy of Management Journal*, *43*, 2, 178–190.

Wright, L. and Smye, M. (1996) *Corporate abuse: How lean and mean robs people and profits*. New York: MacMillan.

3 Bully/victim problems in school

Basic facts and an effective intervention programme[1]

Dan Olweus

Bullying among schoolchildren is certainly a very old phenomenon. The fact that some children are frequently and systematically harassed and attacked by other children has been described in literary works, and many adults have personal experience of it from their own school days. Though many are acquainted with the bully/victim problem, it was not until fairly recently – in the early 1970s – that the phenomenon was made the object of more systematic research (Olweus, 1973a, 1978). For a number of years, these efforts were largely confined to Scandinavia. In the 1980s and early 1990s, however, bullying among schoolchildren attracted attention in other countries such as Japan, Ireland, Great Britain, The Netherlands, Australia, Canada and the USA. There are now clear indications of an increasing societal as well as research interest into bully/victim problems in several parts of the world.

What is meant by bullying?

A much-used definition of *bullying or victimisation* is the following: *A student is being bullied or victimised when he or she is exposed, repeatedly and over time, to negative actions on the part of one or more other students* (e.g. Olweus, 1986a, 1993a, 1996b). It is a negative action when someone intentionally inflicts, or attempts to inflict, injury or discomfort upon another – basically what is implied in the definition of aggressive behaviour in the social sciences (Olweus, 1973b). Negative actions can be carried out by physical contact, by words, or in other ways, such as by making faces or mean gestures, spreading rumours, and intentional exclusion from a group. Although children or youth who engage in bullying very likely vary in their degree of awareness of how the bullying is perceived by the victim, most or all of them probably realise that their behaviour is at least somewhat painful or unpleasant to the victim.

In order for the term bullying to apply, there should also be an *imbalance in strength* (*an asymmetric power relationship*). The student who is exposed to the negative actions has difficulty in defending himself or herself and is somewhat helpless against the student or students who harass. The actual and/or perceived imbalance in strength or power may come about in several different ways. The target of bullying may actually be physically weaker, or may simply perceive himself or herself as physically or mentally weaker than the perpetrator. Or there may be a difference in numbers, with several students ganging up on a single victim. A somewhat different kind of imbalance may be achieved, when the 'source' of the negative actions is difficult to identify or confront as in social exclusion from the group, backtalking or when a student is being sent anonymous mean notes. In line with this reasoning, we do not talk about bullying when there is a conflict or aggressive interchange between two persons of approximately the same physical or mental strength.

In this context, it is also natural to consider briefly the relationship between bullying and teasing. In the everyday social interactions among peers in school, there occurs a good deal of (also recurrent) teasing of a playful and relatively friendly nature – which in most cases cannot be considered bullying. On the other hand, when the repeated teasing is of a degrading and offensive character, and, in particular, is continued in spite of clear signs of distress or opposition on the part of the target, it certainly qualifies as bullying. Here it is thus important to try to distinguish between malignant and more friendly, playful teasing, although the line between them is sometimes blurred and the perception of the situation may to some extent depend on the perspective taken: that of the target or of the perpetrator(s).

In this definition, then, bullying is characterised by the following *three criteria*: (1) It is aggressive behaviour or intentional 'harm-doing' (2) which is carried out 'repeatedly and over time' (3) in an interpersonal relationship characterised by an imbalance of power. One might add that the bullying behaviour often occurs without apparent provocation. This definition makes it clear that bullying can be considered a form of abuse, and I sometimes use the term *peer abuse* as a label for the phenomenon. What sets it apart from other forms of abuse such as child abuse and spouse abuse is the context in which it occurs and the relationship characteristics of the interacting parties.

It may also be useful to distinguish between *direct bullying/victimisation* – with relatively open attacks on the victim – and *indirect bullying/victimisation* in the form of intentional exclusion from a group, backtalking and the spreading of malicious rumours. (See Olweus, 1996b, 1999a for a more extended discussion of the term bullying including its relationship to aggression and violence.)

Basic facts about bully/victim problems

Prevalence

On the basis of our survey of more than 130,000 Norwegian students with my Bully/Victim Questionnaire (Olweus, 1983a), one can estimate that some 15 per cent of the students in elementary and secondary/junior high schools (grades 2–10, roughly corresponding to ages 7 through 16) in Norway were involved in bully/victim problems with some regularity – as bullies, victims or bully/victims (Olweus, 1985, 1987, 1991, 1993a, 1994). This percentage represents one student out of seven, or about 84,000 students (in 1983). Approximately 9 per cent, or 52,000 students, were victims, and 41,000, or 7 per cent, bullied other students. Some 9,000 students were both victim and bully (1.6 per cent of the total of 568,000 students or 17 per cent of the victims). A total of some 5 per cent of the students were involved in more serious bullying problems (as bullies or victims or bully/victim), occurring about once a week or more frequently. As the prevalence questions in the Questionnaire refer to only part of the autumn term, there is little doubt that the figures presented actually underestimate the number of students involved in such problems during a whole year.

It is apparent, then, that bullying is a considerable problem in Norwegian schools, a problem that affects a very large number of students. Data from other countries (in large measure collected with my Bully/Victim Questionnaire), such as Sweden (Olweus, 1992b), Finland (Lagerspetz *et al.*, 1982), Ireland (O'Moore and Brendan, 1989), Great Britain (Smith, 1991; Whitney and Smith, 1993), the USA (Perry *et al.*, 1988), Canada (Ziegler and Rosenstein-Manner, 1991), The Netherlands (Haeselager and van Lieshout, 1992; Junger, 1990), Japan (Hirano, 1992), Spain (Ruiz, 1992), and Australia (Rigby and Slee, 1991), indicate that this problem certainly exists outside Norway as well, and with similar or even higher prevalence rates (see also Olweus and Limber, 1999, and Smith *et al.*, 1999).

There are many more boys than girls who bully others, and a relatively large percentage of girls report that they are mainly bullied by boys. Also, there is a somewhat higher percentage of boys who are victims of bullying. Although direct bullying is thus a greater problem among boys, there occurs a good deal of bullying among girls as well. Bullying with physical means is less common among girls, however; girls typically use more subtle and indirect ways of harassment such as slandering, spreading rumours, intentional exclusion from the group and manipulation of friendship relations (e.g. depriving a girl of her 'best friend'). Such forms of bullying may be more difficult to detect for adults. We have also found that it is the younger and weaker students who are most exposed to bullying. Although most bullying occurs among students at the same grade level, a good deal of bullying is also carried out by older students towards younger ones. (More details about bullying in different grades and among boys and girls, etc. are given in Olweus, 1993a.)

There is a good deal of evidence to indicate that the behaviour patterns involved in bully/victim problems are fairly stable over time: This means that being a bully or a victim is something that is likely to continue for substantial periods of time, unless systematic efforts (from adults) are made to change the situation (Olweus, 1977, 1978).

Three common 'myths' about bullying

A common view holds that bully/victim problems are a consequence of large classes and/or schools: the larger the class or the school, the higher the level of bully/victim problems. Closer analysis of this hypothesis, making use of the Norwegian survey data from more than 700 schools and several thousand classes (with great variations in size) resulted in the conclusion that the size of the class or the school appears to be of negligible importance for the relative frequency or level of bully/victim problems in the class or the school (Olweus, 1993a).

Second, in the general debate it has been commonly maintained that bullying is a consequence of competition and striving for grades in school. More specifically, it has been argued that the aggressive behaviour of the bullies towards their environment can be explained as a reaction to failures and frustrations in school. Also, this hypothesis failed to receive support from detailed analyses of (longitudinal) data: Though there was an association (of moderate magnitude) between aggressive behaviour and (poor) grades, there was nothing in the results to suggest that the behaviour of the aggressive boys was a *consequence* of poor grades and failure in school (Olweus, 1983b).

Third, a widely held view explains victimisation as 'caused' by external or outward deviations. It is argued that students who are fat, are red-haired, wear glasses or speak with an unusual dialect, etc. are particularly likely to become victims of bullying. This explanation is quite common among students. This hypothesis also received no support in empirical analyses (Olweus, 1978). It was concluded that external deviations play a much smaller role in the origin of bully/victim problems than generally assumed (see also Junger, 1990). In spite of the lack of empirical support for this hypothesis, it seems still to enjoy considerable popularity. Some probable reasons why this is so have been advanced, and the interested reader is referred to this discussion (Olweus, 1978, 1993a).

All of these hypotheses have thus failed to receive support from empirical data. Accordingly, one must look for other factors to find the origins of these problems. The research evidence collected so far and presented in my book *Bullying at school: What we know and what we can do* (1993a), summarised in the next few pages, clearly suggests that personality characteristics/typical reaction patterns, in combination with physical strength or weakness in the case of boys, are quite important for the development of these problems *in individual students*. At the same time, environmental factors such as the teachers' attitudes, routines, and behaviour play a major role in determining the extent to which the problems will manifest themselves *in a larger unit* such as a classroom or a school (see Olweus, 1993a).

What characterises the typical victims?

A relatively clear picture of both the typical victims and the typical bullies has emerged from research (Olweus, 1973a, 1978, 1981a, 1984; Björkqvist *et al.*, 1982; Boulton and Smith, 1994; Farrington, 1993; Lagerspetz *et al.*, 1982; Perry *et al.*, 1988). By and large, this picture seems to apply to both boys and girls, although it must be noted that clearly less research has so far been done on bullying among girls. Typical victims are more anxious and insecure than students in general. Further, they are often cautious, sensitive and quiet. When attacked by other students, they commonly react by crying (at least in the lower grades) and through withdrawal. In addition, victims suffer from low self-esteem, they have a negative view of themselves and their situation. They often look upon themselves as failures and feel stupid, ashamed and unattractive.

The victims are lonely and abandoned at school. As a rule, they do not have a single good friend in their class. They are not aggressive or teasing in their behaviour, however, and, accordingly, one cannot explain the bullying as a consequence of the victims themselves being provocative to their peers (see below). These children often have a negative attitude towards violence and the use of violent means. If they are boys, they are likely to be physically weaker than other boys (Olweus, 1978).

I have labelled this type of victim *the passive or submissive victim*, as opposed to the far less common type described below. In summary, it seems that the behaviour and attitude of the passive/submissive victims *signal to others that they are insecure and worthless individuals who will not retaliate if they are attacked or insulted.* A slightly different way of describing the passive/submissive victims is to say that they are characterised by an *anxious or submissive reaction pattern combined* (in the case of boys) with *physical weakness.*

In-depth interviews with parents of victimised boys indicated that these boys tended to be characterised by a certain cautiousness and sensitivity at an early age (Olweus, 1993b). Boys with such characteristics (perhaps combined with physical weakness) are likely to have had difficulty in asserting themselves in the peer group and may have been somewhat disliked by their age mates. There are thus good reasons to believe that these characteristics contributed to making them victims of bullying (see also Schwartz *et al.*, 1993). At the same time, it is obvious that repeated harassment by peers must have considerably increased their anxiety, insecurity and generally negative evaluation of themselves. Some common characteristics of victims may thus be both a 'cause' and a consequence of their victimisation. It must also be pointed out that, at least in some situations, there may be a certain element of arbitrariness in the selection of targets of victimisation.

As mentioned earlier, there is also another clearly smaller group of victims, *the provocative victims*, who are characterised by a combination of both anxious and aggressive reaction patterns. These students often have problems with concentration, and behave in ways that may cause irritation and tension around them. Some of these students can be characterised as hyperactive. It is not uncommon for their behaviour to provoke many students in the class, thus resulting in negative reactions from a large part of, or even the entire, class. The dynamics of bully/victim problems in a class with provocative victims differ in part from problems in a class with passive victims (Olweus, 1978).

A follow-up study of two groups of boys who had or had not been victimised by their peers in school showed that the former victims (mostly of the passive/submissive type) were much more likely to be depressed and had poorer self-esteem as young adults, at age 23 (Olweus, 1993b). The pattern of findings clearly suggested that this was a consequence of the earlier, persistent victimisation which thus had left its scars on their minds.

What characterises the typical bullies?

A distinctive characteristic of typical bullies is their aggression toward peers – this is implied in the definition of a bully. But bullies are often aggressive towards adults as well, both teachers and parents. Generally, bullies have a more positive attitude towards violence and the use of violent means than students in general. Further, they are often characterised by impulsiveness and a strong need to dominate others in a negative way. They have little empathy with victims of bullying. If they are boys, they are likely to be physically stronger than boys in general, and the victims in particular.

A commonly held view among psychologists and psychiatrists is that individuals with an aggressive and tough behaviour pattern are actually anxious and insecure 'under the surface'. The assumption that the bullies have an underlying insecurity has been tested in several of my own studies, also using 'indirect' methods such as stress hormones (adrenaline and noradrenaline) and projective techniques. There was nothing in the results to support the common view; they rather pointed in the opposite direction: the bullies had unusually little anxiety and insecurity, or were roughly average on such dimensions (Olweus, 1981a, 1984, 1986; see also Pulkkinen and Tremblay, 1992). They did not suffer from poor self-esteem.

These conclusions apply to the bullies as a group. The results do not imply that there cannot be individual bullies who are both aggressive and anxious. It should also be emphasised that there are students who participate in bullying but who do not usually take the initiative – these may be labelled *passive bullies*, *followers* or *henchmen* (see Olweus, 2001b, for a more detailed discussion of this issue in the context of a 'bullying circle'). A group of passive bullies is likely to be fairly mixed, and may also contain insecure and anxious students (Olweus, 1978).

Several studies have found bullies to be of average or slightly below average popularity (Björkqvist *et al.*, 1982; Lagerspetz *et al.*, 1982;

Olweus, 1973a, 1978; Pulkkinen and Tremblay, 1992). Bullies are often surrounded by a small group of two to three peers who support them and who seem to like them (see also Cairns *et al.*, 1988). The popularity of the bullies decreases, however, in the higher grades, and is considerably less than average in grade 10 (around age 16). Nevertheless, the bullies do not seem to reach the low level of popularity that characterises the victims.

In summary, typical bullies can be described as having an *aggressive reaction pattern combined*, in the case of boys, with *physical strength*.

As regards the possible psychological sources underlying bullying behaviour, the pattern of empirical findings suggests at least three, partly interrelated motives (in particular for male bullies). First, the bullies have a strong need for power and dominance; they seem to enjoy being 'in control' and subduing others. Second, considering the family conditions under which many of them have been reared (below), it is natural to assume that they have developed a certain degree of hostility towards the environment; such feelings and impulses may make them derive satisfaction from inflicting injury and suffering upon other individuals. Finally, there is an 'instrumental component' to their behaviour. The bullies often coerce their victims to provide them with money, cigarettes, beer and other things of value. In addition, it is obvious that aggressive behaviour is in many situations rewarded in the form of prestige.

Bullying can also be viewed as a *component of a more generally antisocial and rule-breaking ('conduct-disordered') behaviour pattern*. From this perspective, it is natural to predict that youngsters who are aggressive and bully others run a clearly increased risk of later engaging in other problem behaviours such as criminality and alcohol abuse (e.g. Loeber and Dishion, 1983; Olweus, 1979). In my follow-up studies we have found strong support for this view. Approximately 60 per cent of boys who were characterised as bullies in grades 6–9 had been convicted of at least one officially registered crime by the age of 24. Even more dramatically, as much as 35–40 per cent of the former bullies had three or more convictions by this age, while this was true of only 10 per cent of the control boys (those who were neither bullies nor victims in grades 6–9). Thus, as young adults, the former school bullies had a fourfold increase in the level of relatively serious, recidivist criminality, as documented in official crime records (Olweus, 1993a).

It may be mentioned, that the former victims had an average or somewhat below average level of criminality in young adulthood.

Development of an aggressive reaction pattern

In light of the characterisation of bullies as having an aggressive reaction pattern – that is, displaying aggressive behaviour in many situations – it becomes important to examine the question: what kind of rearing and other conditions during childhood are conducive to the development of an aggressive reaction pattern? Very briefly, the following four factors have turned out to be particularly important (based chiefly on my research with boys, Olweus 1980; see also Loeber and Stouthamer-Loeber, 1986):

- the basic emotional attitude of the primary caretaker(s) toward the child during early years;
- parental permissiveness towards aggressive behaviour by the child;
- the use of power-assertive child-rearing methods such as physical punishment and violent emotional outbursts; and
- an active and hot-headed temperament in the child. (See Olweus, 1980, 1993a for more details.)

Group mechanisms

When several students jointly engage in bullying another student, certain group mechanisms are likely to be at work. Due to space limitations, some such important mechanisms are only listed here: (1) social 'contagion'; 2) weakening of control or inhibitions against aggressive tendencies; (3) 'diffusion of responsibility'; and (4) gradual cognitive changes in the perceptions of bullying and of the victim (see Olweus, 1978, 1993a for more details).

A question of fundamental human rights

The victims of bullying form a large group of students who are often neglected by the school. We have shown that many of these youngsters are the targets of harassment for long periods of time, often for many years (Olweus, 1977, 1978). It does not require much imagination to understand what it is to go through the school years in a state of more or less permanent anxiety and insecurity, and with poor self-esteem. It is not surprising that the victims' devaluation of themselves sometimes becomes so overwhelming that they see suicide as the only possible solution.

Bully/victim problems in school really concern some of our basic values and principles. For a long time, I have argued that *it is a fundamental human right for a child to feel safe in school and to be spared the oppression and repeated, intentional humiliation implied in bullying.* No student should be afraid of going to school for fear of being harassed or degraded, and no parent should need to worry about such things happening to his or her child.

As early as 1981, I proposed the introduction of a law against bullying at school (Olweus, 1981b). At that time, there was little political support for the idea. In 1994, however, this suggestion was followed up by the Swedish Parliament, with a new school law article including formulations that are very similar to those expressed above. In addition, the associated regulations place responsibility for realisation of these goals, including the development of an intervention programme against bullying for the individual school, with the principal. At present, passing of a similar article is being discussed in Norway, and now there seems to be considerable political support for the idea.

Effects of a school-based intervention programme

The first evaluation

Against this background, it is appropriate briefly to describe the effects of the Olweus Bullying Prevention Program, developed and evaluated over a period of almost 20 years (see, e.g., Olweus, 1986a, 1993a, 1999b; Olweus and Limber, 1999).

The first evaluation of the effects of the intervention programme was based on data from approximately 2,500 students originally belonging to 112 grade 4–7 classes (modal ages were 11–14 years at the start of the project) in forty-two primary and secondary/junior high schools in Bergen, Norway. The subjects of the study were followed over a period of 2.5 years, from 1983 to 1985 (see, e.g., Olweus, 1991 for details).

The *main findings* of the analyses can be summarised as follows:

- There were marked reductions – by 50 per cent or more – in bully/victim problems for the periods studied, with eight and twenty months of intervention, respectively. By and large, these reductions were obtained for 'direct bullying' (where the victim is exposed to relatively open attacks), 'indirect bullying' (where the victim is isolated and excluded from the group, involuntary loneliness) and 'bullying others'. The results generally applied to both boys and girls and to students from all grades studied.
- There was also a clear reduction in general antisocial behaviour such as vandalism, fighting, pilfering, drunkenness and truancy.
- In addition, we registered marked improvement as regards various aspects of the 'social climate' of the class: improved order and discipline, more positive social relationships, and a more positive attitude to schoolwork and the school. At the same time, there was an increase in student satisfaction with school life.

In the majority of comparisons for which reductions were reported above, the differences between base line and intervention groups were quite marked and highly significant. Detailed analyses of the quality of the data and the possibility of alternative interpretations of the findings led to the following general statements (Olweus, 1991): It is very difficult to explain the results obtained as a consequence of (1) underreporting by the students, (2) gradual changes in the students' attitudes to bully/victim problems, (3) repeated measurement, and (4) concomitant changes in other factors, including general time trends.

In addition, a clear 'dosage-response' relationship has been established in preliminary analyses at the class level (which is the natural unit of analysis in this case): those classes that showed larger reductions in bully/victim problems had implemented three presumably essential components of the intervention programme (including establishment of class rules against bullying and use of regular class meetings) to a greater extent than those with smaller changes (Olweus and Alsaker, 1991). This finding provides corroborating evidence for the hypothesis that the changes observed were due to the intervention programme.

All in all, it was *concluded that the changes in bully/victim problems and related behaviour patterns were likely to be mainly a consequence of the intervention programme and not of some other 'irrelevant' factor*. It was also noted that self-reports, which were implicated in most of these analyses, are probably the best data source for the purposes of this study. At the same time, largely parallel results were obtained for two peer rating variables and for teacher ratings of bully/victim problems at the class level; for the teacher data, however, the effects were somewhat weaker.

Basic principles

The intervention programme is built on *a set of four key principles* derived chiefly from research on the development and modification of the problem behaviours concerned, in particular aggressive behaviour. It is thus important to try to create a school (and ideally, also a home) environment characterised by (1) warmth, positive interest and involvement from adults, on the one hand, and (2) firm limits to unacceptable behaviour, on the other. Third (3), in cases of violations of limits and rules, non-hostile, non-physical sanctions should be consistently applied. Implied in the latter two principles is also a certain degree of monitoring and surveillance of the students' activities in and out of school (Patterson 1982, 1986). Finally (4), adults both at school and home are supposed to act as authorities at least in some respects.

As regards the role of adults, the intervention programme is based on an authoritative (NB. *not* authoritarian) adult–child interaction or child-rearing model (cf., e.g., Baumrind, 1967), in which the adults are encouraged to take responsibility for the children's total situation, not only their learning, but also their social relationships.

These principles were 'translated' into a number of specific measures to be used at the *school*, *class* and *individual* levels. It is considered important to work on all of these levels. Space limitations prevent a description of the various measures, but such an account can be found in *Bullying at school: What we know and what we can do* (Olweus, 1993a) and a new teacher handbook *Olweus' core program against bullying and antisocial behavior* (Olweus, 2001b). Table 3.1 lists a set of core components which are considered, on the basis of statistical analyses and our experience with the programme, to be particularly important in any implementation of the programme.

With regard to implementation and execution, the programme is mainly based on a utilisation of the existing social environment: teachers and other school personnel, students and parents. Non-mental-health professionals thus play a major role in the desired *restructuring of the social environment*. 'Experts' such as school psychologists, counsellors and social workers serve important functions as planners and co-ordinators in counselling teachers and parents (groups), and in handling more serious cases.

Additional evaluations

Results similar to those reported in the first evaluation study were obtained in 'The New Bergen Project against Bullying' (from 1997 to1999) comprising some 3,500 students from thirty schools (Olweus and Limber, 1999), and in another new intervention project (from 1999 to 2000) comprising ten Oslo (capital of Norway) schools with about 1,900 students. Results from the latter project using an 'extended selection cohorts design with contiguous cohorts' (see Cook and Campbell, 1979; the same design as used in the first evaluation study) are presented in Figures 3.1 and 3.2. These graphs show the results for students in grades 5–7 with modal ages between 11 and 13 years. Positive, although some-

Table 3.1 Overview of the Olweus Bullying Prevention Program

General prerequisites
- Awareness and involvement on the part of adults

Measures at the school level
- questionnaire survey
- school conference day
- effective supervision during break times
- educational teacher discussion groups
- formation of co-ordinating group

Measures at the class level
- class rules against bullying
- regular class meetings with students
- meetings with parents of the class

Measures at the individual level
- serious talks with bullies and victims
- serious talks with parents of involved students
- development of individual intervention plans

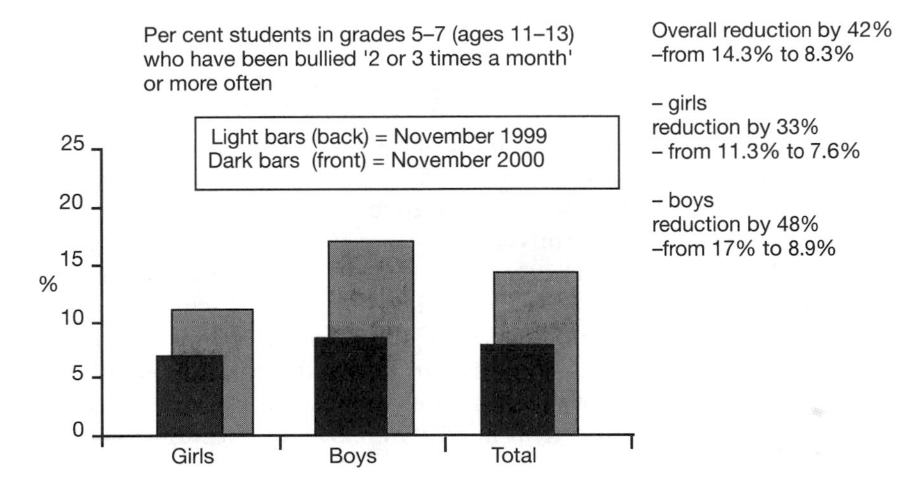

Figure 3.1 The effects of the intervention programme on 'being bullied' (being exposed to direct as well as indirect bullying)

Notes: Light bars (back) show base line data (at Time 1, before intervention) for grade 5, 6 and 7 (or year 12, 13 and 14) cohorts, whereas the dark bars (front) display data for corresponding cohorts at Time 2 (one year later), when they had been exposed to the intervention programme for approximately eight months

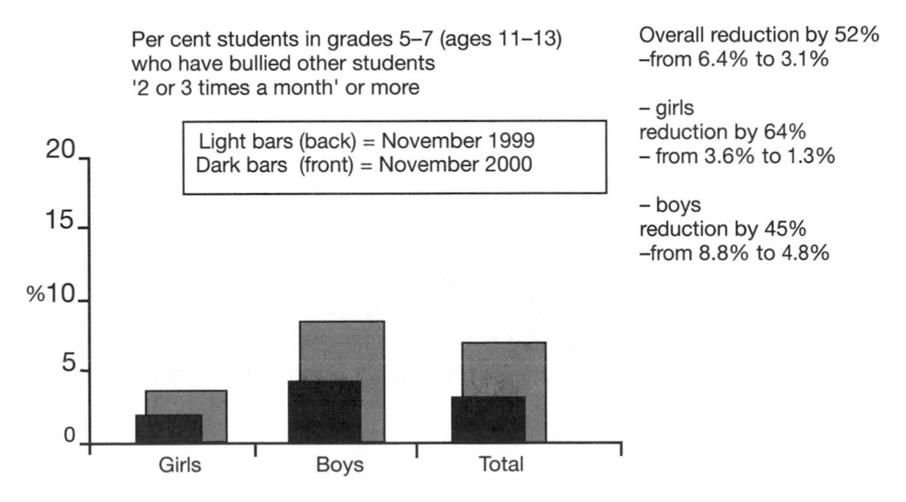

Figure 3.2 The effects of the intervention programme on 'bullying other students'

Notes: See Figure 3.1 for explanation

what weaker, effects have also been reported in partial replications in Great Britain (Smith and Sharp, 1994), Germany and the USA (see Olweus and Limber, 1999).

Possible reasons for the effectiveness of this intervention programme have been discussed in some detail (Olweus, 1992a). They include a *change of the 'opportunity' and 'reward structures' for bullying behaviour* (resulting in fewer opportunities and rewards for bullying). It is also generally emphasised that bully/victim problems can be seen as an excellent *entry point* for dealing with a variety of problems that plague today's schools. Furthermore, one can view the programme from the perspective of planned organisational change (with quite specific goals), and in this way link it with work on school effectiveness and school improvement. It may also be pointed out that the programme in many ways represents what is sometimes called 'a whole school policy approach to bullying' in the English literature. It consists of a set of routines, rules and strategies of communication and action for dealing with existing and future bullying problems in a school.

The need for evidence-based intervention programmes

As bully/victim problems have gradually been placed on the official school agenda in many countries, a number of suggestions about their handling and prevention have been proposed. Some of these suggestions and approaches seem poorly conceived or even counterproductive, whereas others appear to be meaningful and potentially useful. There is a great problem, however, in that most of these approaches or programmes have either failed to document positive results or have never been exposed to systematic research evaluation. Therefore it is difficult to know what actually works and what does not. As demonstrated in other contexts, adult user satisfaction with a programme is certainly not a sufficient criterion.

The situation is well illustrated by the following facts. Recently, a US expert committee systematically evaluated more than 400 presumably violence- (or problem behaviour) preventing programmes according to certain minimum-level criteria (see Elliott, 1999). Only ten of the programmes satisfied the specified criteria, one of which was my Bullying Prevention Program (four of the programmes were school based). These so-called 'blueprint' or model programmes are now being implemented in a number of sites with financial support from the US Department of Justice (OJJDP).

A similar evaluation by an officially appointed committee was recently made in Norway as well. In this case, fifty-five programmes designed to counteract and/or prevent 'problem behaviour' and in use in Norwegian schools were evaluated. Only one programme, my Bullying Prevention Program, was recommended for further use without reservations.

A new national initiative against bullying in school

Against this background, the Norwegian government (via the Ministry of Education and Research and the Ministry of Children and Family Affairs) decided in the year 2000 to introduce the Olweus Bullying Prevention Program over an extended period as an offer to all comprehensive schools in Norway. My colleagues and I are now in the process of realising these ambitious plans, building up an organisational structure for dissemination of the programme and quality control of its implementation in schools. A key element of the new initiative is the establishment of 'educational teacher discussion groups' in each school. These groups receive training and supervision from special instructors (the first cohort of instructors began training in August 2001), who in turn are trained and supervised by members of the 'Olweus Group against Bullying and Antisocial Behavior' at the University of Bergen. In this way, it will be possible to reach out to a considerable number of schools in a relatively short period of time.

Final words

The basic message of our research findings and experience is quite clear: *with a suitable intervention programme, it is definitely possible to reduce dramatically bully/victim problems in school, and related problem behaviours.*[2]

Notes

1 The research programme reported on in this article was supported by grants from the Department for Children and Family Affairs (BFD), the Research Council of Norway (NFR, NAVF), the Johann Jacobs Foundation, Switzerland, and in earlier phases, from the Norwegian Ministry of Education (KUD), which is gratefully acknowledged.

2 The intervention 'package' consists of the book *Bullying at school: What we know and what we can do* (Olweus, 1993a; this book is sold in bookstores or through direct order from the publisher: Blackwell, 108 Cowley Road, Oxford OX4 1JF, UK, or its North American division: Blackwell, 238 Main Street, Cambridge, MA 02142, USA), *Olweus' core program against bullying and antisocial behavior: A teacher handbook* (Olweus, 2001b), the *Revised Olweus Bully/Victim Questionnaire* (Olweus, 1996b) with accompanying PC program, and a video cassette on bullying (Olweus and Limber, 1999). More information about the intervention programme can be obtained from the author on the e-mail address: Olweus@psyhp.uib.no.

Bibliography

Baumrind, D. (1967) Child care practices anteceding three patterns of preschool behavior. *Genetic Psychology Monographs*, *75*, 43–88.

Björkqvist, K., Ekman, K. and Lagerspetz, K. (1982) Bullies and victims: Their ego picture, ideal ego picture and normative ego picture. *Scandinavian Journal of Psychology*, *23*, 307–313.

Boulton, M. J. and Smith, P. K. (1994) Bully/victim problems among middle school children: Stability, self-perceived competence, and peer acceptance. *British Journal of Developmental Psychology*, *12*, 315–329.

Cairns, R. B., Cairns, B. D., Neckerman, H. J., Gest, S. D. and Gariépy, J. L. (1988) Social networks and aggressive behavior: Peer support or peer rejection? *Developmental Psychology*, *24*, 815–823.

Cook, T. D. and Campbell D. T. (1979) *Quasi-experimentation: Design and analysis issues for field settings*. Chicago: Rand McNally.

Elliott, D. S. (1999) Editor's introduction. In D. Olweus and S. Limber, *Blueprints for violence prevention: Bullying Prevention Program*. Boulder, CO: Institute of Behavioral Science, University of Colorado.

Farrington, D. (1993) Understanding and preventing bullying. In M. Tonry (ed.), *Crime and justice: A review of research*, vol. 17. Chicago: University of Chicago Press.

Haeselager, G. J. T. and van Lieshout, C. F. M. (1992) *Social and affective adjustment of self- and peer-reported victims and bullies*. Paper presented at the European Conference on Developmental Psychology, Seville, Spain.

Hirano, K. (1992) *Bullying and victimization in Japanese classrooms*. Paper presented at the European Conference on Developmental Psychology, Seville, Spain.

Junger, M. (1990) Intergroup bullying and racial harassment in the Netherlands. *Sociology and Social Research*, *74*, 65–72.

Lagerspetz, K. M., Björkqvist, K., Berts, M. and King, E. (1982) Group aggression among school children in three schools. *Scandinavian Journal of Psychology*, *23*, 45–52.

Loeber, R. and Dishion, T. (1983) Early predictors of male delinquency: A review. *Psychological Bulletin*, *94*, 69–99.

Loeber, R. and Stouthamer-Loeber, M. (1986) Family factors as correlates and predictors of conduct problems and juvenile delinquency. In M. Tonry and N. Morris (eds), *Crime and justice*, vol. 7. Chicago: University of Chicago Press.

Olweus, D. (1973a) *Hackkycklingar och översittare. Forskning om skolmobbning*. Stockholm: Almqvist and Wicksell.

—— (1973b) Personality and aggression. In J. K. Cole and D. D. Jensen (eds), *Nebraska Symposium on Motivation 1972*. Lincoln: University of Nebraska Press.

—— (1977) Aggression and peer acceptance in adolescent boys: Two short-term longitudinal studies of ratings. *Child Development*, *48*, 1301–1313.

—— (1978) *Aggression in the schools. Bullies and whipping boys*. Washington, DC: Hemisphere Press (Wiley).

—— (1979) Stability of aggressive reaction patterns in males: A review. *Psychological Bulletin*, *86*, 852–875.

—— (1980) Familial and temperamental determinants of aggressive behavior in adolescent boys: A causal analysis. *Developmental Psychology*, *16*, 644–660.

—— (1981a) Bullying among school-boys. In N. Cantwell (ed.), *Children and violence*. Stockholm: Akademilitteratur.

—— (1981b) Vad skapar aggressiva barn? In A. O. Telhaug and S. E. Vestre (eds), *Normkrise og oppdragelse*. Oslo: Didakta.

—— (1983a) *The Olweus Bully/Victim Questionnaire*, mimeo. Bergen: Research Center for Health Promotion (HEMIL), University of Bergen.

—— (1983b) Low school achievement and aggressive behavior in adolescent boys. In D. Magnusson and V. Allen (eds), *Human development. An interactional perspective*. New York: Academic Press.

—— (1984) Aggressors and their victims: Bullying at school. In N. Frude and H. Gault (eds), *Disruptive behavior in schools*. New York: Wiley.

—— (1985) 80,000 barn er innblandet i mobbing. *Norsk Skoleblad*, *2*, 18–23.

—— (1986a) *Mobbning. Vad vi vet och vad vi kan göra*. Stockholm: Liber.

—— (1986b) Aggression and hormones: Behavioral relationship with testosterone and adrenaline. In D. Olweus, J. Block and M. Radke-Yarrow (eds), *Development of antisocial and prosocial behavior*. New York: Academic Press.

—— (1987) Bully/victim problems among schoolchildren. In J. P. Myklebust and R. Ommundsen (eds), *Psykologprofesjonen mot år 2000*. Oslo: Universitetsforlaget.

—— (1991) Bully/victim problems among schoolchildren: Basic facts and effects of a school based intervention program. In D. Pepler and K. Rubin (eds), *The development and treatment of childhood aggression*. Hillsdale, NJ: Erlbaum.

—— (1992a) Bullying among schoolchildren: Intervention and prevention. In R. D. Peters, R. J. McMahon, V. L. Quincy (eds), *Aggression and violence throughout the life span*. Newbury Park, CA: Sage.

—— (1992b) *Mobbning i skolan: Vad vi vet och vad vi kan göra*. Stockholm: Almqvist and Wiksell.

—— (1993a) *Bullying at school: What we know and what we can do*. Oxford, UK, and Cambridge, MA: Blackwell Publishers. (This book has been published in a number of other languages as well.)

—— (1993b) Victimization by peers: Antecedents and long-term outcomes. In K. H. Rubin and J. B. Asendorf (eds), *Social withdrawal, inhibition, and shyness in childhood*. Hillsdale, NJ: Erlbaum.

—— (1994) Annotation: Bullying at school: Basic facts and effects of a school based intervention program. *Journal of Child Psychology and Psychiatry*, *35*, 1171–1190.

—— (1996a) Bully/victim problems in school. *Prospects*, *26*, 331–359.

—— (1996b) *The Revised Olweus Bully/Victim Questionnaire*, mimeo. Bergen: Research Center for Health Promotion (HEMIL), University of Bergen.

—— (1999a) *Sweden*. In P. K. Smith, Y. Morita, J. Junger-Tas, D. Olweus, R. Catalano and P. Slee (eds), *The nature of school bullying: A cross-national perspective* (pp. 7–27). London: Routledge.

—— (1999b) *Norway*. In P. K. Smith, Y. Morita, J. Junger-Tas, D. Olweus, R. Catalano and P. Slee (eds), *The nature of school bullying: A cross-national perspective* (pp. 28–48). London: Routledge.

—— (2001a) Peer harassment. A critical analysis and some important issues. In J. Juvonen and S. Graham (eds), *Peer harassment in school* (pp. 3–20). New York: Guilford Publications.

—— (2001b) *Olweus' core program against bullying and antisocial behavior: A teacher handbook*. Bergen: Research Center for Health promotion (Hemil Center).

Olweus, D. and Alsaker, F. D. (1991) Assessing change in a cohort longitudinal study with hierarchial data. In D. Magnusson, L. R. Bergman, G. Rudinger and B. Törestad (eds), *Problems and methods in longitudinal research* (pp. 107–132), New York: Cambridge University Press.

Olweus, D. and Limber, S. (1999) *Blueprints for violence prevention: Bullying Prevention Program*. Boulder, CO: Institute of Behavioral Science, University of Colorado.

O'Moore, M. and Brendan, H. (1989) Bullying in Dublin schools. *Irish Journal of Psychology*, *10*, 426–441.

Patterson, G. R. (1982) *Coercive family process*. Eugene, Oregon: Castalia Publishing Co.

—— (1986) Performance models for antisocial boys. *American Psychologist*, *41*, 432–444.

Patterson, G. R. and Stouthamer-Loeber, M. (1984) The correlation of family management practices and delinquency. *Child Development*, *55*, 1299–1307.

Perry, D. G., Kusel, S. J. and Perry, L. C. (1988) Victims of peer aggression. *Developmental Psychology*, *24*, 807–814.

Pulkkinen, L. and Tremblay, R. E. (1992) Patterns of boys' social adjustment in two cultures and at different ages: A longitudinal perspective. *International Journal of Behavioral Development*, *15*, 527–553.

Rigby, K. and Slee, P. (1991) Victims in school communities. *Journal of the Australasian Society of Victimology*, 25–31.

Ruiz, R. O. (1992) *Violence in schools. Problems of bullying and victimization in Spain*. Paper presented at the European Conference on Developmental Psychology, Seville, Spain.

Schwartz, D., Dodge, K. and Coie, J. (1993) The emergence of chronic peer victimization in boys' play groups. *Child Development*, *64*, 1755–1772.

Smith, P. K. (1991) The silent nightmare: Bullying and victimization in school peer groups. *The Psychologist*, *4*, 243–248.

Smith, P. K. and Sharp, S. (1994) *School bullying: Insights and perspectives*. London and New York: Routledge.

Smith, P. K., Morita, Y., Junger-Tas, J., Olweus, D., Catalano, R. and Slee, P. (eds) (1999) *The nature of school bullying: A cross-national perspective*. London: Routledge.

Ziegler, S. and Rosenstein-Manner, M. (1991) *Bullying at school: Toronto in an international context* (Report No. 196). Toronto: Toronto Board of Education, Research Services.

Whitney, I. and Smith, P. K. (1993) A survey of the nature and extent of bullying in junior/middle and secondary schools. *Educational Research*, *35*, 3–25.

4 Sexual harassment research in the United States

John B. Pryor and Louise F. Fitzgerald

What is sexual harassment?

In the United States, the concept of workplace sexual harassment developed first as a legal concept and subsequently as a concept empirically studied by social and behavioural scientists. In the US, workplace sexual harassment is considered a form of gender-based discrimination under Federal Law. In 1980, the US Equal Employment Opportunity Commission issued a set of guidelines that have become the cornerstone of legal and policy definitions of sexual harassment throughout the United States. These guidelines describe two general types of sexual harassment: (1) unwelcome sex or gender-related behaviour that creates a *hostile environment* and (2) *quid pro quo* behaviours, where the unwelcome behaviour becomes a term or condition of employment or advancement. The EEOC Guidelines emphasise the importance of examining the specifics of each case in determining whether sexual harassment has taken place. This point seems to acknowledge the importance of contextual factors in judging whether behaviours should be considered to be sexually harassing.

Many subsequent legal cases have since refined and added to the legal understanding of sexual harassment. It is generally understood that unwelcome sexual or gender-related behaviour must be sufficiently severe or pervasive so as to alter the work environment in order to cross the threshold of sexual harassment in a legal sense. While sexual harassment often represents a pattern of behaviour in workplaces where it exists, it is possible for single episodes of behaviour to cross a threshold of severity to be considered sexual harassment in a legal sense – particularly instances of blatant *quid pro quo* sexual harassment. Legal precedents also uphold the notion that victims sometimes may voluntarily participate in behaviours that they nevertheless find unwelcome or offensive (*Meritor Savings Bank v. Vinson*, 1986).

From a legal standpoint, intentions for sexually harassing behaviour are superfluous. The US legal understanding of sexual harassment focuses upon potential employment consequences for victims. Sexual harassment is illegal because it is thought to represent a barrier for equal employment.

Whether the perpetrator intended the behaviour to be offensive or not is not the point of legal deliberations. That the behaviour occurred and was unwelcome are the main considerations. What are considered to be unwelcome sexual or gender-related behaviours obviously varies from person to person and across circumstances. Women may often have perspectives about unwelcome behaviour that differ from those of men (*Ellison v. Brady*, 1991). US courts recognise that they must try to consider the subjective perspective of the victim when determining whether unwelcome behaviour has taken place. While psychological damage may sometimes result from sexual harassment, US courts do not require plaintiffs to prove psychological damage to establish that they have been sexually harassed. Plaintiffs are only required to show that the behaviour would be considered offensive or abusive by a reasonable person who shares the perspective of the victim (*Harris v. Forklift Systems, Inc.*, 1993). A hostile work environment created by sexual or gender-related behaviour can be considered sexual harassment even for co-workers who are not directly targeted for these behaviours (*Broderick v. Ruder*, 1990). So bystanders potentially have legal recourse under US law. US Federal Law makes no assumptions about the gender of the perpetrators or the targets of sexually harassing behaviours. Males and females can fill both roles. Same-sex and other-sex harassment are both possible (*Oncale v. Sundowner Offshore Services, Inc.*, 1998).

Perhaps the key legal issue that distinguishes sexual harassment from bullying is that plaintiffs in sexual harassment law suits in the US must prove that the harassment is somehow based upon gender. For a behaviour to be based upon gender implies that men and women are treated differently (MacKinnon, 1979). In practice, this is often difficult to prove because actual differential comparisons of how men and women are treated are often lacking in sexual harassment cases. Attorneys often appeal to people's general beliefs about how men and women are typically treated to try to argue that a behaviour was based upon gender.

Sexual harassment from a social scientific perspective

The developing legal understanding of sexual harassment has influenced not only how sexual harassment has been studied, but also the selection of topics to be studied in conjunction with sexual harassment. For example, social scientists' selections of potential antecedents and consequences of sexual harassment for study have been influenced by the legal literature. The direction of influence has also gone the other way as well. For example, the legal understanding of a 'reasonable person standard' has been informed by research in the social sciences on gender differences in the interpretation of potentially sexually harassing behaviours (Blumenthal, 1998; Burns, 1995). As in the studies of bullying, much of the research concerning sexual harassment is based upon victim surveys.

Some of the earliest systematic surveys of sexual harassment were conducted by the US Merit Systems Protection Board (USMSPB, 1981). The MSPB scientifically selected sample consisted of Federal workers from all across the US. The operational definition of sexual harassment used in this survey was 'uninvited and unwanted sexual attention on the job'. Respondents were asked to indicate the frequency with which they had experienced several kinds of 'unwanted sexual attention' in the last twenty-four months. A similar survey of Federal workers was repeated in 1987 and 1994. The 1994 survey added 'stalking' to the list of unwanted behaviours. Remarkably, these three surveys found very similar incidence rates. A breakdown of the percentages of women and men reporting each of the specific behaviours is presented in Table 4.1.

One finding that was replicated in all three of the MSPB surveys concerns the role of organisational power. Contrary to popular stereotypes about sexual harassment being just something that supervisors do to underlings, a large majority of the respondents describing their most significant experience reported that the perpetrators were co-workers or other employees with no supervisory authority over the respondent. Another

Table 4.1 Percentages of women and men reporting sexual harassment experiences

Women	*1980*	*1987*	*1994*
Sexual teasing, jokes, remarks	33	35	37
Sexual looks, gestures	28	28	29
Deliberate touching, cornering	26	26	24
Pressure for dates	15	15	13
Suggestive letters, calls, materials	9	12	10
Pressure for sexual favours	9	9	7
Stalking	NA	NA	7
Actual /attempted rape, assault	1	0.8	4
Any type	42	42	44
Men	*1980*	*1987*	*1994*
Sexual teasing, jokes, remarks	10	12	14
Sexual looks, gestures	8	9	9
Deliberate touching, cornering	7	8	8
Pressure for dates	3	4	4
Suggestive letters, calls, materials	3	4	4
Pressure for sexual favours	2	3	2
Stalking	NA	NA	2
Actual /attempted rape, assault	0.3	0.3	2
Any type	15	14	19

Source: From the 1980, 1987 and 1995 US Merit Systems Protection Board Surveys of Federal workers.

interesting finding is that the most frequently described behaviour in the most significant incident was 'unwanted sexual teasing, jokes, remarks, or questions'. Either this behaviour or 'unwanted sexually suggestive looks or gestures' or some combination of these two behaviours accounted for the vast majority of respondents' most significant experiences.

Another early victim survey of sexual harassment was conducted by Gutek (1985). This telephone survey used random digit dialling to collect data concerning the social-sexual behaviours at work experienced by 827 women and 405 men in Los Angeles County. Gutek asked respondents whether they had ever experienced each of the following behaviours at work: (1) sexual comments from a man (or a woman if the respondent was male) that he meant to be insulting; (2) sexual looks or gestures from a man that he meant to be insulting; (3) sexual touching from a man; (4) requests to go out with a man as a part of a job (with the understanding that refusal would hurt the woman's job); and (5) requests to engage in sexual relations with a man as a part of a job (with the understanding that refusal would hurt the woman's job). The percentage of women reporting these behaviours were as follows: 12.2 per cent, 9.1 per cent, 15.3 per cent, 2.8 per cent and 1.8 per cent, respectively. The percentage of men reporting these behaviours were as follows: 12.6 per cent, 12.3 per cent, 20.9 per cent, 2.7 per cent and 1 per cent, respectively.

Obviously, the MSPB surveys found more sexually harassing behaviours reported than did Gutek's (1985) survey. Such differences could have been the result of differences in the samples. Another possibility is that wording differences in the survey questions could account for some of these differences. For example, the MSPB survey asked about 'unwanted sexually suggestive looks or gestures' while the Gutek survey asked about 'sexual looks or gestures from a man that he meant to be insulting'. Thus, some of Gutek's questions required respondents to report only behaviours that they thought were intentionally hostile. Another potentially crucial difference is that Gutek's survey explicitly asked women about behaviours where men were the perpetrators, and asked men about behaviours where women were the perpetrators. In the MSPB surveys, respondents were told to report 'unwanted sexual attention on the job from either sex'. So, some of the reported behaviour in the MSPB surveys may have been same-sex. While the MSPB surveys consistently found more women reporting unwanted sexual behaviours at work than men, Gutek (1985) found slightly more men than women reporting three out of five of the behaviours in her survey. There are many other points of methodological divergence between these two surveys that make them difficult to compare. These comparisons illustrate a general problem in the sexual harassment literature: the absence of a common method of measuring the incidence of sexual harassment. Fortunately, some progress has been made on this issue in recent times.

Many subsequent sexual harassment surveys have modelled the questions used in the original MSPB surveys (e.g. Martindale, 1992). One problem common to the MSPB surveys, Gutek's (1985) survey and many other victim surveys of sexual harassment (Culbertson *et al.*, 1993) is the use of collections of discrete behavioural questions without reference to underlying constructs. This makes it difficult to pose fundamental psychometric questions about the reliability of what is being measured in these surveys. There have been several recent attempts to develop more psychometrically sound instruments for measuring the incidence of sexual harassment in victim surveys (Dekker and Barling, 1998; Fitzgerald *et al.*, 1995; Murdoch and Nichol, 1995). Perhaps the most extensive research on instrument development concerns the Sexual Experiences Questionnaire (SEQ). The SEQ was originally developed to assess sexual harassment experiences in an academic context (Fitzgerald *et al.*, 1988), but has subsequently been adapted for use in business settings (Fitzgerald *et al.*, 1997) and military settings (Fitzgerald *et al.*, 1999; Rosen and Martin, 1998). Variants of the SEQ have also been used in cross-cultural research (Gelfand *et al.*, 1995; Wasti *et al.*, 2000).

The SEQ assesses victims' experiences of three general forms of sexually harassing behaviour: sexual coercion, unwanted sexual attention and gender harassment. Sexual coercion includes the use of threats or bribes to solicit sexual involvement. Unwanted sexual attention includes repeated requests for dates and persistent attempts to establish unwanted sexual relationships. With regard to gender harassment, Gelfand *et al.* (1995) stated: 'This category encompasses a wide range of verbal and nonverbal behaviors not aimed at sexual co-operation; rather, they convey insulting, hostile, and degrading attitudes about women' (p. 8). In the SEQ, each of these three types of sexually harassing behaviours is measured by multiple behavioural items. This allows for standard assessments of construct reliability. Respondents are asked to rate the frequency with which they have experienced these behaviours, usually within a circumscribed time period. Another important methodological aspect of the SEQ is that the term 'sexual harassment' is not mentioned in any of the behavioural questions. Fitzgerald and her colleagues suggest that this term should not be mentioned until after the SEQ behavioural questions. The logic of this juxtaposition is that signalling to participants that these questions are 'about sexual harassment' might potentially bias responses. Respondents may have varying personal definitions of what it means to be sexually harassed. In some cases, respondents might even fail to respond to specific behavioural questions if they have already judged that the questions are not relevant to their experiences.

Research has shown consistently that gender harassment is the most common form of sexually harassing experience, followed by unwanted sexual attention, and then by sexual coercion. This same rank order pattern holds across many organisational and cultural settings (Fitzgerald *et al.*, 1995). In comparing this to a legal analysis of sexual harassment, sexual coercion often constitutes *quid pro quo* sexual harassment, where sexual or gender-related behaviour is a term or condition of employment or advancement. Unwanted sexual attention and gender harassment often constitute what is termed hostile environment sexual harassment, where the sexual or gender-related behaviour creates an intimidating, offensive or hostile work environment. Sometimes unwanted sexual attention may be argued to constitute a form of *quid pro quo* sexual harassment, if toleration of such behaviour becomes a term or condition of employment (Fitzgerald *et al.*, 1995).

Research has found that most targets of sexually harassing behaviour are women and most perpetrators are men. However, a sizeable minority of men have been found to report being the targets of sexually harassing behaviour across the many surveys in which men have been asked about sexual harassment. Recent survey research by Waldo and his colleagues (Waldo *et al.*, 1998) has revised the SEQ so as to be more gender-neutral in content, and has added items that focus upon harassment regarding the enforcement of traditional gender roles – a type of harassment that is often directed towards male targets. One conclusion drawn from research using this revised SEQ is that men may be exposed to behaviour that is potentially sexual harassment far more than previously estimated. Another conclusion is that men are much more likely to be the targets of same-sex harassment than women.

Recent research has developed yet another version of the SEQ that was designed specifically to be used in military settings with samples of both men and women: the SEQ-DoD (SEQ-Department of Defense; Fitzgerald *et al.*, 1999). In one section of this survey, respondents were asked to describe the details of the 'one situation that had the greatest effect'. Among other things, respondents were asked to rate the extent to which they appraised the behaviours in this their most significant experience as annoying, offensive, disturbing and threatening. Fitzgerald and her colleagues have argued that sexual harassment represents a form of work-related stress (Fitzgerald *et al.*, 1997). In a general stress/coping framework (Lazarus, 1993), the primary appraisal of an event is an important psychological step in determining the degree of stress it induces and the coping resources needed to deal with it. Palmieri *et al.* (2000) suggest that appraisal measures such as those used in the SEQ-DoD may be useful in discerning whether a survey respondent actually experienced sexual harassment. Much of the previous research on sexual harassment has used a behavioural item-counting method in establishing whether people have experienced sexual harassment (e.g. Fitzgerald *et al.*, 1988; Gutek, 1985; Martindale, 1992; USMSPB, 1981, 1988, 1995). The question posed was simply whether respondents indicated that they experienced some behaviour or not. One problem with this method of determining 'who counts?' is that some of the experiences indicated by respondents might have been relatively innocuous – for example, the occasional sexual joke. It is likely that survey estimates of the incidence of sexual harassment that do not query whether the respondents found the behaviours in question offensive or somehow bothersome (i.e. the subjective interpretation or appraisal of the behaviour) have overestimated the degree to which respondents have experienced sexual harassment. One might speculate that the solution to this conundrum would be simply to ask people whether they considered the behaviours they experienced to be 'sexual harassment'. However, many researchers have argued that this self-labelling approach is inherently unreliable because of the often fuzzy understanding that lay people have of what can be or should be considered 'sexual harassment'. Also, there is empirical evidence that people who do not label their experiences as 'sexual harassment' nevertheless experience psychological distress, negative job attitudes and other consequences at rates comparable to those who do label them as 'sexual harassment' (Magley *et al.*, 1999; Munson *et al.*, 2001). The solution proposed by Palmieri *et al.* (2000) is a two-step criterion: (1) did the respondent experience any of the SEQ-DoD behaviours, and (2) did they indicate a negative appraisal of the behaviours (i.e. were they annoyed, offended, disturbed and/or threatened)? By this criterion, approximately 53 per cent of the women sampled could be counted as having experienced sexual harassment.[1] This approach for determining the incidence of sexual harassment from a survey sample seems to hold the promise of helping the social scientific and legal understandings to achieve a closer convergence.

Social antecedents of sexual harassment in the workplace

Research reveals several 'risk factors' associated with the occurrence of sexual harassment and other forms of gender-related discrimination. One set of risk factors involves the nature of the job – sometimes referred to as the 'job gender context' (Fitzgerald *et al.*, 1997). The absolute numbers of men and women who interact at work have an impact upon the opportunities of male/female sexual harassment. Sheer inter-gender 'contact' has been found to be an important predictor of sexual harassment (Gruber, 1998; Gutek *et al.*, 1990). In addition, traditionally masculine jobs have been found to pose higher risk for sexual harassment problems for women (Mansfield *et al.*, 1991; Ragins and Scanduri, 1995). Research in military settings has also found that the sexual harassment of women is more likely to occur in work groups where men numerically dominate, and in jobs where women are 'gender pioneers' (i.e. one of the first women in the occupation or job) (Niebuhr, 1997). Having a male immediate supervisor predicts sexual harassment for women (USMSPB, 1995).

Another set of risk factors centres around social norms or organisational climate factors (Gruber, 1998). Some organisations through their stated policies, enforced sanctions and effective leadership actively and successfully discourage sexual harassment. Other organisations through their vague policies, indifference to instances and even sometimes the bad examples set by leaders actually encourage or condone sexual harassment. Men are more likely to harass women sexually at work when management is perceived as tolerating or condoning such behaviour. For example, a study conducted at a US Federal agency (Pryor *et al.*, 1995; Pryor *et al.*, 1993) found that women were more likely to experience sexual harassment in offices where the men believed that local management discouraged complaints about sexual harassment. In studies of military personnel also reported by Pryor *et al.* (1995), the degree to which women on a base or post reported experiences of sexual harassment was related to the degree to which men on the base or post perceived the local commander as tolerating or condoning such behaviour. These studies illustrate that management in an organisation or people in authority in a small group can play important roles in influencing the local social norms regarding sexual harassment. Recent research using data from a service-wide survey of sexual harassment in the US Military found that an organisational climate tolerant of sexual harassment directly contributes to the occurrence of sexual harassment among military personnel (Williams *et al.*, 1999). Perceived military efforts related to the implementation of sexual harassment policies (e.g. thoroughly investigating complaints, enforcing penalties against harassers and making honest efforts to stop harassment) were more strongly related to the incidence of harassment than training efforts or providing resources to victims. Parallel to the findings reported by Pryor and his colleagues, these perceived implementation efforts largely concerned the actions taken by local management. In a study of a sample of women employed at a large, regulated utility company, Fitzgerald *et al.* (1997) found that individual perceptions of organisational tolerance were significantly related to the incidence of harassment. Perceptions of organisational tolerance were operationalised as beliefs concerning the degree of risk to a female who would report harassment, the likelihood that complaints would be taken seriously, and the degree to which a harasser would be punished. Again, these social climate factors seem to emanate from managerial stances on sexual harassment, implying that leadership within an organisation can importantly influence the incidence of sexual harassment.

Gruber (1998) and others have demonstrated that organisational policies can play an important role in preventing sexual harassment. The US Equal Employment Opportunity Commission first issued guidelines on sexual harassment in 1980. By 1987, some surveys indicated that as many as 97 per cent of US companies had sexual harassment policies (National Council for Research on Women, 1995). The passage of Title IX of the Education Amendments of 1972 applied Title VII standards to educational settings. This prohibited discrimination on the basis of sex in educational programmes or activities that receive Federal financial aid. By the mid-1980s, 80 per cent of US public institutions of higher education had policies about sexual harassment (Robertson *et al.*, 1988). Today the presence of sexual harassment policies seems near universal at US institutions of higher education (Kelley, 2000). Research on the introduction of sexual harassment policies in institutions of higher education found that while official complaints about sexual harassment typically increased following the introduction of a policy (Robertson *et al.*, 1988), prevalence surveys showed that the rates at which female students reported experiencing sexually behaviours from instructors and staff declined (Williams *et al.*, 1992). It is not surprising that complaints would rise following the introduction of a sexual harassment policy since most policies outline the procedures for filing complaints. That the prevalence of sexual harassment would decline after a policy has been issued is related not just to the presence of the policy, but the enforcement or implementation of the policy (e.g. effectively investigating complaints and sanctioning harassers). As Williams *et al.* (1999) have found, even when a policy purports zero tolerance of sexual harassment as in the US Armed Forces, the failure of local management to implement the policy can create a climate at risk for sexually harassing behaviour. From a social norm standpoint, policies and their implementation are inputs to an overall organisational climate. The dissemination of policies is a way in which an organisation communicates to employees the desired social norms for employee behaviour. When management fails to enforce policies against sexual harassment, this communicates that the 'real' norms are different from the 'stated' norms.

One way to think of the research on social norms is that it shows that men who want to harass sexually are more likely to do so in social settings where they feel they can get away with it (Pryor and Meyers, 2000). This notion is illustrated by recent research in which men have been questioned about whether they have engaged in sexually harassing behaviour and the social norms they perceive as relevant for such behaviour. In a survey of male Canadian university faculty and staff, Dekker and Barling (1998) asked respondents whether they had engaged in several specific sexually harassing behaviours with a member of the opposite sex at work in the past three months. Men who admitted to having engaged in these behaviours were also likely to see the organisational climates where they worked as supporting such behaviour. Thus, the tendency to admit sexual harassment was related to perceptions that there were few company sanctions for such behaviour and that the workplace was highly sexualised.

In summary, sexual harassment is more likely to occur in a male-dominated job gender context and in a social climate that tolerates or condones such behaviour. When we observe sexual harassment that has taken place repeatedly over a long period of time, it is reasonable to assume that a climate of tolerance has existed. Minimally, the organisation has shown indifference about sexual harassment. When victims complain and nothing is done, the harasser is encouraged to continue such behaviour. Finally, there is also good evidence that not all men are would-be harassers. Individual proclivities for sexual harassment are more likely to translate into sexually harassing behaviours when those with such proclivities find themselves in a climate of organisational tolerance (Dekker and Barling, 1998; Pryor and Meyers, 2000).

Characteristics of harassers and harassing behaviours

The lion's share of research and theory on harassers has focused upon the most common perpetrators and victims – men who sexually harass women. In examining the psychological characteristics of men who sexually harass women, two things are important to keep in mind. First, the term 'sexual harassment' is used to refer to a variety of different types of behaviour. So it is unlikely that a single psychological profile would characterise all harassers. Second, sexually harassing behaviour has been shown to be more likely to occur in some social settings than others (Dekker and Barling, 1998; Pryor *et al.*, 1995). So even though some men may have stronger proclivities for sexually harassing behaviours than others, some social situations may enable or even encourage such behaviours. Research has found that even men with proclivities for sexually harassing behaviours may not act upon these proclivities when the social environment encourages professional behaviour (Dekker and Barling, 1998; Perry *et al.*, 1998; Pryor *et al.*, 1993). This framework for understanding the antecedents of sexually harassing behaviour is called a Person X Situation analysis. Pryor *et al.* (1993) have suggested that both person factors (relatively stable personality and attitudinal characteristics of particular men) and situational factors (organisational climate factors) may importantly contribute to repeated sexually harassing behaviour. Below potential proclivities for different forms of sexual harassment are discussed.

Gender harassment

By all accounts, gender harassment is the 'garden variety' of sexually harassing behaviour, that is the type that occurs most often and the type that is most likely to occur in isolation of the others (Fitzgerald *et al.*, 1999). Gender harassment is not sexual behaviour, per se, in that its aim does not seem to be to gain sexual access to the target. Rather it may be better construed as sexist behaviour or behaviour that puts down or offends those targeted. Pryor and Whalen (1996) theorised that individual differences in sexism may contribute to male propensities for gender harassment. Recent research by Pryor and his students (Pryor *et al.*, 1999; Hahn *et al.*, 1999) has confirmed this prediction. Manipulated situational factors, such as having one's sense of masculinity threatened by being outperformed by a woman on a masculine task and being exposed to a film depicting the sexist treatment of women, also contributed to men's propensities to exhibit sexist verbal behaviours in these studies.

Dekker and Barling (1998) recently applied a similar analysis of person and situation factors in a study of Canadian university faculty and staff. Male respondents were asked whether they had engaged in gender harassment behaviours with a member of the opposite sex at work in the past three months. Dekker and Barling found that men who admitted to having performed various forms of gender harassment were likely to possess what Dekker and Barling called inappropriate sexual harassment beliefs – for example, the belief that women exaggerate sexual harassment problems (a person factor). Men who admitted to having engaged in gender harassment were also likely to see the organisational climates where they worked as supporting such behaviour. Thus, the tendency to admit gender harassment was related to perceptions that there were few company sanctions for such behaviour and that the workplace was highly sexualised (situational factors).

Unwanted sexual attention

Pryor and Whalen (1996) also theorised that a common contributor to men's persistence in making unwanted sexual advances towards women might simply be a lack of social sensitivity. Research by Abbey (1982, 1987) and others (see Stockdale, 1993) suggests that men often interpret women's friendly behaviours toward men as conveying more sexual intent than women intend to convey. Ridge and Reber (in press) suggest that men's erroneous initial expectations about sexual intentions may lead them to engage in a dynamic behavioural confirmation process that exacerbates their misperceptions of women's mutual sexual intent.

Dekker and Barling (1998) also found evidence for person and situation factors as potential contributors to unwanted sexual behaviours. Men with weak perspective-taking skills, adversarial sexual beliefs or inappropriate sexual harassment beliefs were likely to engage in sexualised harassment. Such behaviours were also more likely when company sanctions were perceived as few. In addition, the presence of fewer company sanctions tended to exacerbate men's proclivities to engage in sexualised harassment such that men with weak perspective-taking skills, inappropriate sexual harassment beliefs or adversarial sexual beliefs were even more likely to engage in sexualised harassment when company sanctions were perceived as few.

Sexual coercion

Since 1987, a great deal of research has explored male proclivities for sexually exploitative or coercive behaviours such as a willingness to use a social power differential to enlist sexual co-operation (for a review, see

Pryor and Meyers, 2000). Based upon a variety of empirical studies, a psychological profile of men who might be likely to use social power for sexually exploiting women would include the following: (1) proclivities for rape and other forms of sexual aggression; (2) a strong desire to dominate women; (3) psychological justification of male dominance; (4) viewing sex in non-relationship terms, for example as a way to escape boredom, achieve social recognition or indulge appetites; (5) a traditional view of gender roles; (6) weak perspective-taking skills; and (7) a personality that combines low Conscientiousness (especially high impulsiveness) and low Openness (narrow-mindedness) (Gutek and Done, 2001; Pryor and Whalen, 1996).

Research has found that whether men with proclivities for sexually exploitative behaviour choose to behave in this way in situations where they have power over women seems to be related to situation factors such as differences in organisational climate or local social norms (Pryor *et al.*, 1995). For example, research has found that these men are strongly influenced by the behaviours of their peers and by the behaviours of people in authority in their sexually harassing behaviours. When such sexually coercive behaviours are modelled or condoned by peers and authority figures, these men are also more likely to enact them.

In summary, there is a building body of research that supports the Person X Situation Model for understanding the occurrence of different types of sexually harassing behaviour. Certain personal characteristics of men seem to present risks for sexually harassing behaviour. Social situation factors can enable or disable, and encourage or discourage such behaviour. So far, this analysis has been applied only to male-perpetrator/female-target behaviours.

Victim characteristics

There has not been a great deal of research on personal characteristics of victims of sexual harassment. Rosen and Martin (1998) found that victims of sexual harassment in the US Military were more likely than non-victims to have a history of childhood sexual abuse. Whether would-be harassers target these people because they appear more vulnerable or whether the victims of previous abuse are more likely to report sexual harassment on surveys is not known. Clearly more research needs to be done in this area. Experimental research by Dall'Ara and Maass (1999) has found that male harassers may target women more for gender harassment (sending them electronic pornographic material) when the women are perceived to be egalitarian than when they are perceived to have traditional gender role orientations. Men thus may harass women they see as threatening male hegemony.

Consequences of sexual harassment in the workplace

Most research on the consequences of sexual harassment has focused upon female targets of this behaviour. Research has found two broad classes of negative psychosocial reactions that women have to sexual harassment at work: (1) psychosocial reactions relevant to women's personal lives, and (2) psychosocial reactions relevant to women's professional lives (Gutek and Koss, 1993). Pryor (1995) found that the degree to which harassed women experience negative psychosocial reactions was related to various factors. Harassed women who thought that complaining to management was ineffective were more likely to experience both of these types of negative psychosocial reactions. This characteristic sense of futility or helplessness when one's complaints are unheeded seems to add to the stress sexually harassed women experience. Pryor (1995) also found that the degree to which harassed women experienced negative emotional reactions from sexual harassment was related to the duration of the harassment and the organisational status of the harasser. Long-term harassment from someone who has the power to do you harm seems particularly stressful. Because the negative impact of stress is cumulative to some degree, long-term sexual harassment is likely to be particularly harmful.

The experience of sexual harassment at work is considered a form of *work-related stress* (Fitzgerald *et al.*, 1995). Several studies have found that many forms of sexual harassment such as exposure to sexist put-downs or unwanted sexual remarks are correlated with subsequent psychological problems for women (e.g. Glomb *et al.*, 1997). Across studies, there appears to be a consistent association between experiencing sexual harassment and symptoms of post-traumatic stress disorder (PTSD), anxiety and depression, and alcohol misuse (Dansky and Kilpatrick, 1997; Firth-Cozens, 1991; Murdoch and Nichol, 1995; Richman *et al.*, 1992; Rospenda *et al.*, 2000; Schneider *et al.*, 1997; Wolfe *et al.*, 1998). Recent research by Schneider *et al.* (2001) has established a causal connection between experiences of gender harassment and physiological measures of stress. Women who were subjected to sexist remarks from a male co-worker showed significant levels of autonomic physiological activity consistent with the pattern of cardiac and vascular reactions that people generally display in 'threat' situations. Such cardiovascular reactivity to stress has been related to coronary heart disease and depressed immune functioning. This suggests that women who are continually exposed to sexist behaviour may be at a greater risk for long-term health consequences.

Evidence of psychological stress has also been found among women who were aware of the sexual harassment of one of their female co-workers, but not themselves the targets of harassment (Glomb *et al.*, 1997). As Gruber (1992) and others have argued, the organisational environment that tolerates harassment is likely to be stressful to all women who work there. Women who witness or become aware of others' experiences of sexual harassment may reasonably worry about whether they may be next in line for such treatment.

Recent research by Fitzgerald and her colleagues (Fitzgerald *et al.*, 1997) examined the work-related outcomes of sexual harassment for women in the context of an integrated model of the antecedents and consequences of sexual harassment in an organisational setting. This research differentiated work-related outcomes into work withdrawal (i.e. attempts to avoid work tasks, such as absenteeism and tardiness) and job withdrawal (e.g. turnover and retirement intentions) (Hanisch and Hulin, 1990, 1991). A structural equation analysis indicated that sexual harassment was related to work withdrawal through its impact upon job satisfaction. Sexual harassment was related to job withdrawal through its impact upon health conditions and health satisfaction. Similar findings were reported by Schneider *et al.* (1997) using different samples. Both work and job withdrawal reflect a woman's responses to an aversive work environment. Essentially, these behaviours represent attempts to avoid or escape the social context in which sexual harassment takes place.

In summary, the experience of sexual harassment has been found to have a negative impact upon both the professional and personal lives of targeted women, damaging both their achievement opportunities and their psychological well-being. Specifically, experiences of sexual harassment are related to several stress-induced psychological conditions. Sexual harassment also has been found to predict several work-related outcomes such as work and job withdrawal.

Training/education programmes and sexual harassment

There seems to be a general perception among workers that training/education programmes have some positive impact upon sexual harassment. For example, 63 per cent of the respondents in the MSPB (USMSPB, 1995) survey of Federal workers indicated that the training provided by their agencies helps reduce or prevent sexual harassment. All of the Federal Agencies surveyed provide sexual harassment training programmes. Eighty-seven per cent of the supervisors and 77 per cent of the non-supervisors who completed this survey said that they had indeed received training. However, one of the big gaps seems to be in the scientific evaluation of these training programmes. Evaluations of training consisted almost entirely of post-training course evaluations. These are, in essence, gauges of how participants feel about the training experience. The MSPB reports that none of the Federal Agencies they surveyed had ever conducted evaluations of whether the sexual harassment training offered in their organisations had any effect upon the particular problems it was designed to address. In analyses of an earlier MSPB survey of sexual harassment in the Federal workplace, a more comprehensive attempt was made to explore the possible connections between the estimates of training efforts of various Federal Agencies and their sexual harassment rates (USMSPB, 1988). These analyses revealed no consistent correlations between training efforts and agency incidence of sexual harassment. More questions about the effectiveness of training arise from recent analyses conducted of data from the 1995 DoD survey of US military personnel. While the majority of military personnel rated sexual harassment training as at least moderately effective (Bastian *et al.*, 1996), survey measures assessing the thoroughness of the specific training experiences of survey respondents or of the training experiences offered at the respondents' duty stations were not predictive of sexual harassment rates (Williams *et al.*, 1999).

Organisation-sponsored training/education programmes aimed at sexual harassment prevention are sometimes embedded in 'diversity training' programmes (Hemphill and Haines, 1998) or 'awareness training' programmes (Meyer *et al.*, 1981). A review of the research literature by Myers (1995) concluded that the effectiveness of training or other interventions that focus on changing diversity-related attitudes and behaviours has not been clearly demonstrated. Recent research examined factors related to human resources (HR) professionals' impressions of diversity training success. In a mail survey of 785 SHRM members (Society for Human Resource Management), Rynes and Rosen (1995) found that respondents' holistic evaluations of diversity training success correlated with mandatory management attendance in the training, the use of post-training evaluations (i.e. evaluations of the training other than trainee reactions at the close of training), the HR professionals' perception that the company used a broad definition of diversity, the extent to which managers were explicitly rewarded for increasing the diversity in their units, and the HR professionals' perception that top management supported diversity training. Finally, male respondents to the survey tended to perceive diversity training as less successful than female respondents.

Over the years, there have been a number of experimental or quasi-experimental studies examining the effectiveness of short-term training upon harassment-related attitudes and perceptions (for a review, see Pryor and Meyers, 2000). The majority of these investigations have involved samples of undergraduates. Some of these studies seemed to suffer from potential experimenter demand biases (e.g. Beauvais, 1986). Research by York *et al.* (1997) found that a frequently used training technique, asking participants to analyse written cases, can affect undergraduates' sensitivity to sexually harassing behaviours portrayed in training films. Standard sexual harassment training films have been found to have similar effects (e.g. Blakely *et al.*, 1998). Training has also been shown to have an impact of gender differences in reactions to sexual harassment. Moyer and Nath (1998) examined whether and how men and women differed in their ability to discriminate descriptions of sexually harassing behaviour from those of non-sexually harassing behaviour, and whether any gender differences in judgements could be reduced by training. Training involved learning about legal definitions of sexual harassment and, in some conditions, seeing examples portrayed in a film. Without training, women were found to be both more biased (they more often judged sexual harassment when it was not there) and more accurate (they better discriminated harassment from non-harassment) than men. After training, Moyer and Nath found a convergence in women's and men's judgements, which was due to men adopting a response bias similar to that of women.

One goal of training might be to reduce the likelihood that men at risk for sexually harassing behaviours would actually perform them. Research by Perry *et al.* (1998) seems to represent one of the only studies that has sought to establish that training can have an impact upon such higher risk men. Perry and her colleagues exposed undergraduate volunteers to a popular, commercially available sexual harassment training film (Meyer, 1992). This research found that men with strong proclivities for sexually harassing behaviours as measured by the Likelihood to Sexually Harass (LSH) Scale (Pryor, 1998) in a pre-training test showed gains in knowledge about sexual harassment after viewing the film and displayed fewer sexual advances to a woman they were assigned to teach a golf lesson. The Perry *et al.* (1998) study suggests that these cognitive/educational changes may be followed by short-term changes in behaviour. Interestingly, when Perry and her colleagues re-measured LSH scores in a post-training test, they found that men's scores did not differ significantly from pre-training measures. This suggests that exposure to the training film might have had an impact upon participants by making salient anti-harassment norms rather than by changing men's inherent proclivities for sexual harassment.

Other training/education studies have focused primarily upon samples of potential victims (Maurizio and Rogers, 1992). For example, Barak (1994) found that participating in a cognitive-behavioural workshop can increase women's understanding of sexual harassment and enhance their skills in coping with it. Barak's evaluation of victim-focused training is unique in that it included exclusively female groups and focused upon coping skills. Most training that transpires in organisational settings does not segregate workers by gender. However, most organisationally based training does include discussion of some coping mechanisms such as filing formal complaints, and some training includes discussion of informal resolutions to harassment problems. The practicality of segregating workers into groups by gender for sexual harassment training while maintaining a climate of gender equality remains to be seen. Indeed, there seems to be a potential to intensify a 'we/they' sense among men and women in the workplace that could actually exacerbate harassment problems (see Pryor and Whalen, 1996).

In summary, while workers who attend training/education programmes seem to feel that such programmes have a positive impact upon sexual harassment in the workplace, research is mixed about such a connection. The various experimental studies reviewed showed that short-term, training/education programmes can have demonstrable effects upon how people (especially men) think about sexual harassment, and how men with harassment proclivities behave. While limited in scope, these experimental results are encouraging. Certainly, such short-term changes are necessary if training is ultimately to have an impact on workplace behaviours.

Conclusions

For US researchers, sexual harassment is generally considered to be a form of gender-based discrimination. Sexually harassing behaviours are more likely in organisational environments that tolerate or condone such behaviour. While some individuals seem more likely to commit these behaviours than others, even those with proclivities for sexually harassing behaviour seem reluctant to harass sexually when the local social norms discourage such behaviour. Experiences of sexually harassing behaviour are potentially sources of stress for workers. Such negative behaviours undermine the physical and mental health of workers and reduce their productivity. As we continue into the twenty-first century, hopefully research will lead to a better understanding of how to encourage a healthy and productive workplace where each person is treated with dignity and respect – where work is enjoyed and is perceived as the fulfilment of human potential and not as an ordeal to be dreaded.

Notes

1 Palmieri *et al.* did not include respondents who only experienced sexist behaviour in this count. The rationale for excluding this group is that DoD policy guidelines consider such gender discrimination as separate from sexual harassment. Also, Palmieri and his colleagues did not report how this rational-empirical algorithm for determining 'how counts?' would affect counts for men.

Bibliography

Abbey, A. (1982) Sex differences in attributions for friendly behavior: Do males misperceive females' friendliness? *Journal of Personality and Social Psychology*, *42*, 830–838.

—— (1987) Misperceptions of friendly behavior as sexual interest: A survey of naturally occurring incidents. *Psychology of Women Quarterly*, *11*, 173–194.

Barak, A. (1994) A cognitive-behavioral educational workshop to combat sexual harassment in the workplace. *Journal of Counseling and Development*, 72, 595–602.

Bastian, L. D., Lancaster, A. R. and Reyst, H. E. (1996) *Department of Defense 1995 sexual harassment survey*. Arlington, VA: Defense Manpower Data Center.

Beauvais, K. (1986) Workshops to combat sexual harassment: A case of studying changing attitudes. *Signs*, *12*, 130–145.

Blakely, G. L., Blakely, E. H. and Moorman, R. H. (1998) The effects of training on perceptions of sexual harassment allegations. *Journal of Applied Social Psychology*, *28*, 71–83.

Blumenthal, J. A. (1998) The reasonable woman standard: A meta-analytic review of gender differences in perceptions of sexual harassment. *Law and Human Behavior*, *22*, 33–57.

Broderick v. Ruder, 685 F.Supp. 1269, 46 EPD 37,963 (DDC 198–8).

Burns, S. E. (1995) Issues in workplace sexual harassment law and related social science research. *Journal of Social Issues*, *51*, 193–207.

Culbertson, A. L., Rosenfeld, P. and Newel, C. E. (1993) *Sexual harassment in the Active-Duty Navy: Findings from the 1991 Navy-wide survey*. San Diego, CA: Navy Personnel Research and Development Center.

Dall'Ara, E. and Maass, A. (1999) Studying sexual harassment in the laboratory: Are egalitarian women at higher risk? *Sex Roles*, *41*, 681–704.

Dansky, B. S. and Kilpatrick, D. G. (1997) Effects of sexual harassment. In W. O'Donohue (ed.), *Sexual harassment: Theory, research, and treatment* (pp. 152–174). Boston: Allyn and Bacon.

Dekker, I. and Barling, J. (1998) Personal and organizational predictors of workplace sexual harassment of women by men. *Journal of Occupational Health Psychology*, *3*, 7–18.

Ellison v. Brady, 924 F. 2d 872 (CA9 1991).

Equal Employment Opportunity Commission. (1980) Guidelines and discrimination because of sex (Sec. 1604.11) *Federal Register*, *45*, 74676–74677.

Firth-Cozens, J. (1991) Sources of stress in women junior house officers. *British Medical Journal*, *328*, 89–91.

Fitzgerald, L. F., Drasgow, F. and Magley, V. J. (1999) Sexual harassment in the armed forces: A test of an integrated model. *Military Psychology*, *11*, 3, 329–343.

Fitzgerald, L. F., Gelfand, M. J. and Drasgow, F. (1995) Measuring sexual harassment: Theoretical and psychometric advances. *Basic and Applied Social Psychology*, *17*, 4, 425–445.

Fitzgerald, L. F., Hulin, C. and Drasgow, F. (1995) The antecedents and consequences of sexual harassment in organizations: An integrated model. In G. Keita and J. Hurrell (eds), *Job stress in a changing workforce: Investigating gender, diversity, and family issues* (pp. 55–73). Washington, DC: American Psychological Association.

Fitzgerald, L. F., Drasgow, F., Hulin, C. L., Gelfand, M. J. and Magley, V. (1997) Antecedents and consequences of sexual harassment in organizations: A test of an integrated model. *Journal of Applied Psychology*, *82*, 578–589.

Fitzgerald, L. F., Magley, V. J., Drasgow, F. and Waldo, C. R. (1999) Measuring sexual harassment in the military: The Sexual Experiences Questionnaire (SEQ–DoD). *Military Psychology*, 11, 3, 243–263.

Fitzgerald, L. F., Shullman, S. L., Bailey, N., Richards, M., Swecker, J., Gold, Y., Ormerod, A. J. and Weitzman, L. (1988) The incidence and dimensions of sexual harassment in academia and the workplace. *Journal of Vocational Behavior*, *32*, 152–175.

Gelfand, M. J., Fitzgerald, L. F. and Drasgow, F. (1995) The structure of sexual harassment: A confirmatory analysis across cultures and settings. *Journal of Vocational Behavior*, *47*, 2, 164–177.

Glomb, T. M., Richman, W. L, Hulin, C. L., Drasgow, F., Schneider, K. T. and Fitzgerald, L. F. (1997) Ambient sexual harassment: An integrated model of antecedents and consequences. *Organizational Behavior and Human Decision Processes*, *71*, 3, 309–328.

Gruber, J. E. (1998) The impact of male work environments and organizational policies on women's experiences of sexual harassment. *Gender and Society*, *12*, 3, 301–320.

—— (1992) A typology of personal and environmental sexual harassment: Research and policy implications for the 1990s. *Sex Roles*, *26*, 447–464.

Gutek, B. A. (1985) *Sex and the workplace*. San Francisco: Jossey-Bass.

Gutek, B. A. and Done, R. S. (2001) Sexual harassment. In R. Unger (ed.), *Handbook of the psychology of women and gender* (pp. 367–387). New York: John Wiley and Sons.

Gutek, B. A. and Koss, M. P. (1993) Change women and changed organizations: Consequences of and coping with sexual harassment. *Journal of Vocational Behavior* (Special Issue), *42*, 1–21.

Gutek, B. A., Cohen, A. G. and Konrad, A. M. (1990) Predicting social-sexual behavior at work: A contact hypothesis. *Academy of Management Journal*, *33*, 560–577.

Hahn, E., Pryor, J. B., Hitlan, R. and Olson, M. (1999) *Gender harassment conceptualized as a form of hostile outgroup discrimination against women.* Paper presented at the June meeting of the American Psychological Society, Boulder, CO.

Hanisch, K. A. and Hulin, C. L. (1990) Job attitudes and organizational withdrawal: An examination of retirement and other voluntary withdrawal behaviors. *Journal of Vocational Behavior*, *37*, 60–78.

—— (1991) General attitudes and organizational withdrawal: An evaluation of a causal model. *Journal of Vocational Behavior*, *39*, 110–128.

Harris v. Forklift Systems, Inc., 115 S. Ct. 367 (1993)

Hemphill, H. and Haines, R. (1998) Combat harassment by starting with mind (establish a zero-tolerance policy). *Business Insurance*, *32*, 16.

Kelley, M. (2000) Sexual harassment in the 1990s. *The Journal of Higher Education*, *71*, 548–568.

Lazarus, R. S. (1993) Coping theory and research: Past, present, and future. *Psychosomatic Medicine*, *55*, 234–247.

MacKinnon, C. A. (1979) *The sexual harassment of working women: A case of sex discrimination*. New Haven, CT: Yale University Press.

Magley, V. J., Hulin, C. L., Fitzgerald, L. F. and DeNardo, M. (1999) Outcomes of self-labeling sexual harassment. *Journal of Applied Psychology*, *84*, 390–402.

Mansfield, P. K., Koch, P. B., Henderson, J., Vicary, J. R., Cohn, M. and Young, E. W. (1991) The job climate for women in traditionally male blue-collar occupations. *Sex Roles*, *25*, 63–79.

Martindale, M. (1992) Sexual harassment in the military: 1988. *Sociological Practice Review*, *2*, August, 200–216.

Maurizio, S. J. and Rogers, J. L. (1992) Sexual harassment and attitudes in rural community care workers. *Health Values*, *16*, 40–45.

Meritor Savings Bank v. Vinson, 106 S. Ct. 2399 (1986).

Meyer, A. (1992) Getting to the heart of sexual harassment. *HR Magazine*, *37*, 82–84.

Meyer, M. C., Berchtold, I. M., Oestreich, J. L. and Collins, F. J. (1981) *Sexual harassment*. New York: Petrocelli Books, Inc.

Moyer, R. S. and Nath, A. (1998) Some effects of brief training interventions on perceptions of sexual harassment. *Journal of Applied Social Psychology*, *28*, 4, 333–356.

Munson, L. J., Miner, A. G. and Hulin, C. (2001) Labeling sexual harassment in the military: An extension and replication. *Journal of Applied Psychology*, *86*, 293–303.

Murdoch, M. and Nichol, K. (1995) Women veterans' experiences with domestic violence and sexual harassment while they were in the military. *Archives of Family Medicine*, *4*, May, 411–418.

Myers, J. G. (1995) Enhancing the effects of diversity awareness training: A review of the research literature. *FAA Office of Aviation Medicine Reports*, *10*, March, 1–25.

National Council for Research on Women (1995) *Sexual harassment: Research and resources*. New York: National Council for Research on Women.

Niebuhr, R. E. (1997) Sexual harassment in the military. In W. O'Donohue (ed.), *Sexual harassment: Theory, research, and treatment* (pp. 250–262). Boston: Allyn and Bacon.

Oncale v. Sundowner Offshore Services, Inc., et al. 118 S. Ct. 998 (1998).

Palmieri, P., Harned, M., Collinsworth, L., Fitzgerald, L.F. and Lancaster, A. (2000) *Who counts? A rational-empirical algorithm for determining the incidence of sexual harassment in the military*. University of Illinois-Urbana/Champaign: DMDC Technical Report.

Perry, E. L., Kulik, T. and Schmidtke, J. M. (1998) Individual differences in the effectiveness of sexual harassment awareness training. *Journal of Applied Social Psychology*, *28*, 8, 698–723.

Pryor, J. B. (1995) The psychosocial impact of sexual harassment on women in the US Military. *Basic and Applied Social Psychology*, *17*, 581–603.

—— (1998) The Likelihood to Sexually Harass Scale. In C. M. Davis, W. H. Yarber, R. Bauserman, G. Schreer and S. L. Davis (eds), *Sexuality-related measures: A compendium* (pp. 295–298). Beverly Hills, CA: Sage.

Pryor, J. B. and Meyers, A. B. (2000) Men who sexually harass women. In L. B. Schlesinger (ed.), *Serial offenders: Current thought, recent findings, unusual syndromes*. Boca Raton, FL: CRC Press LLC.

Pryor, J. B. and Whalen, N. J. (1996) A typology of sexual harassment: Characteristics of harassers and the social circumstances under which sexual harassment occurs. In W. O'Donohue (ed.), *Sexual harassment: theory, research, and treatment* (pp. 130–151). Needham Heights, MA: Allyn and Bacon.

Pryor, J. B., Giedd, J. L. and Williams, K. B. (1995) A social psychological model for predicting sexual harassment. *Journal of Social Issues*, *51*, 69–84.

Pryor, J. B., LaVite, C. and Stoller, L. (1993) A social psychological analysis of sexual harassment: The person/situation interaction. *Journal of Vocational Behavior* (Special Issue), *42*, 68–83.

Pryor, J. B., Hitlan, R., Olson, M. and Hahn, E. (1999) *Gender harassment: Some social psychological studies of the antecedents*. Invited address at the Meeting of the Midwestern Psychological Association, Chicago.

Ragins, B. R. and Scandura, T. A. (1995) Antecedents and work-related correlates of reported sexual harassment: An empirical investigation of competing hypotheses. *Sex Roles*, *32*, 7–8, 429–455.

Richman, J. A., Flaherty, J. A. and Rospenda, K. M. (1992) Mental health consequences and correlates of reported medical student abuse. *Journal of the American Medical Association*, *267* (5 February), 692–694.

Ridge, R. D. and Reber, J. S. (in press) I think she's attracted to me: The effect of men's beliefs on women's behavior. *Basic and Applied Social Psychology*.

Robertson, C., Dyer, C. E. and Campbell, D. (1988) Campus harassment: sexual harassment policies and procedures at institutions of higher learning. *Signs: Journal of Women in Culture and Society*, *13*, 792–812.

Rosen, L. N. and Martin, L. (1998) Childhood maltreatment history as a risk factor for sexual harassment among US Army soldiers. *Violence and Victims*, *13*, 3, 269–286.

Rospenda, K. M., Richman, J. A., Wislar, J. S. and Flaherty, J. A. (2000) Chronicity of sexual harassment and generalized work-place abuse: Effects on drinking outcomes. *Addiction*, *95*, 1805–1820.

Rynes, S. and Rosen, B. (1995) A field survey of factors affecting the adoption and perceived success of diversity training. *Personnel Psychology*, *48*, 247–270.

Schneider, K. T., Swan, S. and Fitzgerald, L. F. (1997) Job-related and psychological effects of sexual harassment in the workplace: Empirical evidence from two organizations. *Journal of Applied Psychology*, *82*, 3, 401–415.

Schneider, K. T., Tomaka, J., Palacios-Esquivel, R. and Goldsmith, S. D. (2001) Women's cognitive, affective, and physiological reactions to a male co-worker's sexist behavior. *Journal of Applied Social Psychology*, *31*, 1995–2018.

Stockdale, M. S. (1993) The role of sexual misperceptions of women's friendliness in an emerging theory of sexual harassment. *Journal of Vocational Behavior*, *42*, 84–101.

United States Merit Systems Protection Board (1981) *Sexual harassment in the Federal workplace: Is it a problem?* Washington, DC: US Government Printing Office.

—— (1988) *Sexual harassment in the Federal Government: An update*. Washington, DC: US Government Printing Office.

—— (1995) *Sexual harassment in the Federal Government: Trends, progress, continuing challenges*. Washington, DC: US Government Printing Office.

Waldo, C. R., Berdahl, J. L. and Fitzgerald, L. F. (1998) Are men sexually harassed? If so by whom? *Law and Human Behavior*, *22*, 59–79.

Wasti, S. A., Bergman, M. E., Glomb, T. M. and Drasgow, F. (2000) Test of the cross-cultural generalizability of a model of sexual harassment. *Journal of Applied Psychology*, *85*, 766–778.

Williams, J. H., Fitzgerald, L. F. and Drasgow, F. (1999) The effects of organizational practices on sexual harassment and individual outcomes in the military. *Military Psychology*, *11*, 303–328.

Williams, E. A., Lam, J. A. and Shively, M. (1992) The impact of a university policy on the sexual harassment of female students. *Journal of Higher Education*, *63*, 50–64.

Wolfe, J., Sharkansky, E., Read, J., Dawson, R., Martin, J. and Oimette, P. C. (1998) Sexual harassment and assault as predictors of PTSD symptomatology among US female Persian Gulf War military personnel. *Journal of Interpersonal Violence*, *13*, 40–57.

York, K. M., Barblay, L. and Zajack, A. (1997) Preventing sexual harassment: The effect of multiple training methods. *Employee Responsibilities and Rights Journal*, *10*, 277–289.

Part 2

The evidence

5 Empirical findings on bullying in the workplace

Dieter Zapf, Ståle Einarsen, Helge Hoel and Maarit Vartia

Introduction

This chapter aims at summarising some descriptive empirical findings of bullying in the workplace. We will start with the frequency and the duration of bullying, the number, gender and status of bullies and victims, distribution of bullying across branches and the use of various categories of bullying. The empirical basis of this chapter is restricted to studies carried out in Europe.

The phenomenon of bullying, which includes being exposed to persistent insults or offensive remarks, persistent criticism, personal or even physical abuse, has been labelled 'mobbing at work' in some Scandinavian and German countries (Leymann, 1996) and 'bullying at work' in many English-speaking countries (Liefooghe and Olafsson, 1999). Typically, a victim is constantly teased, badgered and insulted, and perceives that he or she has little recourse to retaliate in kind. Bullying may take the form of open verbal or physical attacks on the victim, but may also take the form of more subtle acts, such as excluding or isolating the victim from his or her peer group (Einarsen *et al.*, 1994; Leymann, 1996; Zapf *et al.*, 1996). The following definition of bullying or mobbing seems to be widely agreed upon (Einarsen *et al.*, this volume; cf. Einarsen, 2000; Einarsen and Skogstad, 1996; Leymann, 1993b; Zapf, 1999a):

> Bullying at work means harassing, offending, socially excluding someone or negatively affecting someone's work tasks. In order for the label bullying (or mobbing) to be applied to a particular activity, interaction or process it has to occur repeatedly and regularly (e.g. weekly) and over a period of time (e.g. about six months). Bullying is an escalating process in the course of which the person confronted ends up in an inferior position and becomes the target of systematic negative social acts. A conflict cannot be called bullying if the incident is an isolated event or if two parties of approximately equal 'strength' are in conflict.
>
> (Einarsen *et al.*, this volume, p. 15)

It should be noted that other authors use less strict definitions with regard to the timeframe (less than six months) and the frequency of the bullying behaviour (less often than once a week) (cf. Einarsen, 2000; Einarsen *et al.*, this volume; Hoel *et al.*, 1999; Keashly and Jagatic, this volume; Zapf, 1999a).

The frequency of bullying

Especially for practical reasons it is important to know how frequently bullying actually occurs in organisations, because efforts to develop measures against this phenomenon would depend on it. However, it is not easy to provide reliable numbers. The problem is that the frequency of bullying depends very much on how it is measured (cf. Hoel *et al.*, 1999). The measures, again, are influenced by the general understanding of what bullying is.

Some researchers administered questionnaires and fixed a cut-off point (e.g. Björkqvist *et al.*, 1994). Respondents scoring higher than the cut-off point were considered to be victims of bullying. Usually, these studies report a prevalence rate as high as 10–17 per cent bullying (cf. Table 5.1). Other researchers used a strategy developed by Leymann (1996) which we will call the 'Leymann criterion': Here, the Leymann Inventory of Psychological Terrorization (LIPT), a questionnaire of forty-five items (Leymann, 1990, 1996), or a similar questionnaire such as the Negative Acts Questionnaire (NAQ) (Einarsen and Raknes, 1997) is administered. To be considered a bullying victim, the response to at least one item or to one general item on the frequency of bullying actions should be: 'at least once a week', and the duration of bullying should be at least six months. Studies using this strategy report frequencies of bullying between 3 and 7 per cent (Table 5.1). However, extremely high numbers are also reported in some studies using this method (Niedl, 1996). Others have asked directly: 'Have you been bullied during the last six months?' (e.g. Rayner, 1997). This, typically, leads to a comparatively high amount of bullying of 10–25 per cent, because people will also say that they have been bullied when only occasional minor negative acts have occurred. Yet another strategy is to give a precise definition of what bullying is, for example the definition given above, and ask respondents whether they consider themselves victims of bullying according to the definition. This strategy leads to 1–4 per cent bullying (e.g. Einarsen and Skogstad, 1996; Hoel and Cooper, 2000; Leymann, 1996; Mackensen von Astfeld, 2000; Piirainen *et al.*, 2000). Yet, in studies employing both measures of exposure to specific behaviours as well as measures of overall feelings of being victimised according to a given definition, the latter seem to provide the lowest number of victims (Mikkelsen and Einarsen, 2001; Salin, 2001).

Taking all studies in Table 5.1 together, a percentage of between 1 and 4 per cent serious bullying has emerged in Europe, in the sense of the above-

Table 5.1 International studies on the frequency of workplace bullying (see also Hoel *et al.*, 1999; Niedl, 1995; Zapf, 1999a)

Country	*Authors*	*Sample*	*No.*	*Definition*	*Prevalence*
Austria	Niedl (1995)	Hospital employees	368	1b+3a	26.6 % in sample; 7.8 % of the population
		Research institute employees	63	1b+3a	17.5 % in sample; 4.4 % of the population
Denmark	Hogh and Dofradottir (2001)	Randomised sample	1857	5	2 %
	Mikkelsen and Einarsen (2001)	Course participants at the Royal Danish School of Educational Studies	99	1b+3a+4	4: 2 %; 1b 3a: 14 % (7.8 % for a more stringent criterion)
		Hospital employees	236	1b+3a+4	4: 3 % now and then; 1b 3a: 16 % (2 %)
		Manufacturing company	224	1b+3a+4	4: 4.1 % now and then; 1b 3a: 8 % (2.7 %)
		Department store	215	1a+3a+4	4: 0.9 %; 1b 3a: 25 % (6.5 %)
Finland	Björkqvist *et al.* (1994)	University employees	338	1a+2	16.9 %
	Salin (2001)	Random sample of business professionals holding a university degree	385	1b+4	1.6 %; 8.8 % occasionally; 1b and 3a: 24.1 %
	Vartia (1996)	Municipal employees	949	4	10.1 %
	Vartia and Hyyti (2002)	Prison officers	896	1a+4	20 %; 11.8 % bullied several times a month
	Piirainen *et al.* (2000)	Representative of employed	1991	4	4.3 %
	Kivimäki *et al.* (2000)	Hospital staff	5655	4	5.3 %

Table 5.1 continued on next page

Table 5.1 continued

Country	*Authors*	*Sample*	*No.*	*Definition*	*Prevalence*
Germany	Minkel (1996)	Employees of a rehabilitation clinic	46	1b+3a	8.7 %
	zur Mühlen *et al.* (2001)	Communal administration	552	1b+3a	10.0 %
	Mackensen von Astfeld (2000)	Administration within federal armed forces	511	1b+3a	10.8 %
		Administration	1989	1b+3a	2.9 %
Ireland	O'Moore (2000)	Random national sample of 4425	1009	4	16.9 % occasionally; 6.2 % frequently
Hungary	Kaucsek and Simon (1995)	Army	323	1b+3a	5.6 %
		Bank employees	41	1b+3a	4.9 %
		Bank inspectors	43	1b+3a	2.5 %
The Netherlands	Hubert *et al.* (2001)	Mixed production office business		4	4.4 %
		Financial institutions; stacked sample	427	3a+4	1 %
	Hubert and van Veldhoven (2001)	Sample including the following branches: 'industry', 'education', 'health care', '(local) government and public administration', 'trade' (included is repairing of consumer articles), 'business services', 'financial institutions', 'construction industry', 'transport' (included are storage and communication), 'public utilities' and 'service organisations on environmental, cultural and recreational issues	3011 66764	2+5	2.2 % mean of 4 items referring to aggressive and unpleasant situations often or always

Country	*Authors*	*Sample*	*No.*	*Definition*	*Prevalence*
Norway	Einarsen and	14 different samples; total	7787	1a+4	Weekly 1.2 % (yes, by and then: 3.4 %);
	Skogstad	Health and welfare managers	344		8.6 % occasional bullying
	(1996)	Psychologists' union	1402		0.3 % (12.0 %)
		Employers' Federation	181		0.6 % (2.3 %)
		University	1470		0.6 % (2.3 %)
		Electricians' union	480		0.7 % (2.8 %)
		Health-care workers	2145		0.8 % (3.1 %)
		Industrial workers	485		1.1 % (2.2 %)
		Graphical workers' union	159		1.3 % (6.5 %)
		Teachers' union	554		1.9 % (8.9 %)
		Trade and Commerce	383		2.4 % (2.0 %)
		Union of hotel /restaurant workers	172		2.9 % (4.3 %)
		Clerical workers and officials	265		2.9 % (4.1 %)
	Matthiesen	Nurses and assistant nurses	99		3.9 % (3.9 %)
	et al. (1989)	Teachers	84		10.3 %
	Einarsen *et*	Representative sample from a county	745		6 %
	al. (1998)				3 %. 8.4 % with previous experience
Portugal	Cowie *et al.* (2000)	International organisation	221	4	33.5 %

Table 5.1 continued on next page

Table 5.1 continued

Country	*Authors*	*Sample*	*No.*	*Definition*	*Prevalence*
Sweden	Leymann (1993a, 1993b)	Representative of employed except self-employed	2438	1b+3a	3.5 %
	Leymann and Tallgren (1993)	Steelworks employees	171	1b+3a	3.5 % (probably lower because of dropouts)
	Leymann (1993a)	Sawing factory	120	1b+3a	1.7 %
	Leymann *et al.* in Leymann (1993b)	Nursery schools	37	1b+3a	16.2 %
	Leymann (1992)	Handicapped employees; non-profit organisation	179	1b+3a	8.4 %; 21.6 % handicapped; 4.4 % not handicapped
	Lindroth and Leymann (1993)	Nursery school teachers	230	1b+3a	6 %
UK	Rayner (1997)	Part time students	581	1c+4	53 %
	UNISON (1997)	Public sector union members	736	1a+4	14 %; 1c+4: 50 %
	Quine (1999)	National Health Service	1100	3b	38 % persistently bullied within last 12 months
	Cowie *et al.* (2000)	International organisation	386	4	15.4 %
	Hoel *et al.* (2001)	Representative sample	5288	1a+3a+4	1.4 %; 3b: 10.6 %

Notes:
1 denotes duration of acts: 1a within the last 6 months; 1b over 6 months; 1c ever in the career
2 denotes type of acts included in judgements
3 denotes frequency of acts: 3a at least weekly; 3b less frequently than weekly
4 denotes victims label themselves as bullied based on a definition
5 denotes approximate criterion

given definition. For somewhat less severe cases (bullying less often than weekly), frequencies between 8 and 10 per cent have been repeatedly reported (e.g. Einarsen and Skogstad, 1996; Hoel *et al.*, 2001). Moreover, the data on less strict criteria on bullying (e.g. Rayner, 1997) as well as studies which simply assess if negative social behaviour at work occurs at all (e.g. Hoel and Cooper, 2000; Hubert and van Veldhoven, 2001; Mikkelsen and Einarsen, 2001; Zapf, 2001) suggest that in many organisations, up to 20 per cent or more of the employees are occasionally exposed to negative social acts, such as being yelled at, teased or humiliated, which do not fall under the stringent definition of bullying but which do mean that these employees are exposed to intensive social stressors at work.

There are various reasons why data on bullying may be exaggerated (e.g. Hoel *et al.*, 1999). Victims may, of course, be much more motivated to respond to questionnaires on bullying than employees who are not concerned (e.g. Niedl, 1996). However, a Norwegian study showed that only 10 per cent more victims were found in the original sample than in a follow-up study among those who did not respond in the original survey (Einarsen and Raknes, 1991). Moreover, some researchers measured observed bullying by asking if the respondents had observed someone in their workplace being subjected to bullying. For example, when all prison workers were studied in Finland, 23 per cent felt themselves to have been bullied and 46 per cent had observed bullying in their own workplace (Vartia and Hyyti, 1999). The fact that observed bullying is higher than self-reported bullying may also support the view that bullying is not exaggerated by the victims.

On the other hand, there are also reasons for underestimating the frequency of bullying because people may not want to admit their status as victims as well as the fact that they are in an inferior position and are not able adequately to cope with the situation. Three per cent serious bullying in a 1,000-employee organisation means that around thirty people are involved at any point in time. Given that not only these thirty people, but usually also a larger group of bullies as well as bystanders are, in one way or another, negatively affected by the bullying situation, we consider this a serious number.

The duration of bullying

In many European countries, meanwhile, even minor conflicts and arguments in everyday life may be described as 'mobbing' or 'bullying'. Therefore, the duration of bullying is an important criterion to differentiate between bullying and everyday conflicts in organisations. Studies reporting on the duration of bullying are summarised in Table 5.2. The studies show that bullying is a long-lasting conflict. Looking at some large representative samples in Sweden (Leymann, 1996) and Norway (Einarsen and Skogstad, 1996), the average duration of bullying was 15

Table 5.2 Average duration of workplace bullying in months

Study	*No.*	*Duration in Months*
Finland (Salin, 2001)	34	32
Germany (DAG-Study, Zapf, 1999a)	56	47
Germany (Gießen Study, Zapf, 1999a)	50	40
Germany (Halama and Möckel, 1995)	183	40
Germany (Konstanz Study, Zapf, 1999a)	87	46
Germany (Stuttgart Study, Zapf *et al.*, 1996)	188	29
Germany (communal administration, zur Mühlen *et al.* 2001)	55	34
Germany (army administration, zur Mühlen *et al.*, 2001)	55	24
Ireland (O'Moore, 2000)	248	41
Norway (Einarsen and Skogstad, 1996)	268	18
Sweden (Leymann, 1996)	85	15
Switzerland (von Holzen-Beusch *et al.*, in press)	28	36

and 18 months respectively. Among bullied Finnish prison officers, 66 per cent of the females and 53 per cent of the males had been bullied for more than two years (Vartia and Hyyti, 2002). In the study by Hoel and Cooper (2000), 39 per cent of the victims had been bullied for more than two years. Among victims in a Finnish municipal institution 29 per cent had been bullied for 2–5 years and as many as 30 per cent for over five years (Vartia, 2001). In studies of victims only, the average duration was much higher, with a mean of more than three years (e.g. Leymann and Gustafsson, 1996; Zapf, 1999a). This difference is probably due to method: If one tries to find bullying victims and to contact them by advisory telephones and self-help groups, etc., one will receive a self-selected sample of more severe bullying cases. In any case, these numbers underscore that bullying is not a short episode but a very long-lasting process that 'wears down' its victims, mostly lasting much longer than one year.

Gender effects of bullying

A frequently asked question in the public is whether there are gender effects of bullying. Although data exist on the gender of bullies and victims (see Table 5.3) there has not been much theorising or in-depth research on this issue (Vartia and Hyyti, 2002).

Table 5.3 The gender distribution of bullying victims (%)

Sample	*Gender of bullying victims*			*Total sample*[a]		
	Men	*Women*	*No.*	*Men*	*Women*	*No.*
Austria (Niedl, 1995, hospital)	37	63	98	39	61	368
Austria (Niedl, 1995, research institute)	18	82	11	45	55	63
Denmark (Mikkelsen and Einarsen, 2002)	9	91	118			
Finland (Björkqvist *et al.*, 1994)	39	61	70	48	52	338
Finland (Salin, 2001)	33	67	6	43	57	376
Finland (Vartia, 1993)	33	67	95	36	64	949
Finland (Vartia and Hyyti, 2002)	86	14	145	87	13	896
Finland (Kivimäki *et al.*, 2000)	12	88	302	12	88	5655
Finland (Nuutinen *et al.*, 1999)	35	65	84	48	52	751
Germany (Bielefeld study, Zapf, 1999a)	35	65	99			
Germany (DAG-Study, Zapf, 1999a)	32	68	56			
Germany (Dick and Dulz, 1994)	26	74	200			
Germany (Gießen Study, Zapf, 1999a)	30	70	50			
Germany (Halama and Möckel, 1995)	25	75	183			
Germany (Konstanz Study, Zapf, 1999a)	44	56	86			
Germany (Mackensen von Astfeld, 2000)	41	59	115	41	59	1972
Germany (Zapf *et al.*, 1996)	38	62	183			
Ireland (O'Moore *et al.*, 1998)	30	70	30			
Italy (Ege, 1998)	49	51	301			
Norway (Einarsen and Skogstad, 1996)	46	54	96	44	56	7787
Norway (Matthiesen and Einarsen, 2001)	23	77	85			
Sweden (Leymann and Gustafsson, 1996)	31	69	64			
Sweden (Leymann and Tallgren, 1993)	50	50	24	53	47	171
Sweden (Leymann, 1993b)	45	55	85	51	49	2438
Switzerland (von Holzen Beusch *et al.*, in press)	32	68	28			
UK (Rayner, 1997)	53	47	581	52	48	1121
UK (Quine, 1999)	18	82	418	16	84	1091
UK (Hoel and Cooper, 2000)	48	52	553	52	48	5288

Note:
[a] Samples with missing data in these columns comprise victims only

Gender of the victims

One can argue that there is some relation between female socialisation and the victim role because women are said to be educated to be less self-assertive and less aggressive, and tend to be more obliging than men (Björkqvist, 1994). Consequently, women are even less able than men to defend themselves when bullying is starting. Moreover, for various reasons, women hold less powerful positions in organisations. For example, they are less often in managerial or supervisor positions (Davidson and Cooper, 1992). Data is shown in Table 5.3. In most samples, the victims are about one-third men and two-thirds women. However, several Scandinavian samples (Einarsen and Skogstad, 1996; Leymann, 1996; Leymann and Tallgren, 1993) and the UK studies by Hoel and Cooper (2000) and Rayner (1997) show a more balanced picture.

In the research institute sample of Niedl (1995), the university sample of Björkqvist *et al.* (1994) and in a sample of police officers (Nuutinen *et al.*, 1999) women appeared more often among the victims in comparison to their number in the overall sample in which the victims were identified. However, in most other samples (Einarsen and Skogstad, 1996; Kivimäki *et al.* 2000; Leymann, 1996; Leymann and Tallgren, 1993; Niedl, 1995, hospital sample; Rayner, 1997; Vartia, 1993) the men/women ratio of the victims corresponded to the respective ratio in the overall sample, suggesting that the over-representation of women in some samples (i.e. in the samples of Niedl, 1995, hospital) is due to the over-representation of women in the respective population. That is, these authors studied bullying in women-dominated sectors such as the service sector, and especially the social services and health sector. All in all, there seems to be little evidence that women are more at risk because of a specific female socialisation.

Moreover, it is known from research on stress and health, that women are more prepared to take part in studies on issues where personal problems or weaknesses play a role (Kasl and Cooper, 1987). Finally, bullying is more often a top-down process than the other way round (see below). Women are more often in subordinate positions which may also contribute to their increased risk of becoming a victim of bullying.

An explanation for the higher risk associated with women in the police sample (Nuutinen *et al.*, 1999) may lie in their salient role in a male-dominated organisation. Minority groups who differ from the main groups in salient characteristics carry a higher risk of being socially excluded from the group (Schuster, 1996). Women may be seen as intruders in the male-dominated cultures of researchers, business professionals or the police force.

Gender of the bullies

In the studies by Zapf (1999a), altogether (N = 209) 26 per cent of the bullies were men only, 11 per cent were women only, and in 63 per cent of all cases men and women were among the bullies. Einarsen and Skogstad (1996) report that 49 per cent of the victims were men, 30 per cent were women, and that in 21 per cent of all cases the bullies were both men and women. The respective numbers in the study by Mackensen von Astfeld (2000) were: 32 per cent men, 27 per cent women and 37 per cent both. In the study by Rayner (1997) two-thirds of the bullies were men. All in all, men seem to be clearly over-represented among the bullies in most studies (an exception is UNISON, 1997). This result corresponds to similar findings in research on bullying at school (Olweus, 1994). Bullying, at least in part, includes forms of direct aggression, such as shouting or humiliating somebody. There is substantial empirical evidence that this kind of aggression is much more typical for men than for women, who prefer forms of indirect aggression such as social exclusion or spreading rumours (Björkqvist, 1994). Moreover, supervisors play a dominant role in bullying (see below). Men are over-represented in supervisor positions. This may all explain why more men are among the bullies.

Finally, Leymann (1993a, 1993b) reported that women are more often bullied by other women, and that men are more often bullied by other men, which he explained in terms of the segregation of the labour market. Similar results were found by Einarsen and Skogstad (1996), Hoel *et al.* (2001), Mackensen von Astfeld (2000), Niedl (1995), Rayner (1997), and Zapf (1999a). Whereas women are sometimes exclusively bullied by men, it is very seldom that men are exclusively bullied by women. This finding may be explained by the different power positions of men and women in organisations.

The number of bullies

Although bullying can be a conflict between two people, some victims report that the entire organisation is bullying them. Data on the number of bullies in various studies are summarised in Table 5.4. In most studies, in between 20 and 40 per cent of all cases there was only one person bullying the victim, whereas in between 15 and 25 per cent there were more than four bullies involved. In the German victims studies (Zapf, 1999a) bullying by only one person was much rarer. In fact, in these studies, in more than 50 per cent of all cases more than four bullies were involved. These differences may be explained as follows: as described above, samples consisting of bullying victims usually consist of more serious bullying cases, which, for example, show a longer mean duration of the bullying conflict. There is some evidence, that bullying becomes more and more serious the longer it lasts. Studies by Einarsen and Skogstad (1996) and Zapf and Gross (2001) showed that bullying occurred more often the longer it lasted. In the study by Zapf (1999a), the duration of bullying correlated positively with the number of bullies. The average duration of bullying of those who were bullied by only one

Table 5.4 The number of bullies (%)

Bullies	*No.*	*1 bully*	*2–4 bullies*	*More than 4 bullies*
Austria (Hospital, Niedl, 1995)	82	20	52	28
Austria (Research institute, Niedl, 1995)	11	55	27	18
Germany (DAG Study, Zapf, 1999a)	55	9	35	56
Germany (Gießen Study, Zapf 1999a)[b]	50	10	50	40
Germany (Konstanz Study, Zapf, 1999a)	78	9	32	59
Germany (Mackensen von Astfeld, 2000)	115	38	46	16
Hungary (Army, Kauscec and Simon, 1995)	18	23	62	14
Ireland (O'Moore, 2000)	248	62	38	0
Ireland (O'Moore *et al.*, 1998)	30	63	33	3
Italy (Ege, 1998)	301	20	46	34
Norway (Einarsen and Skogstad, 1996)[a]	392	42	43	15
Sweden (Leymann and Tallgren, 1993)	24	43	50	7
Sweden (Leymann, 1993b)	85	34	43	23

Notes:
[a] The third category of this study was '4 and more bullies'
[b] The middle category of this study was '2 –5 bullies'

person was twenty-eight months, for those who were bullied by two to four and more than four people, it was thirty-six months and fifty-five months respectively. These data suggest that it is getting more and more difficult to remain a neutral bystander the longer bullying lasts. Therefore more and more people may become involved as bullies in the course of time. This may explain the higher mean number of bullies in the pure victim samples which show a higher mean duration of bullying.

Some studies, especially the British ones (Hoel and Cooper, 2000; Rayner, 1997), report that many victims share their experience with other colleagues. For example, in the study by Hoel and Cooper, as much as 55 per cent of the bullying victims reported that they shared their experience with other work colleagues, and 15 per cent reported that everyone in the work group was bullied. Similar results were reported in the UNISON studies (1997, 2000). In other countries such as Austria (Niedl, 1995) or Germany (Zapf, 1999a, 1999b), this is reported only occasionally. This may be a country-specific phenomenon; however, it may have to do with the definition of bullying. The more stringent the definition of bullying, the more likely it is that it involves only one victim. While a perpetrator may occasionally bully everyone in the work group for months and years, it seems much more unlikely that he or she can bully to such an intensity that everyone in the work group is exposed to bullying at least on a weekly basis.

The organisational status of bullies and victims

The relationship between bullying and organisational status has so far received limited attention (Hoel *et al.*, 2001). Organisational status in this respect refers to the formal position within the organisational hierarchy.

The status of the victim

Relatively little has been reported about the status of the victim. Einarsen and Raknes (1997), in a study of male employees at a Norwegian engineering plant, found no difference between the experience of negative behaviours for workers, on the one hand, and supervisors/managers, on the other. Similar results were found by Hoel *et al.* (2001), who present the most elaborated study on this issue. They found similar numbers of bullying for workers, supervisors, and middle and senior management. A representative sample of Finnish employees showed that upper white-collar employees experienced bullying somewhat more often than lower white-collar employees or workers (Piirainen *et al.*, 2000). Salin (2001), however, found less bullying at the higher levels of the organisation. Hoel *et al.* report some interesting interaction effects with gender: Whereas male workers and supervisors were bullied more than women, this was the other way round at the management level. The largest differences occurred for the senior management level, where 16 per cent of the female senior managers reported having been bullied. This finding may be due to the salience of women at this male-dominated hierarchical level and may reflect common prejudice against women in leadership positions (see Davidson and Cooper, 1992).

All in all, the findings of Hoel *et al.* (2001) question a common assumption in various European countries that the weak and defenceless, in terms of organisational status, become the primary victims of bullying. Rather, there seem to be similar risks at all organisational status levels. Supervisors and senior managers may also experience a power imbalance relative to colleagues and superiors.

The status of the bully

By contrast, the issue of perpetrator status has received considerable attention. Interestingly, the findings vary across countries. Leymann (1993b) introduced 'mobbing' as the definition of a lasting conflict among colleagues. Yet even in his study, there were only marginally more colleagues among the bullies than there were supervisors. However, taking the Scandinavian studies as a whole, these identified people in superior positions as offenders in approximately equal numbers to peers, with only a small number bullied by a subordinate (Einarsen and Skogstad, 1996; Leymann, 1992, 1993b). In contrast, British studies have consistently identified people in superior positions as bullying in an overwhelming majority of cases (Cowie *et al.*, 2001; Hoel *et al.*, 2001; Rayner, 1997). Studies in other European countries (e.g. Niedl, 1995; Zapf, 1999a) were in between. In part, they topped the British numbers with regard to supervisor bullying (e.g. Zapf, 1999a, DAG-study); however, the number of colleagues among the bullies was much higher in these studies than the British ones. For example, in the British UNISON (1997) study, 84 per cent of respondents were bullied by their managers as opposed to 16 per cent by a colleague. Hoel *et al.* (2001), however, found a much higher percentage of co-worker bullying (see Table 5.5). In Britain, there seems to be a common perception of bullying predominantly as a process whereby a worker is being bullied

Table 5.5 Position of the bullies (%)

Bullies	*No.*	*Supervisors*	*Colleagues*	*Subordinates*
Austria (Hospital, Niedl, 1995)	76	75	55	13
Austria (Research institute, Niedl, 1995)	11	73	45	0
Finland (Björkqvist *et al.*, 1994) observed bullying	137	55	32	12
Finland (Vartia, 1993)	98	59	70	4
Finland (Vartia and Hyyti, 2002)	145	43	55	5
Germany (DAG Study, Zapf, 1999a)	56	91	71	16
Germany (Gießen Study, Zapf 1999a)	50	82	78	18
Germany (Konstanz Study, Zapf, 1999a)	86	67	80	16
Germany (Mackensen von Astfeld, 2000)	115	62	53	4
Germany (Stuttgart Study, Zapf *et al.*, 1996)	119	82	78	18
Germany (Communal administration, zur Mühlen *et al.* 2001)	55	58	37	27
Germany (Army administration, zur Mühlen *et al.* 2001)	55	41	67	9
Ireland (O'Moore, 2000)	248	70	32	9
Ireland (O'Moore *et al.*, 1998)	30	93	7	0
Italy (Ege, 1998)	301	87	40	2
Norway (Einarsen and Skogstad, 1996)	489	54	54	15
Portugal (Cowie *et al.*, 2000)	74	45	71	0
Sweden (Leymann, 1993b)	85	47	54	9
Switzerland (von Holzen Beusch *et al.*, in press)	28	85	59	26
UK (Rayner, 1997)	582	71	12	3
UK (UNISON, 1997)	103	84	16	0
UK (Quine 1999)	421	54	34	12
UK (Cowie *et al.*, 2000)	59	71	17	0
UK (Hoel *et al.*, 2001)	553	75	37	7

by someone in a managerial capacity (Hoel *et al.*, 2001), a fact which seems to have been amplified both by the media and by British trade unions.

Occasionally, gender differences are reported. For example, Vartia and Hyyti (2002) found that women were more often bullied by co-workers, whereas men were more often bullied by immediate supervisors or managers.

Einarsen (2000), referring to Hofstede (1993), argued that some cultural differences between the Nordic and the central European countries may explain some of these differences. Hofstede's studies suggest that low power differentials and feminine values prevail in the Scandinavian countries. The abuse of formal power is much more sanctioned in such countries. Power differences between immediate supervisors and their colleagues are small, hence producing similar numbers of perpetrators for supervisors and colleagues.

Generally, superiors are seldom bullied by subordinates. In particular, there are only a very few cases reported where superiors were exclusively bullied by their subordinates. Usually, subordinates bully a superior together with other supervisors or managers. The reason for this is, of course, that it is not easy to overcome the formal power of a superior using informal power. It is possible if the superior is isolated (which points at tensions or conflicts within management), but it is almost impossible if the superior is backed up by superiors at the same level and/or by senior management. One can certainly say that only isolated superiors, who have lost the support of their colleagues and of higher management, carry the risk of becoming the victims of subordinates' bullying.

With regard to negative supervisory behaviour there is still a lack of knowledge. As stated by Ashforth (1994), leadership study has focused almost exclusively on the positive aspects of leadership, that is on factors that contribute to job satisfaction and performance – obviously on the implicit assumption that negative leadership behaviour can be equated with a lack of positive behaviour such as task orientation or consideration. Although some of the common leadership questionnaires, such as the leader behaviour description questionnaire LBDQ (Fleishman, 1953), contain some items similar to those which appear in workplace bullying questionnaires, negative leadership behaviour has not really been investigated in this tradition. Humiliating, yelling or threatening somebody is not simply the absence of consideration or employee orientation. Bullying by superiors is, therefore, a leadership research issue of its own right.

The frequency of bullying in various sectors

In this section some findings on the frequency of bullying in various sectors are summarised. Leymann (1993a, 1993b) reported an over-representation of bullying in the educational (approximately 2:1) and administrative

(1.5:1) sectors, and an under-representation in the trade, production and health sectors. The prevalence of bullying in Swedish public administration was 1 per cent higher than the average of 3.5 per cent (Leymann, 1993a). However, in other studies Leymann also found a high level of occurrence in the health sector. In the study by Leymann and Gustafsson (1996), public administration, the social and health sectors, as well as religious organisations showed greater prevalence, whereas trade and industry were under-represented. Niedl (1995), Piirainen *et al.* (2000) and Vartia (1993, 1996) also report high levels of bullying in the health and social sector. Similarly, in the studies by Einarsen and Skogstad (1996), the highest rate of frequent bullying (weekly or more often) was found among clerical and officials (3.9 per cent) and within trade and commerce (3.5 per cent). For occasional bullying the results were different. Here, in contrast to Leymann, there were significantly fewer respondents from public companies who reported bullying than from private enterprises. The highest prevalence rate was found among industrial workers, where 17.4 per cent reported having been occasionally bullied during the last six months. Bullying was also frequent among those who did graphical work, and hotel and restaurant workers. The lowest rate of bullying was found among psychologists and university employees.

In Germany, analyses based on almost 400 victims of serious bullying (Zapf, 1999a) showed that employees of the health and social sector had a seven-fold risk of being bullied. Other risk factors were: public administration, 3.5:1; educational sector, 3:1. Moreover, there was an increased risk of being bullied in the banking and insurance sectors. In contrast, the risk was relatively low in the traffic, trade and farming industries, in the hotel and restaurant sector, as well as in the building industry. Hubert and van Veldhoven (2001) found increased risks of aggressive and unpleasant behaviour in service organisations, in industry and in education. Salin (2001) reported more frequent bullying in the public sector than in the private sector, and Piirainen *et al.* (2000) in the municipal sector rather than the private sector or the civil service.

Taking the studies together, a higher risk of being bullied is reported for the social and health, public administration and education sectors, which all belong to the public sector. There may be various reasons which explain the differences between sectors. First, one may assume that bullying is less frequent in small family enterprises such as occur frequently in the hotel and restaurant business as well as in the building sector. Here personal relationships can be expected to develop between employees and between employers. If severe conflicts arise, one party may leave the 'family'. Moreover, in these and other areas, short-term job contracts prevail; conflicts lasting several years are almost impossible because the employees would find it relatively easily to terminate their job.

On the other hand, in many European countries, for example Germany, Norway and Sweden, working in public administration means having a secure, lifelong job which usually compensates for a somewhat lower than average salary. In this case, it is much more complicated to give up one's job when bullying occurs, because this would involve giving up the high job security which is among the most important aspects of these jobs. Frequently the specific knowledge gained in such jobs cannot be applied in the private sector. Moreover, it is often difficult to change one's job within the public sector because it is all one organisation. An example would be the case of a bullied police officer. In a single organisation, rumours spread fast and the officer's potential new superior might receive biased information, and, to be on the safe side, reject the bullied officer's application (cf. Leymann, 1993b).

Yet another aspect may be inherent in the very nature of the job itself. Some jobs in the service sector, and in particular in the social and health sector, require a high level of personal involvement, i.e. a form of emotional labour (Hochschild, 1983; Zapf, 2002) which means sensing and expressing emotions and building up personal relationships. In other jobs, such as manufacturing work, a much more instrumental job attitude may suffice. The higher the level of personal involvement, the more personal information is available, and the more possibilities for being attacked exist. Moreover, it is much more difficult to evaluate these jobs. Again this offers possibilities for attacking someone. If a production worker is accused of doing a bad job, he or she can easily defend him or herself, in contrast with a teacher or a nurse who may have much greater difficulties proving that he or she is doing a good job.

All in all, looking at the distribution of bullying across sectors, bullying seems to be primarily a problem among white-collar workers, service employees and employees in supervisory positions, and less a problem among blue-collar workers, although some studies (e.g. Einarsen and Raknes, 1997) show that bullying does exist there as well.

Categories of bullying

The final question of this chapter is: Is bullying a homogeneous construct or are there typical types or categories of bullying which can be differentiated? Homogeneity of bullying would imply that all bullying actions show similar frequencies, have similar causes and consequences, and occur under the same circumstances (Zapf *et al.*, 1996). Leymann (1996) differentiated between five classes of bullying behaviour, which he referred to as the manipulation of: (1) the victim's reputation; (2) the victim's possibilities of communicating with co-workers; (3) the victim's social relationships; (4) the quality of a person's occupational and life situation; and (5) the victim's health. In an empirical study, Leymann (1992) found factors which he labelled as negative communication: humiliating behaviour, isolating behaviour, frequent changes of tasks to punish someone, and violence or threat of violence. Using factor analyses, Zapf *et al.* (1996)

found seven factors in two samples: 'organisational measures' consisting of behaviours initiated by the supervisor or aspects directly related to the victim's tasks. 'Social isolation' is related to informal social relationships at work. The third factor is related to individual attributes of the victim and the victim's private life. 'Physical violence' includes two items of sexual harassment as well as general physical violence or threat of violence. 'Attacking the victim's attitudes' is related to political, national and religious attitudes. The factor 'verbal aggression' consists of items related to verbal attacks. Finally, there was a factor consisting of two items related to spreading rumours. Other studies in several countries utilising the LIPT found similar factor structures (Niedl, 1995, Vartia, 1991,1993; Mühlen *et al.*, 2001).

Factor analysis of the Negative Acts Questionnaire (NAQ) (Einarsen and Raknes, 1997) identified five factors, four of which appears to overlap with attacking the private person, social isolation, work-related measures and physical violence. Based on a revised version of the NAQ applied to a random sample of 5,288 UK employees, Einarsen and Hoel (2001) suggest two factors: personal bullying and work-related bullying.

Taking these studies together, it appears that 'organisational measures' affecting the victims' tasks and competencies, 'social isolation', 'attacking the private person', 'verbal aggression' and 'spreading rumours' are typical categories of bullying (Leymann, 1992; Niedl, 1995; Vartia, 1993; Zapf *et al.*, 1996), whereas 'attacking attitudes' and 'physical violence' are found in some studies, but occur only occasionally in the context of bullying. In the shipyard study by Einarsen and Raknes (1997) physical violence was reported by 2.4 per cent, whilst in the various studies reported by Zapf (1999a) physical aggression occurred in between 3.6 and 9.1 per cent of the bullying cases. Thus the results underline that, in the first instance, bullying is primarily a form of psychological rather than physical aggression.

'Rumours', in contrast, seem to be among the most used behaviours, probably because of their subtle and indirect nature which can be applied without revealing one's identity. In countries where most of the bullying appears to come from superiors, work-related bullying seems to be the most frequently applied behaviour (Hoel and Cooper, 2000). However, contextual factors will influence which behavioural category is most frequent.

Correlational analyses of overall samples (e.g. Niedl, 1995) show that the bullying categories are very highly correlated. This means that if people are bullied they tend to experience a large number of bullying behaviours from different behavioural categories. With regard to gender-specific bullying categories, Leymann and Tallgren (1993) report that women used slander and making someone look a fool, whereas men preferred social isolation. Mackensen von Astfeld (2000) found that women used significantly more strategies affecting communication, social relationships and social reputation, whereas men preferred strategies affecting the victim's work. In a sense, these results correspond to findings regarding schoolyard bullying. Here Björkqvist *et al.* (1992) found that boys used physical aggression more often, whereas girls preferred more indirect strategies such as rumours and social exclusion. In Vartia's (1993) study, women were more often the victims of strategies of indirect aggression such as spreading rumours and social isolation, whereas men were more often the victims of threats and criticism.

Work-related strategies including acts such as being given tasks with impossible targets or deadlines, having one's opinions and views ignored, and being given work clearly below one's level of competence seem to be experienced more often among persons in superior positions (Hoel *et al.*, 2001; Salin, 2001). In the studies reported in Zapf (1999a) and Zapf *et al.* (1996), co-workers used social isolation and attacking the private sphere more often than the supervisors or managers. Bullying was most frequent when both co-workers and supervisors were among the bullies. If only supervisors were bullying, strategies such as social isolation, attacking the private sphere and spreading rumours occurred less often.

One explanation for these findings may be that some categories, such as social isolation and spreading rumours, work only if many people are involved. A single supervisor cannot isolate somebody. For other bullying categories, such as attacking the private sphere, intimate information about the victim is necessary, which may be less often at hand for superiors.

Conclusion

This chapter has summarised some empirical findings of bullying studies in Europe. There has been quite a lot of research in various European countries during the past ten years. Although different definitions and measures were used in these studies, and although there may be some cultural differences, a converging picture emerges showing that between 1 and 4 per cent of employees may experience serious bullying, and between 8 and 10 per cent occasional bullying. Between 10 and 20 per cent (or even higher) of employees may occasionally be confronted with negative social behaviour at work which does not correspond to definitions of bullying but which is stressful for the persons concerned nevertheless. In most countries, there seems to be a tendency for bullying to occur more often in the public sector. Women seem to be more often among the victims, whereas men seem to be more often among the perpetrators. Bullying occurs at all levels of an organisation. Both supervisors and colleagues are among the perpetrators. Most studies report an average duration of bullying much longer than one year. Bullying can be a conflict between two people; however, very often, there is more than one perpetrator. More and more people seem to become involved the longer bullying lasts. Finally, there is some empirical evidence that a variety of bullying behaviour exists. At least some of the differences found in different studies may be due to cultural differences. It is also important to note that overall findings may mask underlying trends with regard to prevalence as well as the nature of experience, for example with respect to gender and occupational status.

Summarising the existing results on workplace bullying shows that great progress has been made during the last decade, which, overall, has led to converging results in the various European countries. However, it is worth bearing in mind that the quality of individual studies varies greatly with respect to sample size, representativeness and response rate. Therefore, more rigorous studies are needed to further substantiate the concept of bullying in the workplace.

Bibliography

Ashforth, B. E. (1994) Petty tyranny in organizations. *Human Relations*, *47*, 755–778.

Björkqvist, K. (1994) Sex differences in aggression. *Sex Roles*, *30*, 177–188.

Björkqvist, K., Lagerspetz, K. M. J. and Kaukiainen, A. (1992) Do girls manipulate and boys fight? Developmental trends in regard to direct and indirect aggression. *Aggressive Behavior*, *18*, 117–127.

Björkqvist, K., Österman, K. and Hjelt-Bäck, M. (1994) Aggression among university employees. *Aggressive Behavior*, *20*, 173–184.

Cowie, H., Jennifer, D., Neto, C., Angula, J. C., Pereira, B., del Barrio, C. and Ananiadou, K. (2000) Comparing the nature of workplace bullying in two European countries: Portugal and the UK. In M. Sheehan, S. Ramsey and J. Patrick (eds), *Transcending the boundaries: Integrating people, processes and systems. Proceedings of the 2000 Conference* (pp. 128–133). Brisbane: Griffith University.

Davidson, M. J. and Cooper, C. L. (1992) *Shattering the glass ceiling* London: Paul Chapman Publishing.

Dick, U. and Dulz, K. (1994) *Zwischenbericht Mobbing-Telefon für den Zeitraum 23.8.93–22.2.1994 (Intermediate report of the mobbing telephone)*. Hamburg: AOK.

Ege, H. (1998) *I numeri del Mobbing. La prima ricera italiana* (The frequency of bullying. The first Italian study). Bologna: Pitagora Editrice.

Einarsen, S. (2000) Harassment and bullying at work: A review of the Scandinavian approach. *Aggression and Violent Behavior*, *4*, 371–401.

Einarsen, S. and Hoel, H. (2001) *The Negative Acts Questionnaire: Development, validation and revision of a measure of bullying at work*. Paper presented at the 9th European Congress of Work and Organizational Psychology in Prague, May.

Einarsen, S. and Raknes, B. I. (1991) *Mobbing I arbeidslivet* (Bullying in work life). Bergen: University of Bergen.

—— (1997) Harassment at work and the victimization of men. *Violence and Victims*, *12*, 247–263.

Einarsen, S. and Skogstad, A. (1996) Prevalence and risk groups of bullying and harassment at work. *European Journal of Work and Organizational Psychology*, *5*, 185–202.

Einarsen, S., Matthiesen, S. B. and Skogstad, A. (1998) Bullying, burnout and well-being among assistant nurses. *Journal of Occupational Health and Safety*, *14*, 563–568.

Einarsen, S., Raknes, B. I., Matthiesen, S. B. and Hellesøy, O. H. (1994) *Mobbing og harde personkonflikter. Helsefarlig samspill pa arbeidsplassen* (Bullying and severe interpersonal conflicts. Unhealthy interaction at work). Soreidgrend: Sigma Forlag.

Halama, P. and Möckel, U. (1995) 'Mobbing'. Acht Beiträge zum Thema Psychoterror am Arbeitsplatz ('Mobbing': Eight contributions to the issue of psychological terror at work). In Evangelischer Pressedienst (ed.), *epd-Dokumentation*, vol. 11/95, Frankfurt am Main: Gemeinschaftswerk der Evangelischen Publizistik.

Hochschild, A. R. (1983) *The managed heart*. Berkeley: University of California Press.

Hoel, H. and Cooper, C. L. (2000) *Destructive conflict and bullying at work*. Manchester: Manchester School of Management (UMIST).

Hoel, H., Cooper, C. L. and Faragher, B. (2001) The experience of bullying in Great Britain: The impact of organisational status. *European Journal of Work and Organizational Psychology*, *10*, 443–465.

Hoel, H., Rayner, C. and Cooper, C. L. (1999) Workplace bullying. In C. L. Cooper and I. T. Robertson (eds), *International review of industrial and organizational psychology*, vol. 14 (pp. 195–230). Chichester: Wiley.

Hofstede, G. (1993) Cultural constraints in management theories. *The Executive*, *7*, 84–91.

Hogh, A. and Dofradottir, A. (2001) Coping with bullying in the workplace. *European Journal of Work and Organizational Psychology*, *10*, 485–495.

Holzen Beusch, E. V., Zapf, D. and Schallberger, U. (in press). Warum Mobbingopfer ihre Arbeitsstelle nicht wechseln (Why the victims of bullying do not change their job). *Zeitschrift für Personalforschung*.

Hubert, A. B. and van Veldhoven, M. (2001) Risk sectors for undesired behaviour and mobbing. *European Journal of Work and Organizational Psychology*, *10*, 415–424.

Hubert, A. B., Furda, J. and Steensma, H. (2001) Mobbing, systematisch pestgedrag in organisaties (Mobbing: systematic harassment in organisations). *Gedrage Organisatie*, *14*, 378–396.

Kasl, S. V. and Cooper, C. L. (eds) (1987) *Stress and health: Issues in research methodology*. New York: Wiley.

Kaucsek, G. and Simon, P. (1995). *Psychoterror and risk-management in Hungary*. Paper presented as poster at the 7th European Congress of Work and Organizational Psychology, 19–22nd April, Györ, Hungary.

Kivimäki, M., Elovainio, M. and Vahtera, J. (2000) Workplace bullying and sickness absence in hospital staff. *Occupational and Environmental Medicine*, *57*, 656–660.

Leymann, H. (1990) *Handbok för användning av LIPT-formuläret för kartläggning av risker för psykiskt vald* (Manual of the LIPT questionnaire for assessing the risk of psychological violence at work). Stockholm: Violen.

—— (1992) *Fran mobbning till utslagning i arbetslivet* (From bullying to exclusion from working life). Stockholm: Publica.

—— (1993a) Ätiologie und Häufigkeit von Mobbing am Arbeitsplatz – eine Übersicht über die bisherige Forschung (Etiology and frequency of bullying in the workplace – an overview of current research). *Zeitschrift für Personalforschung*, 7, 271–283.

—— (1993b) *Mobbing – Psychoterror am Arbeitsplatz und wie man sich dagegen wehren kann* (Mobbing – psychoterror in the workplace and how one can defend oneself). Reinbeck: Rowohlt.

—— (1996) The content and development of mobbing at work. *European Journal of Work and Organizational Psychology*, 5, 165–184.

Leymann, H. and Gustafsson, A. (1996) Mobbing and the development of post-traumatic stress disorders. *European Journal of Work and Organizational Psychology*, 5, 251–276.

Leymann, H. and Tallgren, U. (1990) *Investigation into the frequency of adult mobbing in a Swedish steel company using the LIPT questionnaire*. Unpublished manuscript.

—— (1993) Psychoterror am Arbeitsplatz (Psychological terror in the workplace). *Sichere Arbeit*, 6, 22–28.

Liefooghe, A. P. D. and Olaffson, R. (1999) 'Scientists' and 'amateurs': Mapping the bullying domain. *International Journal of Manpower*, 20, 16–27.

Lindroth S. and Leymann, H. (1993) *Vuxenmobbning mot en minoritetsgrupp av män inom barnomsorgen. Om mäns jämställdhet i ett kvinnodominerat yrke* (Bullying of a male minority group within child-care. On men's equality in a female-dominated occupation). Stockholm: Arbetarskyddstyrelsen.

Mackensen von Astfeld, S. (2000) *Das Sick-Building-Syndrom unter besonderer Berücksichtigung des Einflusses von Mobbing* (The sick building syndrome with special consideration of the effects of mobbing). Hamburg: Verlag Dr Kovac.

Matthiesen, S. B. and Einarsen, S. (2001) MMPI-2-configurations among victims of bullying at work. *European Journal of Work and Organizational Psychology*, 10, 467–484.

Matthiesen, S. B., Raknes, B. I. and Rökkum, O. (1989) Mobbing på arbeidsplassen (Bullying in the workplace). *Tidsskrift for Norsk Psykologforening*, 26, 761–774.

Mikkelsen, G. E. and Einarsen, S. (2001) Bullying in Danish work-life: Prevalence and health correlates. *European Journal of Work and Organizational Psychology*, 10, 393–413.

—— (2002) Basic assumptions and symptoms of post-traumatic stress among victims of bullying at work. *European Journal of Work and Organizational Psychology*, 11, 87–111.

Minkel, U. (1996) *Sozialer Stress am Arbeitsplatz und seine Wirkung auf Fehlzeiten* (Social stress at work and its consequences for sickness absence). Unpublished diploma thesis, University of Konstanz: Social Science Faculty.

zur Mühlen, L., Normann, G. and Greif, S. (2001) *Stress and bullying in two organisations*. Manuscript submitted for publication, University of Osnabrück: Faculty of Psychology.

Niedl, K. (1995) *Mobbing/Bullying am Arbeitsplatz. Eine empirische Analyse zum Phänomen sowie zu personalwirtschaftlich relevanten Effekten von systematischen Feindseligkeiten* (Mobbing/bullying at work. An empirical analysis of the phenomenon and of the effects of systematic harassment on human resource management). Munich: Hampp.

—— (1996) Mobbing and well-being: Economic and personnel development implications. *European Journal of Work and Organizational Psychology*, *5*, 239–249.

Nuutinen, I., Kauppinen, K. and Kandolin, I. (1999) *Tasa-arvo poliisitoimessa* (Equality in the police force). Helsinki: Työterveyslaitos, Sisäasiainministeriö.

Olweus, D. (1994) Annotation: Bullying at school – basic facts and effects of a school based intervention program. *Journal of Child Psychology and Psychiatry*, *35*, 1171–1190.

O'Moore, M. (2000) *Summary report on the national survey on workplace bullying*. Dublin: Trinity College.

O'Moore, M., Seigne, E., McGuire, L. and Smith, M. (1998) Victims of bullying at work in Ireland. *Journal of Occupational Health and Safety: Australia and New Zealand*, *14*, 569–574.

Piirainen, H., Elo, A.-L., Hirvonen, M., Kauppinen, K., Ketola, R., Laitinen, H., Lindström, K., Reijula, K., Riala, R., Viluksela, M. and Virtanen, S. (2000) *Työ ja terveys – haastattelututkimus* (Work and health – an interview study). Helsinki: Työterveyslaitos.

Quine, L. (1999) Workplace bullying in NHS community trust: Staff questionnaire survey. *British Medical Journal*, *3*, 228–232.

Rayner, C. (1997) The incidence of workplace bullying. *Journal of Community and Applied Social Psychology*, *7*, 199–208.

Salin, D. (2001) Prevalence and forms of bullying among business professionals: A comparison of two different strategies for measuring bullying. *European Journal of Work and Organizational Psychology*, *10*, 425–441.

Schuster, B. (1996) Rejection, exclusion, and harassment at work and in schools. *European Psychologist*, *1*, 293–317.

UNISON (1997) *UNISON members' experience of bullying at work*. London: UNISON.

—— (2000) *Police staff bullying report* (No. 1777). London: UNISON.

Vartia, M. (1991) Bullying at workplaces. In S. Lehtinen, J. Rantanen, P. Juuti, A. Koskela, K. Lindström, P. Rehnström and J. Saari (eds), *Towards the 21st century. Proceedings from the International Symposium on Future Trends in the Changing Working Life* (pp. 131–135). Helsinki: Institute of Occupational Health.

—— (1993) Psychological harassment (bullying, mobbing) at work. In K. Kauppinen-Toropainen (eds), *OECD Panel group on women, work, and health* (pp. 149–152). Helsinki: Ministry of Social Affairs and Health.

—— (1996) The sources of bullying – psychological work environment and organizational climate. *European Journal of Work and Organizational Psychology*, *5*, 203–214.

—— (2001) Consequences of workplace bullying with respect to well-being of its targets and the observers of bullying. *Scandinavian Journal of Work Environment and Health*, *27*, 63–69.

Vartia, M. and Hyyti, J. (1999) *Väkivalta vankeinhoitotyössä* (Violence in prison work). Helsinki: Oikeusministeriön vankeinhoito-osaston julkaisuja 1 (English summary).

—— (2002) Gender differences in workplace bullying among prison officers. *European Journal of Work and Organizational Psychology*, *11*, 1–14.

Zapf, D. (1999a). Mobbing in Organisationen. Ein Überblick zum Stand der Forschung (Mobbing in organisations. A state of the art review). *Zeitschrift für Arbeits- and Organisationspsychologie*, *43*, 1–25.

—— (1999b). Organizational, work group related and personal causes of mobbing/bullying at work. *International Journal of Manpower*, *20*, 70–85.

—— (2001) *Social stressors at work*. Invited presentation at the BPS Centenary Conference, Symposium on 'Well-being', 28–31 March, Glasgow.

—— (2002) Emotion work and psychological strain. A review of the literature and some conceptual considerations. *Human Resource Management Review*, *12*, 237–268.

Zapf, D. and Gross, C. (2001) Conflict escalation and coping with workplace bullying: A replication and extension. *European Journal of Work and Organizational Psychology*, *10*, 497–522.

Zapf, D., Knorz, C. and Kulla, M. (1996) On the relationship between mobbing factors, and job content, the social work environment and health outcomes. *European Journal of Work and Organizational Psychology*, *5*, 215–237.

Zapf, D., Renner, B., Bühler, K. and Weinl, E. (1996) *Ein halbes Jahr Mobbingtelefon Stuttgart: Daten und Fakten* (Half a year mobbing telephone Stuttgart: Data and facts). Konstanz: University of Konstanz, Social Science Faculty.

6 Individual effects of exposure to bullying at work

Ståle Einarsen and Eva Gemzøe Mikkelsen

Introduction

Exposure to bullying at work has been classified as a significant source of social stress at work (Zapf, 1999a; Vartia, 2001) and as a more crippling and devastating problem for employees than all other work-related stress put together (Wilson, 1991). Others have claimed work harassment to be a major cause of suicide (Leymann, 1992). Clinical observations have shown effects of exposure to workplace bullying such as social isolation and maladjustment, psychosomatic illnesses, depressions, compulsions, helplessness, anger, anxiety and despair (Leymann, 1990). Although single acts of aggression and harassment occur fairly often in everyday interaction at work, they seem to be associated with severe health problems in the target when they occur on a regular basis (Einarsen and Raknes, 1997; Vartia 2001). To be a victim of intentional and systematic psychological harm, be it real or perceived, by another person seems to produce severe emotional reactions such as fear, anxiety, helplessness, depression and shock (Janoff-Bulman, 1992). Victimisation due to workplace bullying appears to transform employees' perceptions of their work environment and life in general into situations involving threat, danger, insecurity and self-questioning (Mikkelsen and Einarsen, 2002a). According to a number of studies (e.g. Einarsen *et al.*, 1998; O'Moore *et al.*, 1998; Vartia, 2001), this may lead to pervasive emotional, psychosomatic and psychiatric problems in victims. Although this particular type of victimisation has been studied under different labels, such as, for example, 'mobbing', 'emotional abuse at work', 'harassment at work', 'bullying at work', 'mistreatment' and 'victimisation at work', researchers have reached comparable conclusions (Einarsen, 2000): exposure to systematic and prolonged non-physical and non-sexual aggressive behaviours at work is highly injurious to the victim's health.

The aim of this chapter is, first, to review the literature on short-term and long-term effects of exposure to bullying at work, and second, to propose two theoretical models that might help explain the observed relationships between exposure to bullying and victims' self-reported health problems. The background for this presentation is a growing awareness among both researchers and practitioners that employees' health and well-being may be severely affected in situations where they are persistently exposed to tyrannical, offensive or intimidating behaviours by their superiors or colleagues. Indeed, studies (Björkqvist *et al.*, 1994; Einarsen *et al.*, 1999; Leymann and Gustafsson, 1996) have indicated that some victims of bullying develop symptoms analogous to post-traumatic stress disorder (PTSD), a diagnosis commonly used to describe the symptom pattern displayed by victims of highly traumatic events such as natural disasters, violent assaults, rape and torture.

The evidence

Health and well-being

In a theoretical overview on tyrannical leadership, Ashforth (1994) suggests that exposure to a highly aggressive and malicious leadership style may have a range of negative effects on subordinates, such as, for instance, (1) frustration, stress and reactance; (2) helplessness and work alienation; (3) lowered self-esteem and productivity; and (4) low work unit cohesiveness. Correspondingly, a study of the research literature and empirical investigations in the area of bullying at work indicates that bullying may have serious detrimental effects on targets' health and well-being (Einarsen and Hellesøy, 1998). Victims of bullying generally report lowered well-being and lowered job satisfaction, as well as a number of stress symptoms including low self-esteem, sleep problems, anxiety, concentration difficulties, chronic fatigue, anger, depression and various somatic problems (e.g. Brodsky, 1976; Einarsen and Raknes, 1997; Einarsen *et al.*, 1994; Einarsen *et al.*, 1998; Matthiesen *et al.*, 1989; Mikkelsen and Einarsen, 2002a; Niedl, 1996; Vartia, 1996; Zapf *et al.*, 1996). Also, some victims express self-hatred and may suffer from suicidal thoughts (Einarsen *et al.*, 1994). Among thirty Irish victims of bullying, the most commonly reported symptoms were anxiety, irritability and depression (O'Moore, *et al.*, 1998). A similar symptom pattern characterised German (Zapf *et al.*, 1996) and Austrian (Niedl, 1996) victims of bullying. The latter group also displayed many psychosomatic symptoms. An American study (Price Spratlen, 1995) showed that victims of bullying reported lowered self-confidence, self-worth and productivity. Finally, the most common symptoms reported by sixty-seven Swedish victims of bullying were nervousness, insecurity, suspiciousness, bitterness, self-hatred and suicidal thoughts (Thylefors, 1987).

On the basis of clinical observations and interviews with American victims of harassment at work, Brodsky (1976) identified three general reaction patterns. Some victims developed vague physical symptoms such as weakness, loss of strength, chronic fatigue, pains and various aches. Others reacted with depression and symptoms related to depression such as impotence, lack of self-esteem and sleeplessness. Finally, a third group portrayed various psychological symptoms such as hostility, hypersensitivity, loss of memory, feelings of victimisation, nervousness and social withdrawal.

Since most of the above studies involve self-selected samples, typically victims of long-term bullying, it is conceivable that the health problems reported by these victims surpass those of other victims (Einarsen, 2000). However, the relationship between exposure to bullying and victims' health and well-being has also been investigated in community samples of intact work groups as well as among random samples of employees. A study of a random sample of Norwegian assistant nurses (Einarsen, *et al.*, 1998) showed that nurses reporting exposure to bullying portrayed significantly higher levels of burnout, lowered job satisfaction and lowered psychological well-being compared to their non-bullied colleagues. In a study among all 500 male industrial workers in a Norwegian shipyard, significant negative associations were demonstrated between exposure to bullying at work and measurements of psychological health and well-being (Einarsen and Raknes, 1997). In this sample, exposure to bullying explained 23 per cent of the variance in psychological health and well-being. Particularly strong relationships were found between self-reported psychological well-being and exposure to acts involving personal derogation, such as offensive remarks and ridicule. In another survey conducted among 2,200 members of six different labour unions, significant relationships were found between various measurements of experienced bullying at work and psychological, psychosomatic and musculo-skeletal complaints (Einarsen *et al.*, 1996). The strongest correlations were found between exposure to bullying and psychological complaints with measures of experienced bullying predicting 13 per cent of the variance. A total of 6 per cent of the variance in musculo-skeletal complaints and 8 per cent in the variance of psychosomatic complaints could be statistically predicted by measures of exposure to bullying. These findings are very much in line with those of Niedl (1996) and Zapf *et al.* (1996). In the latter studies, highly significant differences in mental health variables were found between bullied and non-bullied respondents. Exposure to personalised attacks was especially strongly correlated with mental health variables (Zapf *et al.*, 1996).

The above findings clearly indicate that bullying at work may negatively affect employee health and well-being. However, since most of these studies are either of a clinical character or have made use of a correlational design, few of them allow for conclusions on cause–effect relationships. Yet, a comprehensive Finnish cohort study among 8,130 employed men and 7,400 employed women showed that both men and women who experienced serious interpersonal conflicts at work had a greater risk of developing psychiatric disease and of being hospitalised in the four to six years following these incidents (Romanov *et al.*, 1996). In a prospective study among 674 male and 4,981 Finnish female hospital employees, the 5 per cent who reported being victimised due to bullying at work subsequently displayed a significantly increased sickness absenteeism rate compared to non-victims (Kivimäki *et al.*, 2000). After having adjusted for sick leaves in the year prior to the survey, victims had a 26 per cent higher risk of medically certified sickness absence and a 16 per cent higher risk of self-certified sickness absence compared to other employees. In this study, sickness absence was collected from the official sickness absence register and adjusted for absence due to maternity leave and sick children.

Furthermore, interview studies show that the victims themselves attribute their health problems to their experiences of being bullied (Kile, 1990; Leymann, 1992). The victims in these studies typically report being normal and healthy prior to their victimisation and that due to exposure to bullying they have subsequently developed severe health problems. In fact, many victims claim that exposure to bullying has ruined their mental and physical health (Mikkelsen, 2001b). Box 6.1 illustrates a victim's subjective experience of the magnitude of her health problems. This narrative demonstrates how some victims may feel that even a questionnaire designed to measure psychological and psychosomatic problems following exposure to traumatic events does not sufficiently capture the degree to which exposure to bullying has damaged their mental and physical health (see also Einarsen, *et al.*, 1999).

However, given the retrospective design of these studies and the use of self-reports, we do not know if the victims who participated were particularly vulnerable prior to being subjected to bullying, perhaps due to an exposure to other distressing life events. Any such exposure might also account for the symptoms reported. Indeed, results of a study of 118 Danish victims of bullying (Mikkelsen and Einarsen, 2002a) showed that

Box 6.1: Example of a female victim's narrative of her health problems when responding to a self-report questionnaire measuring psychological and psychosomatic problems (Einarsen *et al.*, 1999)

This questionnaire did not give me the opportunity to show just how much my health has been ruined. How I went from being a healthy person to a patient with 6 hospitalisations. How my tolerance for stress has changed completely. The slightest stress (such as forgetting where I have put something) makes me stiff in the joints and I start to ache from my ankles right up to my fingers and my neck. My quality of life has been reduced substantially. Answering these questions stirs up many things that I thought were 'buried' and forgotten. After three hours my muscles ache and I have trouble focusing my eyes. Pain and negative feelings are aroused.

many victims had experienced other distressing life events such as, for example, accidents, divorce and bereavements. However, 80.5 per cent of the victims in this study stated that none of these events affected them more negatively than the bullying they had suffered. Nonetheless, those victims who appeared to be most traumatised, as indicated by their scores on a self-report measure of PTSD, also reported feeling more negatively affected by an event other than bullying. Hence, in addition to the bullying, exposure to one or several other stressful life events might have contributed to their severe health problems.

Bullying and post-traumatic stress disorder

In view of the particular symptom constellation displayed by victims of long-term bullying at work, it has been argued that some of them may suffer from post-traumatic stress disorder (PTSD) (Björkqvist *et al.*, 1994; Leymann and Gustafsson, 1996). The PTSD diagnosis refers to a constellation of stress symptoms typically exhibited by victims of exceptionally traumatic events (APA, 2000). The hallmark symptoms of PTSD are re-experiencing, avoidance-numbing and arousal. First, the trauma is relived through repeated, insistent and painful memories of the event(s) or in recurring nightmares. Also, the victims may experience an intense psychological discomfort and/or react physically when exposed to reminders of the trauma. Second, victims with PTSD tend to avoid stimuli related to the traumatic situation(s) and to exhibit a general numbing of responsiveness. For instance, they may have problems remembering the actual event(s) or may exhibit a reduced interest in activities they used to enjoy. Often they feel detached from others and may display limited affect. A third cardinal symptom is hyper-arousal. This may be manifested in, for example, sleeping problems, concentration difficulties, highly tense and irritable behaviour, as well as in exaggerated reactions to unexpected stimuli.

Several studies have provided empirical support for the hypothesis that the symptomatology of many victims of long-term bullying resembles that of PTSD. On the basis of their study among 64 Swedish victims of bullying attending a rehabilitation programme, Leymann and Gustafsson (1996) concluded that 65 per cent of the victims suffered from PTSD. In a study involving 102 victims of long-term bullying at work recruited among members of two Norwegian national associations against bullying, Einarsen *et al.* (1999) found that 75 per cent of the victims portrayed stress symptoms indicating PTSD. Even five years after the bullying had terminated, 65 per cent of the victims reported a symptom pattern indicative of PTSD. Symptom severity scores on the Hopkins Symptoms Check List (Derogatis *et al.*, 1974) showed that 76.5 per cent of the victims scored above the level indicating psychiatric pathology, as opposed to 21.4 per cent for females and 12.4 per cent for males in a control group. In this study, the level of PTSD symptoms was highly related to the intensity of the reported bullying behaviours. Also, the PTSD symptoms were particularly severe if the aggressive behaviours to which the victims were subjected involved personal degradation.

Similar results were found in a sample comprising 118 former and present Danish victims of bullying at work (Mikkelsen and Einarsen, 2002a). Based on the diagnostic criteria outlined in the DSM-IV-TR (APA, 2000), 29 per cent appeared to suffer from PTSD, whereas another 47 per cent failed to fulfil the stressor criterion A1 only; i.e. they did not perceive the bullying as a serious threat to their life or their physical integrity. Moreover, using the Post-traumatic Diagnostic Scale (PDS) (Foa *et al.*, 1997), 61.7 per cent of the 89 victims who met either five or all six DSM-IV-TR criteria for PTSD were found to display moderate to severe or severe levels of PTSD symptoms. Furthermore, 73.6 per cent of these 89 victims displayed a moderate or severe impairment in several important spheres of their lives such as, for example, their relationship with family or friends, leisure activities, household duties or sex life. High levels of reported exposure to specific bullying behaviours, as measured by the Negative Acts Questionnaire, were associated with high levels of PTSD symptoms. Victims who reported exposure to long-term bullying portrayed the highest level of psychological impairment. Finally, results showed that in many cases the victims' psychological problems persisted long after the bullying had ceased. Whereas 90 per cent of the victims bullied at the time of the survey fulfilled five or six PTSD criteria, this was the case for 54 per cent of the victims bullied more than five years earlier.

However, it is important to note that in order to be diagnosed with PTSD, victims must have experienced or witnessed a traumatic event that involved actual or threatened death or serious injury to their own or other people's *physical* integrity (APA, 2000). Also, they must have felt helpless, scared or terrified whilst being victimised. The former stressor criterion A1 poses a problem in relation to stressors such as bullying, which are primarily non-physical. Thus, at the present point in time, most victims of bullying cannot be diagnosed with PTSD (Mikkelsen and Einarsen, 2000a).

Since a revision of the PTSD criterion A1 as outlined in the DSM-IV-TR (APA, 2000) may not be forthcoming, perhaps alternative diagnoses should be considered. Leymann and Gustafsson (1996) found depression and Generalised Anxiety Disorder to be the most common co-morbid diagnoses. Alternatively, Scott and Stradling (1994) have proposed a supplementary diagnosis, Prolonged Duress Stress Disorder (PDSD), which might account for the stress symptoms displayed by victims exposed to prolonged stress of relatively less intensity (see Mikkelsen, 2001b, and Mikkelsen and Einarsen, 2002a for a discussion of the PTSD stressor criterion). Whichever diagnosis one may eventually use, the main issue is to understand just how psychologically destructive bullying can be. Victimisation due to bullying at work may not only ruin employees' mental health, but also their career, social status and thus their way of life. Indeed, the vast majority of the victims in the Danish sample reported that the bullying had damaged their personality and their mental and physical health (Mikkelsen, 2001b). Moreover, nearly all reported a considerable reduction in their quality of life. Therefore, it comes as no surprise that most of these victims perceived their exposure to bullying at work as being the worst thing that had ever happened to them. Clearly, this demonstrates that, at least for some employees, exposure to bullying is highly traumatising (see Mikkelsen, 2001a).

Effects on victim personality

There are several ways in which victim personality may be related to bullying. First, such traits may increase the likelihood of a person displaying behaviours that are socially provocative in turn leading to interpersonal conflict that may escalate into bullying (Einarsen *et al.*, 1994; Olweus, 1993). Second, personality traits may influence the extent to which a person is selected as a target (Coyne *et al.*, 2000; Einarsen, 1999; Zapf, 1999b). Third, personality traits are also likely to play a role in people's perception of being a victim of bullying (see Spector *et al.*, 2000, for a discussion on the role of personality in stress research). Finally, such traits may act as mediating or moderating factors in the relationships between bullying and stress reactions (see Einarsen, 2000).

Empirical research has indeed indicated that variables pertaining to individual differences might affect the degree of reported stress symptoms following exposure to bullying at work. Einarsen *et al.* (1996) showed that self-esteem and social anxiety partially moderated the relationships between bullying and self-report measures of psychological, psychosomatic and musculo-skeletal health complaints. Victims high on social anxiety reported more psychosomatic symptoms than did victims low on social anxiety. Also, Einarsen *et al.* (1996) found that victims low in self-esteem reported more psychological and musculo-skeletal complaints than victims high in self-esteem. Furthermore, a study among Danish factory employees showed that state-negative affectivity partially mediated the relationships between exposure to bullying behaviours and psychological and psychosomatic health complaints, while generalised self-efficacy acted as a weak moderator of the relationship between exposure to bullying behaviours and psychological health complaints (Mikkelsen and Einarsen, 2002b). In a theoretical overview, Mikkelsen (2001a) further proposed that individual variables such as, for example, trait negative affectivity, perceived locus of control, attributional style and coping strategies are likely to influence the extent to which victims of bullying develop severe health problems following exposure to bullying at work.

With regard to victims' supposed causal role in workplace bullying, studies by Coyne *et al.* (2000) as well as Zapf (1999b) have pointed to several personality traits that might predict future victimisation (see also Zapf and Einarsen, this volume). However, other authors have argued that any observed victim personality characteristics must be understood as an effect of their victimisation, rather than its cause (Björkqvist *et al.*, 1994; Leymann and Gustafsson, 1996). Indeed, many of the 'personality traits' commonly found in victims, such as low self-esteem and neuroticism/negative affectivity (Zapf *et al.*, 1996; Vartia, 1996; Mikkelsen and Einarsen, 2002b), are typical psychological sequelae resulting from victimisation either due to bullying or other traumatic events. Even a stable personality trait such as (trait) negative affectivity, which has been found to characterise many victims of bullying (Vartia, 1996; Zapf, 1999b), is likely to be affected as a result of exposure to stressful or traumatic life events such as bullying (Clark *et al.*, 1994). Leymann and Gustafsson (1996) claim that victims with chronic PTSD may develop permanent personality changes that are related either to depression or obsession. According to these authors, victims with a predominantly obsessive personality may develop a generally hostile and suspicious attitude towards their surroundings, and a chronic feeling of nervousness. They also tend to be hypersensitive in regard to perceived injustices and may display a pathological and compulsory identification with the suffering of others. Another compulsory behaviour seen in such victims may be the incessant accounting of their experiences as victims of bullying. Leymann and Gustafsson (1996) strongly argue that these observed victim characteristics must be seen as a sign of the destruction of the victim's personality rather than pre-morbid personality traits.

It may also be the case that such personality changes following exposure to bullying may mediate relationships between exposure to bullying and health problems. First, such personality changes might increase victims' general level of anxiety; second, they may instigate the use of less effective coping strategies, and third, such personality changes may influence the extent to which individuals perceive the environment as being generally hostile (see also Spector *et al.*, 2000). In the case of violence at work, Barling (1996) has argued that victimisation may have an indirect effect on victims' health as exposure to violence increases their negative mood and results in cognitive distraction in addition to instigating fear. These changes may again explain victims' psychological and psychosomatic health problems.

Victims' coping with bullying

Only a few studies have as yet investigated how exposure to bullying may affect victims' attempts at coping with this awkward and often highly unanticipated situation. One interesting finding is that when victims' actual coping strategies are compared with the strategies that non-bullied employees report they would have used if exposed to bullying at work, some noticeable differences appear (Rayner, 1998, 1999). A British study comprising 761 members of a public trade union showed that 73 per cent of the non-bullied respondents claimed that they would confront the bully, while 63 per cent would complain to the manager (Rayner, 1998). Yet, among the victims in this study only 60 per cent actually confronted the bully, whilst only 46 per cent actually complained to the bully's manager. Furthermore, while some 55 per cent and 73 per cent, respectively, of the non-bullied claimed they would go to personnel or to the union, and another 60 per cent claimed they would seek support from their colleagues in order to file a complaint, only 24 per cent of the victims actually went to personnel and only 26 per cent contacted their union. A mere 21 per cent of the victims had sought the support of colleagues with the purpose of filing a complaint. In another study comprising 1,137 British part-time students only 45 per cent of the victims reported having confronted the bully, while 38 per cent stated they had stayed in their job and done nothing (Rayner, 1998). A total of 36 per cent quit their jobs. In a study by Keashly *et al.* (1994), 13.6 per cent of the victims left their jobs due to exposure to bullying. A study among American nurses (Cox, 1987) revealed that at least 18 per cent of staff turnover could be related to this problem. In contrast, only 7 per cent of the non-bullied in Rayner's study (1998) believed they would leave their job if bullied.

In a representative Danish sample, Hogh and Dofradottir (2001) found that people who report being subjected to repeated negative behaviours at work tend to make less use of problem-solving strategies compared to others. These results may be explained by people's tendency only to use such strategies in situations perceived to be controllable. In an interview study among German victims of bullying who had already left their company, Niedl (1996) found support for the notion that the more the situation deteriorates, the less likely victims are to use problem-solving strategies. Initially, constructive problem-solving strategies were used by most of the victims in this study. These strategies included talking to the superior or informing management of the bullying. Inasmuch as this did not improve the situation, an alternative strategy was put into use. The victims tended to increase their loyalty to the company and commenced working even harder. However, since this did not seem to alleviate their situation, victims started to neglect the situation and wait for something to happen. As the situation deteriorated, they typically reduced their commitment and finally left the company. It is most likely that such complete coping failure will lessen victims' self-worth and increase their feelings of vulnerability, which in turn may exacerbate their psychological symptoms (Mikkelsen and Einarsen, 2002a). Zapf and Gross (2001) also found that victims of bullying generally used constructive conflict-solving strategies in the beginning of the conflict, whereafter they subsequently changed their strategy several times, often ending up leaving the organisation. Using interviews these authors also discovered that victims of bullying often recommended to other victims that they leave the organisation and seek social support elsewhere. Zapf and Gross (2001) discovered that victims who succeeded in improving their situation were generally better at recognising escalating behaviours and better able to avoid using them, whilst those victims who fought for justice often contributed to an escalation of the bullying.

Hence, data indicates a general tendency towards victims taking less action than non-bullied employees claim they would do if they were bullied. Moreover, it appears that many victims often fail to put an end to the bullying. In this respect, Leymann (1990, 1992) has claimed that the effect of bullying is one that drains victims' coping resources to such an extent that they are unable to cope with the requirements of daily life.

Effects on the observers

Workplace bullying not only seems to affect the targets but also their colleagues or other bystanders. For instance, in a Norwegian study among 2,215 employees working in seven different occupational sectors, 21 per cent of the respondents reported lowered job satisfaction due to bullying at work, and 27 per cent claimed that bullying reduced productivity in their department (Einarsen *et al.*, 1994). Fourteen per cent perceived bullying as a daily strain. Yet in this study only 9.6 per cent of the respondents claimed they were victims of bullying. In a recent British study, 32 per cent of the participants reported that bullying resulted in reduced efficiency at their workplace, while 28 per cent stated that bullying reduced their own motivation (Hoel and Cooper, 2000). A study among Finnish municipal workers revealed that witnesses of bullying reported more mental stress reactions than workers who had not witnessed anyone being bullied in their department (Vartia, 2001). In a British survey among 761 members of a public sector trade union, 73 per cent of the witnesses of bullying reported increased stress levels, while 44 per cent worried about becoming a target of bullying (Rayner, 1999). Only 16 per cent of the witnesses claimed not to be affected in any way. Hence, the effects of bullying on bystanders may be either direct, such as when they fear being the next target, or indirect, such as when their general well-being is reduced as a result of the hostile and abusive environment in which they work.

Presumably, witnesses of bullying may also suffer due to a real or perceived inability to help the target. In Rayner's study (1999), more than one-third of the witnesses of bullying reported that they wanted to help the victim but did not dare. Rayner (1998, 1999) also found that, generally, the strategies that the non-bullied would have used in order to put an end to the bullying, such as confronting the bully, complaining to the bully's boss or personnel, or even getting support from colleagues to complain, were to no avail once actually implemented by the victims. For example, 41 per cent of the victims who actually confronted the bully reported that this had no effect, while just as many ended up being labelled a trouble-maker. Even more concerning is the fact that complaints put to the bully's manager had no effect in the majority of cases (57 per cent), while 25 per cent of victims experienced being threatened with dismissal following such a complaint. Generally, non-bullied workers claim they would support any victim of bullying within their department. Yet many victims of bullying report that they lack social support at work (Einarsen *et al.*, 1994). One explanation of this finding may be that victims who are particularly distressed may have difficulty relating to and accepting support from others. On the other hand, becoming a witness of bullying may also lead to other kinds of reactions than anticipated. Highly distressed victims may display behaviours thought to be inappropriate within a work context. Consequently, they may break social norms with regard to interpersonal interaction, sometimes resulting in less help and support from colleagues.

Theoretical explanations

As pointed out by Einarsen (1999) and Zapf (1999b), bullying and harassment at work may take various forms. Their intensity may also vary considerably. Although many models or theories may account for the observed relationships between exposure to bullying at work and victims' health problems, few studies have actually been designed to test such models. In fact, no comprehensive model has as yet been forwarded to explain why exposure to bullying may result in severe symptoms of stress (Mikkelsen, 2001a). This chapter will focus on two possible explanations put forward by the present authors: a cognitive explanation (Einarsen *et al.*, 1999; Mikkelsen, 1997, 2001a; Mikkelsen and Einarsen, 2002a) and a socio-biological explanation (Einarsen and Hellesøy, 1998; Einarsen *et al.*, 1999). The two perspectives are not seen as mutually exclusive, but rather as parallel or even complementary.

Explaining a PTSD reaction: a cognitive framework

Intuitively, it may seem strange that people develop symptoms of PTSD when exposed to interpersonal aggression that is typically indirect and often verbal in nature. However, it may not be the external event itself that causes the trauma, but rather the potential effect this event may have on the inner world of the target. According to Janoff-Bulman (1989, 1992) events are traumatic to the extent that they threaten to shatter our most basic cognitive schemas. These core schemas involve fundamental beliefs that the world is benevolent and meaningful, and that we, as individuals, are worthy, decent and capable human beings deserving other people's affection and support (Janoff-Bulman, 1989). Providing us with expectations about ourselves, other people and the world in which we live, these basic schemas or assumptions enable us to operate effectively in our daily lives. Moreover, being fundamentally positive, the assumptions endow us with a sense of invulnerability central to human existence (Lerner, 1970).

When exposed to highly distressing events, suddenly and painfully victims become conscious of the fragility of those basic assumptions on which their lives are founded (Janoff-Bulman and Frieze, 1983). Insofar as we need stability in our conceptual system (Epstein, 1985), such abrupt changes in core schemas are deeply threatening and may result in an intense psychological crisis (Janoff-Bulman, 1992). The conceptual incongruity between the trauma-related information and prior schemas leads to cognitive disintegration (Epstein, 1985; Janoff-Bulman, 1989), which in turn gives rise to stress responses requiring reappraisal and revision of the basic schemas. Hence, victims must rebuild new and more viable core schemas, which account for the experience of being victimised (Janoff-Bulman and Schwartzberg, 1990). However, some victims have difficulty doing so. Instead of resolving the cognitive-emotional crisis forced upon them by the traumatic event, they remain in a chronic state of cognitive confusion and anxiety that is characteristic of PTSD.

This hypothesis was tested empirically by Mikkelsen and Einarsen (2002a) in a group of 118 Danish victims of bullying and a matched non-bullied control group. The results yielded significant group differences on six out of eight basic assumptions. Victims of bullying considered themselves to be less worthy, less capable and unluckier than did the control group. In addition, they perceived the world as less benevolent, other people as less supportive and caring, and the world as less controllable and just. The difference between victims and non-victims were particularly noticeable on the latter assumption.

At least one possible explanation may be forwarded as to why the victims portray these assumptions about the world: many victims consider themselves as competent and resourceful employees (Zapf, 1999a). If victims have had a successful professional career prior to exposure to bullying, then they may with good reason fail to comprehend why they of all people have become targets of repeated allegations of being stupid, useless or ineffective. In Thylefors's (1987) study, victims who were bullied because of a pre-existing conflict at work or due to theft suffered less than other victims, inasmuch as the former could see the link between their own deviant behaviours and the aggression they faced from their co-workers.

However, it may also be the case that the victims who participated in our study (Mikkelsen and Einarsen, 2002a) tended to have negative views of themselves and the world prior to their victimisation. Indeed, such negative views characterise individuals high in negative affectivity (Watson and Clark, 1984). Although individuals high in negative affectivity are prone to experiencing and reporting high levels of stress (Watson and Clark, 1984), the question remains as to whether this personality trait alone can account for the severity of victims' symptoms. If victims' schemas were negative in the first place, exposure to bullying would come as no shock but rather confirm the validity of their schemas. Based on our own personal experience of working with victims of bullying, we have come to believe that some victims may have had unrealistically positive, in some cases even naive, assumptions prior to their victimisation. In a similar manner, Brodsky (1976) claims that some victims appear to have an unrealistic view of their own abilities and resources, and of the demands of their work situation and their tasks. Hence, for these victims, exposure to bullying may be extremely traumatic. If they are unable to rebuild or adjust the assumptions that have previously provided them with a basic feeling of invulnerability, victims of bullying may remain in a constant state of anxiety. In the long run this may then lead to a breakdown in a range of basic physiological processes.

A socio-biological perspective

Although the term 'bullying' appears to connote overt and direct aggressive behaviours, many victims of bullying are also subjected to covert behaviours, such as when they are 'treated like air' or are made to suffer the 'silent treatment'. Indeed, systematic exclusion and rejection from social groups, i.e. social ostracism (Williams, 1997), appears to be a common key feature of bullying. Exposure to social ostracism signals that the target is in danger of being excluded from an important group, in this case the work group. From an evolutionary perspective, there is probably a very basic fear in all human beings of being excluded from interacting with and receiving the attention of important significant others or social groups (Baumeister and Tice, 1990). Indeed, as a social and tribal primate, the survival of human beings depends on their being integrated in a well-functioning social group. Accordingly, from an existential point of view, social exclusion may be life-threatening. At the very least, it symbolises to the target what death is (Williams, 1997). Indeed, many victims describe exposure to bullying as 'psychological drowning' (Einarsen *et al.*, 1999).

Therefore, it may come as no surprise that exposure to social ostracism is associated with extreme anxiety and the breakdown in basic physiological process (Einarsen and Hellesøy, 1998). According to biologists and physiologists, exposure to ostracism leads to a general physiological deregulation by interfering with the immune system and brain functions relating to aggression and depression (Williams, 1997). The ambiguous nature of ostracism combined with its potential extreme consequence results in a situation where even vague perceptions of being ostracised may have strong negative effects on the targets. According to Williams (1997), perceptions of being excluded or rejected from important social relationships threaten four basic social needs.

First of all, ostracism deprives people of a sense of belonging to others. Second, ostracism may threaten the targets' self-esteem by indicating that they are unworthy of other people's love and affection. When the targets are unable to prevent further exclusion, despite whatever prior positive self-image they may have had, this is likely to be replaced by one of weakness and helplessness. Hence, exposure to bullying may in all probability threaten the victim's feelings of being a capable individual, and as such it may damage his or her *desired identity images* (Mikkelsen, 2001a). These desired identity images represent what we would like to be and what we believe we can be within a given context (Schlenker, 1987): for example, a competent employee who contributes significantly to the company and who others find likeable. Due to our fundamental sociality (e.g. Sampson, 1993) our desired identity images may be easily threatened, especially when excluded by someone perceived to be powerful or important (Mikkelsen, 2001a).

Third, ostracism threaten's the target's need to control his or her interactions with others as well as his/her need to control desired outcomes (Williams, 1997). Unlike other forms of aversive behaviours, ostracism involves a unilateral form of aggression in which the target is deprived of the bilateral nature of an interpersonal conflict. If a person's colleagues refuse to communicate with or even respond to him or her, then that person is often powerless to react. In a sense, such a situation communicates that the target has ceased to exist. Consequently, exposure to ostracism also threatens peoples' need for a meaningful existence by reminding them about their fragile and temporary existence. Thus ostracism may be experienced as a kind of social death (Williams, 1997).

In a short-term perspective, a breach in the fulfilment of these four basic needs causes pain, anxiety and worry (Williams, 1997). In the long term, the frustration of these needs may lead to extreme anxiety, depression and in some cases even psychotic reactions. This may account for the desperate, erratic and sometimes highly aggressive behaviours displayed by many victims of bullying (Einarsen *et al.*, 1994). Leymann (1990) describes how exposure to social isolation or social ostracism gradually reduces the victim's ability to cope with the demands of daily living, leading to a situation where he or she displays more and more abnormal behaviours. Such deviant behaviours may in turn reinforce other people's negative attitudes towards the victim, leading to further victimisation. Consequently, the victim's self-esteem and self-confidence may suffer considerably. Combined with the anxiety caused by the ostracism, this might result in him or her developing severe psychological and psychosomatic health problems.

Conclusion

Previous research clearly indicates that there is a relationship between exposure to bullying on the one hand and symptoms of lowered well-being and psychological and somatic health problems on the other. Furthermore, victims themselves are generally convinced that their health problems are linked to their exposure to bullying. Hence, a casual link between exposure to bullying and strain reactions appears plausible, this despite a scarcity of empirical evidence. Therefore, future research needs to test these proposed causal relationships. Such research should also include analyses of the potential mediating and moderating effects of victim personality traits. Moreover, while research suggests that the symptoms portrayed by many victims of long-term bullying is identical to those outlined in the PTSD diagnosis, more empirical and theoretical work needs to be conducted on what makes bullying at work such a potentially traumatic event (see Mikkelsen, 2001a). Also, alternative diagnoses other than PTSD should be considered (see Mikkelsen and Einarsen, 2002a).

In this chapter we have suggested that the symptoms displayed by victims of bullying at work may be explained using either a cognitive or a socio-biological perspective. Future research, derived either from these or other models, should strive to test theoretical-driven hypotheses on the relationship between exposure to bullying and various negative health outcomes among victims. In order to advance our knowledge in this important field of inquiry, future research should generally be more theory driven than has been the case until now. Furthermore, studies need to be conducted which investigate how victims react when exposed to different kinds of bullying behaviours, in addition to how victims and their colleagues cope with bullying at work. Finally, since empirical studies appear to support the hypothesis that exposure to bullying poses a serious strain on victims' and bystanders' health and well-being, the development and evaluation of therapeutic interventions is clearly needed.

Bibliography

American Psychiatric Association (2000) *Diagnostic and statistical manual of mental disorders IV – Text revision*. Washington, DC: American Psychiatric Association.

Ashforth, B. (1994) Petty tyranny in organizations. *Human Relations*, *47*, 7, 755–778

Barling, J. (1996) The prediction, experience, and consequences of workplace violence. In G. R. Van den Bos and E. Q. Bulatao (eds), *Violence on the job*. Washington, DC: American Psychological Association.

Baumeister, R. F. and Tice, D. M. (1990) Anxiety and social exclusion. *Journal of Social and Clinical Psychology*, *9*, 2, 165–195.

Björkqvist, K., Österman, K. and Hjelt-Bäck, M. (1994) Aggression among university employees. *Aggressive Behavior*, *20*, 3, 173–184.

Brodsky C. M. (1976) *The harassed worker*. Toronto: Lexington Books, DC Heath and Co.

Clark, L.A., Watson, D. and Mineka, S. (1994) Temperament, personality and the mood and anxiety disorders. *Journal of Abnormal Psychology*, *103*, 103–116.

Cox, H. (1987) Verbal abuse in nursing: Report of a study. *Nursing Management*, *18*, 47–50.

Coyne, I., Seigne, E. and Randall, P. (2000) Predicting workplace victim status from personality. *European Journal of Work and Organizational Psychology*, *9*, 335–349.

Derogatis, L. R., Lipman, R. S., Rickels, K., Uhlenhuth, E. H. and Covi, L. (1974) The Hopkins Symptom Checklist (HSCL): A self-report symptom inventory. *Behavioral Science*, *19*, 1–15.

Einarsen, S. (1999) The nature and causes of bullying at work. *International Journal of Manpower*, *20*, 16–27.

—— (2000) Harassment and bullying at work: A review of the Scandinavian approach. *Aggression and Violent Behavior: A Review Journal*, *5*, 4, 371–401.

Einarsen, S. and Hellesøy, O. H. (1998) Når samhandling går på helsen løs – helsemesige konsekvenser av mobbing i arbeidslivet (When interaction affects health – health consequences of bullying at work). In *Medicinsk årbog 1998* (pp. 1–11). Cophenhagen: Munksgaard.

Einarsen, S. and Raknes, B. I. (1997) Harassment in the workplace and the victimization of men. *Violence and Victims*, *12*, 247–263.

Einarsen, S., Matthiesen, S. B. and Mikkelsen, E.G. (1999) *Tiden leger alle sår? Senvirkninger af mobbing i arbeidslivet* (Does time heal all wounds? Long-term health effects of exposure to bullying at work). Bergen: University of Bergen.

Einarsen, S., Matthiesen, S. B. and Skogstad, A. (1998) Bullying, burnout and well-being among assistant nurses. *The Journal of Occupational Health and Safety – Australia and New Zealand*, *14*, 6, 563–568.

Einarsen, S., Raknes, B. I., Matthiesen, S. B. and Hellesøy, O. H. (1994) *Mobbing og harde personkonflikter. Helsefarlig samspill på arbeidsplassen* (Bullying and harsh interpersonal conflicts). Bergen: Sigma Forlag.

—— (1996) Bullying at work and its relationships with health complaints – moderating effects of social support and personality. *Nordisk Psykologi*, *48*, 2, 116–137.

Epstein, S. (1985) The implications of cognitive-experiential self-theory for research in social psychology and personality. *Journal for the Theory of Social Behaviour*, *15*, 3, 282–310.

Foa, E. B., Cashman, L., Jaycox, L. and Perry, K. (1997) The validation of a self-report measure of posttraumatic stress disorder: The posttraumatic diagnostic scale. *Psychological Assessment*, *9*, 4, 445–451.

Hoel, H. and Cooper, C. L. (2000) *Destructive conflict at work*. Manchester: Manchester School of Management.

Hogh, A. and Dofradottir, A. (2001) Coping with bullying in the workplace. *European Journal of Work and Organizational Psychology*, *10*, 485–495.

Janoff-Bulman, R. (1989) Assumptive worlds and the stress of traumatic events: Applications of the schema construct. *Social Cognition*, *7*, 113–136.

—— (1992) *Shattered assumptions – towards a new psychology of trauma*. New York: The Free Press.

Janoff-Bulman, R. and Frieze, I. H. (1983) A theoretical perspective for understanding reactions to victimization. *Journal of Social Issues*, *39*, 2, 1–17.

Janoff-Bulman, R. and Schwartzberg, S. S. (1990) Toward a general model of personal change: Applications to victimization and psychotherapy. In C. R. Snyder and D. R. Forsyth (eds), *Handbook of social and clinical psychology: The health perspective* (pp. 488–508). New York: Pergamon.

Keashly, L., Trott, V. and MacLean, L. M. (1994) Abusive behavior in the workplace: A preliminary investigation. *Violence and Victims*, *9*, 341–357.

Kile, S. M. (1990) *Helsefarleg leiarskap. Ein eksplorerande studie* (Health endangering leadership. An exploratory study). Report to the Norwegian Council of Research. University of Bergen: Department of Psychosocial Science.

Kivimäki, M., Elovainio, M. and Vahtera, J. (2000) Workplace bullying and sickness absence in hospital staff. *Occupational and Environmental Medicine*, *57*, 656–660.

Lerner, M. J. (1970) The desire for justice and reactions to victims. In J. Macaulay and L. Berkowitz (eds), *Altruism and behaviour: Social psychological studies of some antecedents and consequences* (pp. 205–229). New York: Academic Press

Leymann, H. (1990) Mobbing and psychological terror at workplaces. *Violence and Victims*, *5*, 119–126.

—— (1992) *Från mobbning till utslagning i arbetslivet* (From bullying to expulsion from working life). Stockholm: Publica.

Leymann, H. and Gustafsson, A. (1996) Mobbing at work and the development of post-traumatic stress disorders. *European Journal of Work and Organizational Psychology*, *5*, 251–275.

Matthiesen, S. B., Raknes, B. I. and Røkkum, O. (1989) Mobbing på arbeidsplassen. (Bullying at work). *Tidskrift for Norsk Psykologforening*, *26*, 761–774.

Mikkelsen, E. G. (1997) *Mobning i arbejdslivet. En krise- og arbejdspsykologisk undersøgelse* (Workplace bullying – an analysis of its trauma related and occupational psychological aspects). Unpublished masters thesis, University of Aarhus: Department of Psychology.

—— (2001a) Mobning i arbejdslivet: Hvorfor og for hvem er den så belastende? (Workplace bullying: Why and for whom is bullying such a strain?). *Nordisk Psykologi*, *53*, 2, 109–131.

—— (2001b) *Workplace bullying: Its prevalence, aetiology and health correlates.* PhD thesis, University of Aarhus: Department of Psychology.

Mikkelsen. E. G. and Einarsen, S. (2002a) Basic assumptions and symptoms of post-traumatic stress among victims of bullying at work. *European Journal of Work and Organisational Psychology*, *11*, in press.

—— (2002b) Relationships between exposure to bullying at work and psychological and psychosomatic health complaints: The role of state negative affectivity and generalised self-efficacy. *Scandinavian Journal of Psychology*, in press.

Niedl, K. (1996) Mobbing and well-being: Economic and personnel development implications. *European Journal of Work and Organizational Psychology*, *5*, 239–249.

Olweus, D. (1993) *Bullying at school. What we know and what we can do.* Oxford: Blackwell.

O'Moore, M., Seigne, E., McGuire, L. and Smith, M. (1998) Victims of bullying at work in Ireland. *The Journal of Occupational Health and Safety – Australia and New Zealand*, *14*, 6, 569–574.

Price Spratlen, L. (1995) Interpersonal conflict which includes mistreatment in a university workplace. *Violence and Victims*, *10*, 4, 285–297.

Rayner, C. (1998) Workplace bullying: Do something! *The Journal of Occupational Health and Safety – Australia and New Zealand*, *14*, 6, 581–585.

—— (1999) From research to implementation: Finding leverage for prevention. *International Journal of Manpower*, *20*, 1 / 2, 28–38.

Romanov, K., Appelberg, K., Honkasalo, M. L. and Koskenvuo, M. (1996) Recent interpersonal conflicts at work and psychiatric morbidity. A prospective study of 15,350 employees aged 24–64. *Journal of Psychosomatic Research*, *40*, 157–167.

Sampson, E. E. (1993) *Celebrating the other. A dialogic account of human nature.* London: Harvester Wheatsheaf.

Schlenker, B. R. (1987) Threats to identity: Self-identification and social stress. In. C. R. Snyder and C. E. Ford (eds), *Coping with negative life events: Clinical and social psychological perspectives* (pp. 273–321). New York: Plenum Press.

Scott, M. J. and Stradling, S. G. (1994) Post-traumatic stress disorder without the trauma. *British Journal of Clinical Psychology*, *33*, 71–74.

Spector, P. E., Zapf, D., Chen, P. Y. and Frese, M. (2000) Why negative affectivity should not be controlled in job stress research: Don't throw out the baby with the bath water. *Journal of Organizational Behavior*, *21*, 79–95.

Thylefors, I. (1987) *Syndabockar. Om utstötning och mobbning i arbetslivet.* (Scapegoats. On exclusion and mobbing at work). Stockholm: Natur och Kultur.

Vartia, M. (1996) The sources of bullying – psychological work environment and organizational climate. *European Journal of Work and Organizational Psychology*, *5*, 203–214.

—— (2001) Consequences of workplace bullying with respect to the well-being of its targets and the observers of bullying. *Scandinavian Journal of Work Environment and Health*, *27*, 1, 63–69.

Watson, D. and Clark, L. A. (1984) Negative affectivity: The disposition to experience aversive emotional states. *Psychological Bulletin*, *96*, 465–490.

Williams, K. (1997) Social ostracism. In R. M. Kowalski (ed.), *Aversive interpersonal behaviors*. New York: Plenum Press.

Wilson, C.B. (1991) US businesses suffer from workplace trauma. *Personnel Journal*, *July*; 47–50.

Zapf, D. (1999a) Mobbing in Organisationen. Ein Überblick zum Stand der Forschung (Bullying in organisations. A state of the art research review). *Zeitschrift für Arbeits- and Organisationspsychologie*, *43*, 85–100.

—— (1999b) Organizational, work group related and personal causes of mobbing/bullying at work. *International Journal of Manpower*, *1*, 70–85.

Zapf, D. and Gross, C. (2001) Conflict escalation and coping with workplace bullying. *European Journal of Work and Organizational Psychology*, *10*, 497–522.

Zapf, D., Knorz, C. and Kulla, M. (1996) On the relationship between mobbing factors, and job content, social work environment and health outcomes. *European Journal of Work and Organizational Psychology*, *5*, 2, 215–237.

7 Organisational effects of bullying

Helge Hoel, Ståle Einarsen and Cary L. Cooper

Introduction

From the very start of research into workplace bullying, attention has been paid to the negative effects the experience may have on victims (see Einarsen and Mikkelsen, this volume). This is not surprising as such effects can be considered part and parcel of the bullying experience (Leymann, 1996). By contrast, much less attention has been paid to a possible relationship between bullying and organisational outcomes. In relation to sexual harassment cases, Pryor (1987) suggested that costs to the organisation include both direct costs relating to sick leave, turnover, reduced productivity among both victims and work groups, and costs in relation to potential litigation. In this respect Leymann (1990) argues that a case of bullying may cost the organisation around $30,000 to $100,000 each year.

Victims of bullying are typically subjected to either direct aggression, in the form of verbal abuse or humiliating and belittling remarks, or more indirect types of behaviour, such as gossiping and rumours, which undermine the personal as well as the professional standing of the target (O'Moore *et al.*, 1998). These behaviours are 'used with the aim or at least the effect of persistently humiliating, intimidating, frightening or punishing the victim' (Einarsen, 2000). Unsurprisingly, the exposure to persistent negative behaviours may have a significant impact on the targets, making them constantly less able to cope with daily tasks and co-operation requirements of the job (Einarsen, 2000). As with other forms of occupational stress (Cooper *et al.*, 1996: Hoel *et al.*, 2001), exposure to bullying is, therefore, likely to manifest itself behaviourally as well as attitudinally. Thus, research has found a relatively strong negative association between exposure to bullying and lowered job satisfaction (Hoel and Cooper, 2000a; Keashly and Jagatic, 2000; Price Spratlen, 1995; Quine, 1999), and commitment (Hoel and Cooper, 2000a).

According to Field (1996), himself a victim of bullying, 'the person becomes withdrawn, reluctant to communicate for fear of further criticism. This results in accusations of "withdrawal", "sullenness", "not co-operating or communicating", "lack of team spirit", etc.' (p. 128).

Research has also identified bullying as being associated with cognitive effects such as concentration problems, insecurity and lack of initiative (Leymann, 1992; O'Moore *et al.*, 1998). Such reactions on the part of the target may in themselves lead to organisational effects, relating to a lack of motivation and creativity, as well as a rise in accidents and mistakes. Whether deemed constructive, negative or maladaptive, many behavioural and attitudinal responses may have a bearing upon the organisation by affecting levels of absenteeism, turnover and productivity as well as team and group performance.

Furthermore, being accused of bullying, whether rightly or wrongly, is also likely to affect the individual and subsequently the organisation (Hoel and Cooper, 2000b). Even where cases are seemingly satisfactorily resolved, there may be a price to pay in terms of organisational upheaval. Moreover, in cases where internal procedures or organisational responses fail to satisfy the need of the complainant, or where organisations are altogether unwilling to acknowledge a complaint, litigation has increasingly become an option, often with negative implications for public relations.

Hence, the aim of this chapter is to review the emerging evidence of the relationship between bullying on the one hand, and negative organisational outcomes on the other, focusing on absenteeism, turnover and productivity, and the dynamic relationship between these outcomes. Also, factors carrying a potential financial cost to the organisation will be identified. The costs accrued in a typical bullying case will be illustrated by means of a case study.

Organisational effects of bullying: the evidence

Absenteeism

From the evidence of the effects of bullying on health, one would intuitively expect that bullying would manifest itself in sickness absenteeism. However, in the studies which have explored this relationship, the association between bullying and sickness absenteeism has been found to be rather weak (Hoel and Cooper, 2000a; Price Spratlen, 1995; UNISON, 1997; Vartia, 2001). For example, in a Norwegian study of members of fourteen trade unions, Einarsen and Raknes (1991) concluded that bullying accounted for a modest 1 per cent of the variation in sickness absenteeism. A stronger relationship between bullying and sickness absenteeism was reported in a recent Finnish study of hospital staff (Kivimäki *et al.*, 2000). In this case the researchers were able to match sickness absence records (certified and self-certified) to company records on absence; they also had access to the sickness certificates register. The risk of medically certified sickness absence was 51 per cent, or 1.5 times higher, for those who had been bullied. When adjusted for a baseline one year prior to the survey, the risk was still 1.2, or 26 per cent more absences. The difference was smaller for self-certified sickness absence but still substantial (16 per cent). These figures are probably an underestimate as many targets may already have been bullied at the time the baseline measurements were obtained. In total, bullying accounted for 2 per cent of total sickness absenteeism, at a cost of £125,000 annually (other related costs not incorporated).

Similarly, Hoel and Cooper (2000a), who also reported a relatively weak relationship in a national study of bullying in the UK, found that victims of bullying took on average seven days more sick leave per year than those who were neither bullied nor had witnessed bullying taking place. Based on a bullying rate of 10 per cent, this would account for 18 million lost working days on a UK basis. Quine (2001), in a UK study of bullying in nurses, reported that 8 per cent had taken time off due to bullying, whilst Vartia (2001), in a Finnish study of municipal employees, reported that 17 per cent had taken time off for the same reason, 10 per cent several times.

In three separate surveys, two covering the UK public sector trade union members, and one covering Australian organisations experiencing restructuring, approximately one-third of targets of bullying reported having taken time off as a result of their experience (McCarthy *et al.*, 1995; UNISON, 1997, 2000). However, among targets taking time off work due to bullying, a substantial number reported prolonged sickness absence, with 29 per cent being absent for more than thirty days and 13 per cent for more than sixty days (UNISON, 2000).

In explaining this seemingly weak relationship, Einarsen and Raknes (1991) refer to the work of Thyholdt *et al.* (1986), who argue that a common finding in health research is that victims typically tend to report more specific symptoms as opposed to general negative health effects such as absenteeism. Another factor is the tendency of people to under-report their own absence (Johns, 1994). Still, and as argued by Einarsen and Raknes (1991), a 1 per cent increase in sickness absences represents a considerable cost to most organisations.

When considering the costs of absenteeism in the broadest terms, it should be noted that the unpredictability and unexpectedness of unscheduled absenteeism may represent a particular problem for organisations, interfering with the normal operation of the organisation, and where applicable, the quality of service provision (Seago, 1996).

According to Steer and Rhodes (1978), absenteeism can be seen as a result of a combination of two factors: the possibility or opportunity to be present at work, and the motivation of the individual to go to work. As far as the first factor is concerned, the presence of illness would be a deciding factor for non-attendance. The motivation for attending work appears to hinge upon two factors: the degree of job satisfaction and the pressure to attend. Pressure to attend work, for its part, may relate to social norms as well as organisational measures, controlling or punishing what may be considered non-legitimate absence.

Following the model of Steer and Rhodes, the relatively strong relationship between bullying and ill health suggests that, for a considerable number of victims, sickness absence may be a necessity and possibly a direct outcome of their health status. Increased health problems associated with perceived bullying may also demotivate individuals to attend through reduced job satisfaction due to bullying. As the causal relationship between job satisfaction and absence may be reversible and dynamic, growing health problems resulting from bullying may gradually affect job satisfaction and motivation to attend, with absence as a possible result. However, real or perceived pressure to attend may act as a negative motivator, forcing targets to go to work even if, strictly speaking, they would be better off absent. But using absence as a coping mechanism to recover from a stressful experience may be considered counter-productive, as the bullying behaviour may simply intensify with their return to work. This supports a recent finding by Zapf and Gross (2001), who revealed that victims who successfully coped with their situation used frequent sickness absence significantly less often than victims who were not successful. In some cases the pressure to attend work is so severe and so strictly enforced that it may be considered to represent a form of bullying in its own right (UNISON, 1997). This disciplining effect may be particularly strong if the individual is made aware that if they are absent, their work colleagues would have to cover for them (Sinclair *et al.*, 1996). This would also be the case where absenteeism is linked to the possibility of job loss, for example in connection with downsizing or organisational restructuring (Voss *et al.*, 2001).

Where absence is seen as deviant, targets may turn up for work to avoid being associated with malingering or disloyalty, even if medically they would benefit from staying at home. In some cases personal guilt in the guise of self-inflicted pressure may be another factor preventing people from taking time off. However, anecdotal evidence suggests that in some incidents, the pressure to attend may lead to 'presenteeism', where targets stay beyond the required number of hours in order to demonstrate their loyalty to the organisation.

Some researchers have reported a gender difference, with a stronger association between bullying and absenteeism found for women than for men (Einarsen and Raknes, 1991; Voss *et al.*, 2001). This finding corresponds with a general finding in health research of women's tendency to report more health complaints and absenteeism (Marmot *et al.*, 1995), reflecting a stronger effect of certain illnesses on women than men (Grunfeld and Noreik, 1996). It may also reflect general exposure to additional stressors such as 'double work' and low earnings. An alternative explanation is that women use absence as a coping mechanism more than men, possibly due to different social norms regarding absenteeism (Einarsen and Raknes, 1991; Johns, 1997).

Turnover

Turnover has been the single organisational outcome receiving most attention in bullying research so far. Several studies have reported that bullying was associated with a greater intention to leave (Hoel and Cooper, 2000a; Keashly and Jagatic, 2000; Quine 1999, 2001). Voluntary turnover was also measured by a US study of abusive supervision, which concluded that employees with an abusive supervisor reported higher turnover than others (Tepper, 2000). In a study of abuse in nursing, at least 18 per cent of the turnover rate was considered to be related to verbal abuse (Cox, 1987). Other studies which asked respondents about their response to bullying found a considerable number, or approximately one in four previously bullied respondents, who reported that they had left their job due to their experience of bullying (Rayner, 1997; UNISON, 1997).

However, most empirical evidence available focuses on 'intention to leave' information as opposed to exit data. Thus Vartia (1993) reported that 46 per cent of all victims were thinking of leaving. The relative unreliability of 'intention to leave' as a measure of exit from the organisation is highlighted by Keashly and Jagatic (2000), in which a much stronger relationship emerged between bullying and 'intention to quit' than between bullying and 'looking for a new job'.

The fact that turnover was the behavioural response most strongly associated with bullying may have several explanations. Leaving the organisation may be seen as a positive coping strategy for some as it removes them from the source of the problem altogether. This also corresponds with the advice victims give other victims (Zapf and Gross, 2001). Others may quit in despair, possibly as a result of prolonged health problems (Einarsen *et al.*, 1994). In other cases victims may be expelled from the organisation or forced out of an otherwise satisfactory job against their will (Leymann, 1996; Zapf and Gross, 2001). There is also anecdotal evidence that organisations may use bullying tactics to get rid of employees considered unproductive or unsuitable, particularly where strong legal labour protection may be present (Einarsen *et al.*, 1994) or where the organisation otherwise would have to pay redundancy money (Lee, 2000).

However, many victims of bullying are reluctant to leave their job before justice has been seen to be done (Kile, 1990). For others labour market considerations and lack of mobility (Tepper, 2000) may prevent them from leaving the organisation. In other cases, where the bullying behaviour is intermittent, the victim may cling to a belief that the problem will go away one day. To sum up, such personal considerations for staying on may help explain why the association between bullying and turnover is not even stronger. However, replacing an individual may also lead to a chain reaction, with implications for other employees as well (Gordon and Risley, 1999), and with possible side-effects for the organisation.

Research has so far for good reason focused on turnover of victims of bullying. However, as organisations start dealing with the problem we are also likely to see a growing number of alleged bullies voluntarily leaving the organisation. Moreover, due to the increasing emphasis on litigation in many countries, organisations may increasingly choose to terminate the contracts of bullies as a preventive mechanism against litigation, with potential for abuse and miscarriage of justice as a likely consequence.

Productivity

Information about the relationship between bullying and productivity is largely anecdotal in nature (e.g. Bassmann, 1992) and often inferred from behavioural implications of bullying (e.g. Field, 1996). For example, Bassmann (1992) points to the likely loss of initiative and creativity when faced with bullying. Similarly, it would make sense to link a reduction in satisfaction, motivation and commitment resulting from bullying to a reduction in performance and productivity. The scarcity of available evidence related to this issue is at least in part a reflection of the difficulties involved in measuring productivity, a fact also recognised in research on stress (Cooper *et al.*, 1996).

However, in a study of Norwegian trade union members, a total of 27 per cent of respondents agreed (totally or partially) with the statement 'bullying at my workplace reduces our efficiency' (Einarsen *et al.*, 1994). The same measure yielded a figure of 32.5 per cent when applied in a later UK study by Hoel and Cooper (2000a). In this study respondents were asked directly, and independently of their experience of bullying, to assess their current performance as a percentage estimate of normal capacity. Using this measure of productivity, a moderate negative correlation was found between self-rated performance and bullying. The results also indicated that the 'currently bullied' group had a relative decrease of approximately 7 per cent in productivity (85 per cent of normal capacity) compared with those who were neither bullied nor had witnessed bullying taking place (who at the time of the survey reported on average 92 per cent of normal capacity (Hoel, Sparks and Cooper, 2001). The figure for those who were bullied in the past was 88 per cent and for 'witnessed bullying' 90 per cent.

We have already pointed to factors such as reduced job satisfaction in explaining a possible negative impact of bullying on productivity. Where reduced commitment or withdrawal are used as a coping strategy, a negative impact on productivity appears to be straightforward. However, even when a strategy of demonstrating loyalty to the organisation is adopted by the target, the net outcome may turn out to be negative as far as productivity and performance are concerned. Eagerness to please and attempts to do the right thing may fail, particularly where health problems and reduced concentration may jeopardise such attempts. Reduced ability to concentrate may itself increase the chance of making mistakes, thereby increasing the possibility of reduced output quality and likelihood of accident. The fact that people may be bullied irrespective of their organisational status or rank, including senior managers (Hoel, Cooper and Faragher, 2001), indicates the possibility of a negative domino effect where the productivity of the entire organisation may be threatened by a bullying scenario in the boardroom.

So far we have considered each of the outcome variables absenteeism, turnover and productivity in isolation. However, in reality there is likely to be a dynamic relationship between them. For example, if work colleagues are not replaced when absent, pressure is likely to mount on their co-workers with more people possibly reaching breaking point, with increased tension among co-workers as a result, possibly reducing productivity, and inflating sickness absence as well as turnover rates. By contrast, in cases where a victim decides to be present whilst strictly speaking, physically and mentally, they would benefit from being off work, they may be unproductive at work due to lack of concentration (Leymann, 1992) or for fear of making mistakes or drawing attention to themselves. This may also affect their relationship with co-workers and supervisors, possibly sharpening rather than reducing the conflict, which again will impact on the total productivity of the work unit.

Effects on bystanders and witnesses

A large number of people report having witnessed bullying taking place (Hoel and Cooper, 2000a; Rayner, 1997). The fact that 'discussing the problem with colleagues' was found to be the response most frequently chosen by targets when faced with bullying (Hoel and Cooper, 2000a) also suggests that a large number of people will be indirectly affected by bullying. Due to a seemingly strong need for the target to seek support for their case, it is also hard to remain uninvolved or neutral in such cases (Einarsen, 1996). With this in mind, it should come as no surprise that there may be a ripple effect involved with bullying (Hoel *et al.*, 1999; Rayner, 1999). According to Vartia (2001) observers of bullying reported higher levels of generalised stress than those who had not experienced or observed bullying taking place.

According to Rayner (1999) in her survey of public sector union members, a substantial number of witnesses (approximately one in five) stated that they considered leaving their organisation as a result of witnessing bullying taking place. In making sense of this figure, she pointed to a presence of a climate of fear within the organisation reflected in victims' endorsement of statements such as 'workers too scared to report' (95 per cent), 'the bully has done this before' (84 per cent) and 'management knew about it' (73 per cent). The fact that no difference emerged between bullied and non-bullied with regard to the above statements lends support to Rayner's interpretation of the figures.

In a recent study of bullying in Great Britain (Hoel and Cooper, 2000a), respondents were divided into four groups: 'currently bullied', 'previously bullied', 'witnessed bullying only' and 'neither experienced nor witnessed bullying'. Those currently exposed to bullying were found to have the worst mental and physical health, the highest sickness absenteeism and 'intention to leave' rates, and the lowest productivity as well as the lowest organisational satisfaction and commitment. The second most affected group were those who were 'bullied in the past', followed by 'witnessed bullying only' and 'neither bullied nor witnessed bullying'. These findings lend support to the idea that bullying may affect third persons to the detriment of the individual as well as the organisation.

With team-working becoming increasingly common, the ripple effect of bullying may turn out to have serious organisational implications (Hoel *et al.*, 1999). Whilst authoritarian or tyrannical behaviour may lead to low work-unit cohesiveness (Ashforth, 1994), and may subsequently increase group tension and the risk of bullying, enforced team-working may increase the likelihood of bullying in the first place, as well as the chance of third parties being drawn into emerging conflict (Zapf *et al.*, 1996). Also, when the complexity of the operation increases and worker interdependence is central to outcome, absence may lead to a lack of familiarity with current work practice, and may increasingly interfere with social interaction within the team (Johns, 1997), and, therefore, the overall productivity of teams.

Other organisational effects

Depending upon variations in national customs and practice, victims may have the opportunity to file a complaint or a grievance, or ask their union or staff association to raise the issue with management. Independently of the practices in place, investigations are time-consuming affairs, often involving a number of people and, therefore, a drain on organisational resources. Where no clear policies and procedures are in place cases can remain unresolved for years, bouncing backwards and forwards as a result of indecision on the part of the organisation, with side-taking and increasing animosity as a likely outcome (Einarsen *et al.*, 1994; Leymann, 1992). However, even in cases where procedures are strictly adhered to and where cases are brought to a conclusion within a reasonable timescale, the process tends to be destructive for all those concerned (Ishmael, 1999). Depending upon the particular situation, it is not unlikely that the organisation would have to suspend the alleged offender during the period of the internal investigation. This is particularly the case where it is considered difficult for the parties to continue to see each other, or when their presence may interfere with the investigation (see also Merchant and Hoel, this volume). Depending upon the outcome of the internal processes, there are likely to be further costs to the organisation. To reduce the impact of the situation on the individual, the victim (and at times the alleged perpetrator) may be in need of support and counselling (Therani, this volume).

Where bullying is prevalent one may, therefore, envisage active involvement on the part of occupational health practitioners. Their roles might include assessing risk, intervention and rehabilitation (Hoel and Cooper, 2000b). Where abuse and interpersonal conflict are endemic, bullying cases may occupy a substantial part of the service's time and resources.

A typical outcome of a bullying complaint is the physical, permanent separation of the perpetrator and victim. Unfortunately, there is evidence that it is the victim who tends to be moved, particularly in cases where the perpetrator is in a superior managerial role (Rayner *et al.*, 2002). Such transfers can be costly as they may encompass replacement costs as well as extra training costs for two or more individuals (Dalton, 1997). And whilst some turnover may be functional to the organisation, forced transfers are, according to Dalton, entirely dysfunctional. As transfers are unlikely to take place without disruption, in this case involving two work groups, they may in their own right have a negative effect on productivity. In some cases a domino effect may emerge, as more than one move may be necessary to accommodate the situation with respect to relevant skills and experience.

The use of the legal system to settle work-related disputes and claims varies between countries and depends upon the national tradition of litigation, and the available legislation as well as national industrial relation practices of individual countries. Reports from some countries indicate that a number of cases involving bullying or inappropriate coercion are resolved by tribunals or in court, and that there is a growing number of inquiries and complaints made to Australian Anti-discrimination Boards (Tidwell, 1998). Such cases currently represent the largest backlog of cases in the UK awaiting a hearing at industrial tribunals (Earnshaw and Cooper, 1996). Litigation may be costly for organisations, with compensation for loss of earning and injury to feelings reaching up to £150,000 in the UK (Dyer, 2002). Similarly, compensation for wrongful termination of contracts may be costly, with such cases reported to average awards of up to $400,000 in California, excluding the cost to employer of out-of-court settlements (Wilson, 1991).

In some cases bullying may also lead to industrial action and unrest as a result of alleged victimisation (Beale, 1999). In one particularly high-profile case in the UK, the union at Ford balloted their members on possible strike action, with allegations that managers tolerated and even encouraged bullying on the shop floor (Guardian, 9 October 1999). In addition to organisational upheaval and likely loss of productivity during the time of unrest, such a collective response to bullying incidents may also harm public relations. This would also be the case for the organisation facing litigation, where allegations and compensation claims may coincide with negative publicity, affecting customer relationships as well as the ability to attract skilled labour in a tight labour market.

Assessing the financial cost of bullying

Some attempts have been made to assess the costs of bullying at corporate as well as national levels. For example, Leymann (1990) calculated that every victimised individual would produce a cost to the organisation of approximately $30,000 to $100,000 annually. Kivimäki *et al.* (2000), in a study of bullying at two Finnish hospitals, estimated that the annual cost of absence from bullying alone accounted for costs of £125,000. Rayner (2000) estimated a cost of approximately £1 million annually in replacement costs alone. Similarly, based on a meta-analysis of workplace bullying research, and taking into consideration prevalence and severity data, Sheehan *et al.* (2001) calculated a cost in the order of $0.6 to $3.6 million per annum for an Australian business with 1,000 employees.

Hoel *et al.* (2001) took this one step further and estimated the costs of bullying at a national level, concluding that costs related to absence and replacement due to bullying alone may account for close to £2 billion annually. However, they acknowledged that any attempt to assess the costs of a complex problem such as bullying was fraught with difficulties, pointing to factors such as questionable quality of data and unclear connections between cause and effect.

We have listed below a number of cost factors which need to be considered when the financial cost is calculated. In this case it is worth bearing in mind that, in general, staff account for 50–80 per cent of organisational costs (Cooper *et al.*, 1996).

- Sickness absenteeism: employers respond to absenteeism in many ways. For example, by finding cover for the absentee by means of voluntary or compulsory cover by colleagues, through overtime or through replacement. However, in most cases the real cost of absenteeism is, by and large, linked to the cost of sick pay, where systems will vary from country to country (Gordon and Risley, 1999).
- Turnover and replacement costs: these primarily comprise recruitment costs (advertising and selection), as well as the cost of training and development. In addition to direct costs arising from these activities, administration (e.g. testing, candidates' travel expenses, termination of contracts, issuing of new contracts) will also have to be included in total costs. Replacement costs will tend to rise according to the experience and skills of the appointee. It should be noted that turnover is not necessarily dysfunctional to the effectiveness of an organisation, as a certain amount of turnover allows the organisation the opportunity to bring in new skills and talents (Gordon and Risley, 1999).

- Impact on productivity and performance: any impact of bullying is likely to be considerable, but largely intangible, and will require some element of 'informed guessing'.
- Grievance/compensation/litigation: practices will vary greatly between countries. It is worth noting that for every case which ends in court, there is likely to be a large number of cases resolved at the level of the organisation. Grievance procedures may give rise to administrative costs in connection with the implementation of investigation and mediation procedures, where such procedures are in place, and these may in many cases be greater than any compensation package. As far as investigations are concerned, costs will include the time of investigators and interviewees, the costs of the hearing as well as the cost of support, such as counselling.
- Loss of public goodwill and reputation: with the high profile of many individual compensation cases, the potential damage to reputation may motivate employers to deal with issues.

A case study of organisational costs

The following case of bullying at work and its costs to the organisation has been drawn from the official account of the case based on investigation reports and reports from the formal hearing. The events took place in a local authority in the UK. The complainant was a graphical designer in charge of a work unit of approximately twenty people. The accused, also a graphical designer and the complainant's subordinate, had applied for the same job as group manager in competition with the complainant eighteen months prior to the first incident of alleged bullying. Whilst the alleged offender, in a meeting with the complainant's line manager, had expressed unhappiness with his work situation, the relationship with his manager (the complainant) was never stated as a cause for dissatisfaction.

The behaviour at the heart of the complaint was described by the complainant as 'continuous undermining and unwillingness to observe instructions, culminating in a "threat of violence"'. According to the complainant, the alleged offender had on several occasions misrepresented him and told 'half-truths' to customers and contractors. On many occasions he had refused to speak to the complainant and had on one occasion told a contractor that he was unwilling to pass on a message and information to his manager. When the complainant had, as a last recourse, resorted to written communication, this was ignored, and on at least one occasion written instructions were ripped up in public. However, by far the most serious incident happened when the alleged offender threw an empty cup and saucer at the complainant, which failed to hit him and smashed into the wall, accompanied by a threat of further actions of this sort.

In response to the complaint, the organisation invoked their procedure for dealing with complaints of bullying. The complaint was looked into by two investigators. Due to the seriousness of the case (threat of violence), which meant that it was automatically considered a health and safety issue, the senior manager decided that a formal investigation should be undertaken. At this stage the negative behaviour on the part of the alleged offender had also escalated, prompting the complainant to ask for a formal investigation. A formal investigation was subsequently initiated involving the immediate line manager, principal officer, staff welfare officer, etc. Four witnesses were interviewed. At this stage the alleged offender was suspended on full pay, as separation of the two parties was considered necessary. The complainant went on medically certified sick leave due to stress and anxiety. Throughout the eight-week-long investigation period, both complainant and alleged offender made extensive use of the counselling service offered by the organisation.

The investigation concluded that, on balance, the respondent was found to have bullied and attempted to ostracise the complainant, and he remained under suspension from work. Due to the seriousness of the matter the offence was likely to be judged as 'gross misconduct', suggesting that dismissal could be a likely outcome. Pleading 'acute illness', the alleged offender failed to turn up at the hearing (involving five independent persons). One day before the second hearing the alleged offender handed in his resignation.

In Table 7.1 are listed the costs involved with the actual case. The figures clearly illustrate how the costs of a bullying case may mount up, representing a considerable drain on an organisation's resources.

Table 7.1 Cost of the case to the organisation

Absence	£6.972
Replacement costs	£7,500
Reduced productivity	?
Investigators' time for grievance investigation	£2,110
Local management line-management time	£1,847
Head office personnel	£2,600
Corporate officers' time (including staff welfare)	£2,100
Cost of disciplinary process (hearing /solicitor)	£3,780
Witness interview costs	£1,200
Transfers	0
Litigation	? (0)
Effects on those indirectly involved	?
Miscellaneous (effects on public relations, etc.)	?
Total costs (minimum)	£28,109

Note: ? = cost unknown or difficult to assess

Conclusions

This chapter has explored a number of potential organisational effects resulting from bullying in the workplace. Whilst we should not forget the human costs involved in bullying, in financial terms we believe that there is evidence to suggest that combating bullying makes good business sense. However, the fact remains that, in empirical terms, the association between bullying and various organisational outcome variables remains relatively weak, or at best moderate. Some factors accounting for this situation have been discussed. Still, correlations of this order indicate an effect of considerable magnitude (Zapf *et al.*, 1996), with a substantially higher risk of negative outcomes such as turnover when bullying is present.

With regard to the relationship between bullying and organisational outcomes, one should also bear in mind that people's reactions to their experience would tend to be complex and idiosyncratic, which is likely to affect overall correlations (Sparks *et al.*, 1997). Use of self-reporting as a way of measuring performance may also weaken the true relationship, as individuals tend to underestimate their own absenteeism and overestimate their own productivity. This relatively weak relationship may also be an artefact of the cross-sectional method normally applied in such studies, where performance is measured entirely at the level of the individual. According to Daniels and Harris (2000), one should bear in mind that a relatively small impact at the individual level may have a substantial aggregated or cumulative effect within an organisation when all behaviours/performance measures are taken into consideration.

Most of the data produced as evidence in this chapter are the outcome of cross-sectional studies and therefore do not allow for inferal of causal relationships. In order to increase our understanding of organisational effects in connection with bullying and their complex interaction, studies of a qualitative nature may represent a way forward, not least to explore the significance to the individual of behaviours such as absenteeism and turnover. Diary studies could be one way forward, though ethical consideration may reduce their effectiveness. Studies involving witnesses or observers could also throw important light on the complexity of the interaction between bullying, on the one hand, and factors such as absenteeism, productivity and turnover, on the other.

Acknowledgement

The authors would like to thank Clive Giddings, Lancashire County Council, for his help with the preparation of the case study material.

Bibliography

Ashforth, B. (1994) Petty tyranny in organizations. *Human Relations*, 47, 755–778.

Bassman, E. (1992) *Abuse in the workplace*. Westport, CT: Quorum Books.
Beale, D. (1999) *Workplace union renewal, resilience and redefining the debate: The case of Royal Mail*. Paper presented to Union Renewal? Workplace Industrial Relations in the Public Sector. Seminar organised by Universities of Northumbria and Hertfordshire, Newcastle-on-Tyne, 8 October.
Cooper, C. L. Liukkonen, P. and Cartwright, S. (1996) *Stress prevention in the workplace: Assessing the costs and benefits for organisations*. Dublin: European Foundation for the Improvement of Living and Working Conditions.
Cox, H. C. (1987) Verbal abuse in nursing: Report of a study. *Nursing Management, November*, 47–50.
Dalton, D. R. (1997) Employee transfer and employee turnover: A theoretical and practical disconnect? *Journal of Organizational Behavior*, *18*, 411–413.
Daniels, K. and Harris, C. (2000) Work, psychological well-being and performance. *Occupational Medicine – Oxford*, *50*, 304–309.
Dyer, C. (2002) Judges curb stress cases: Appeal court rulings scrap compensation awards and put onus on workers to tackle employers. *The Guardian*, 6 February.
Earnshaw, J. and Cooper, C. L. (1996) *Stress and employer liability, law and employment series*. London: Institute of Personnel and Development.
Einarsen, S. (1996). Bullying and Harassment at Work: Epidemiological and Psychosocial Aspects. PhD thesis, Faculty of Psychology, University of Bergen.
—— (2000) Bullying and harassment at work: Unveiling and organizational taboo. In M. Sheehan, S. Ramsay and J. Patrick (eds) *Transcending boundaries: Integrating people, processes and systems* (pp. 7–13). Conference Proceedings, Griffith University, Brisbane, Australia.
Einarsen, S. and Raknes, B. I. (1991) *Mobbing in Arbeidslivet. En undersøkelse av forekomst og helsemessige konsekvenser av mobbing på norske arbeidsplasser* (A study of the prevalence and health effects of mobbing in Norwegian workplaces). University of Bergen: Centre for Occupational Health and Safety Research.
Einarsen, S., Raknes, B. I., Matthiesen, S. B. and Hellesøy, O. H. (1994). *Mobbing og Harde Personkonflikter: Helsefarlig samspill på arbeidsplassen* (Bullying and interpersonal conflict: Interaction at work with negative implications for health). Bergen: Sigma Forlag
Field, T. (1996) *Bully in sight: How to predict, resist, challenge and combat workplace bullying*. Wantage: Wessex Press
Gordon, F. and Risley, D. (1999) *The costs to Britain of workplace accidents and work-related ill health in 1995/6*, 2nd edn. London: HSE Books.
Grunfeld, B. and Noreik, K. (1996) Syke kvinner og uførepensjonerte menn: noen refleksjoner (Sick women and disabled men: Some reflections). *Journal of Norwegian Medical Association*, *116*, 988–989.
Hoel, H. and Cooper, C. L. (2000a) *Destructive conflict and bullying at work*. Unpublished report, UMIST.
—— (2000b) Working with victims of workplace bullying. In H. Kemshall and J. Pritchard (eds), *Good practice in working with victims of violence* (pp. 101–118). London: Jessica Kingsley Publishers.
Hoel, H., Cooper, C. L. and Faragher, B. (2001) The experience of bullying in Great Britain: The impact of organizational status. *European Journal of Work and Organizational Psychology*, 10, 443–465

Hoel, H., Rayner, C. and Cooper, C. L. (1999). Workplace bullying. In C. L. Cooper and I. T. Robertson (eds), *International Review of Industrial and Organizational Psychology* (pp. 195–230), Chichester: John Wiley and Sons.

Hoel, H., Sparks, K. and Cooper, C. L. (2001) *The cost of violence/stress at work and the benefits of a violence/stress-free working environment*. Geneva: International Labour Organisation.

Ishmael, A. (1999) *Harassment, bullying and violence at work*. London: The Industrial Society.

Johns, G. (1994) How often were you absent? A review of the use of self-reported absence data. *Journal of applied Psychology*, *79*, 574–591.

—— (1997) Contemporary research on absence from work: Correlates, causes and consequences. In C. L. Cooper and I. T. Robertson (eds), *International Review of Industrial and Organizational Psychology* (pp. 115–173). Chichester: John Wiley and Sons.

Keashly, L. and Jagatic, K. (2000) *The nature, extent, and impact of emotional abuse in the workplace: Results of a statewide survey*. Paper presented at the Academy of Management Conference, Toronto, Canada.

Kile, S. M. (1990) Helsefarleg Leiarskap: Ein Explorerande Studie (Leadership with negative health implications: An exploratory study). Report from the Norwegian General Science Council, Bergen.

Kivimäki, K., Elovainio, M. and Vathera, J. (2000) Workplace bullying and sickness absence in hospital staff. *Occupational and Environmental Medicine*, *57*, 656–660.

Lee, D. (2000) An analysis of workplace bullying in the UK. *Personnel Review*, *29*, 593–612.

Leymann, H. (1990) Mobbing and psychological terror at workplaces. *Violence and Victims*, *5*, 119–125.

—— (1992) *Vuxenmobbning på svenska arbeidsplatser*. Delrapport 1 om frekvenser (Adult bullying at Swedish workplaces: Report 1 concerning frequencies). Stockholm: Arbetarskyddstyrelsen.

—— (1996) The content and development of mobbing at work. *European Journal of Work and Organizational Psychology*, *5*, 2, 165–184.

Marmot, M., Feeny, A., Shipley, M., North, F. and Syme, S. (1995) Sickness absence as a measure of health status and functioning: From the Whithall II study. *Journal of Epidemiological Community Health*, *49*, 124–130.

McCarthy, P., Sheehan M. and Kearns, D. (1995) *Managerial styles and their effects on employees' health and well-being in organizations undergoing restructuring*. Griffith University, Brisbane, School of Organizational Behaviour and Human Resource Management.

O'Moore, M., Seigne, E., McGuire, L. and Smith, M. (1998) Bullying at work: Victims of bullying at work in Ireland. *Journal of Occupational Health and Safety – Australia and New Zealand*, *14*, 569–574.

Price Spratlen, L. (1995) Interpersonal conflict which includes mistreatment in a university workplace. *Violence and Victims*, *10*, 285–297.

Pryor, J. B. (1987) Sexual harassment proclivities in men. *Sex Roles*, *17*, 269–290.

Quine, L. (1999) Workplace bullying in NHS community trust: Staff questionnaire survey. *British Medical Journal*, *318*, 228–232.

—— (2001) Workplace bullying in nurses. *Journal of Health Psychology*, *6*, 73–84.

Rayner, C. (1997) The incidence of workplace bullying. *Journal of Community and Applied Social Psychology*, *7*, 249–255.

—— (1999) Workplace bullying. PhD thesis, UMIST.

—— (2000) Building a business case for tackling bullying in the workplace: Beyond a cost-benefit analysis. In M. Sheehan, C. Ramsey and J. Patrick (eds), *Transcending boundaries*, proceedings of the 2000 Conference, September, Brisbane.

Rayner, C., Hoel, H. and Cooper, C. L. (2002) *Workplace bullying: What we know, who is to blame, and what can we do?* London: Taylor and Francis.

Seago, J. A. (1996) Work group culture, stress and hostility: Correlations with organisational outcomes. *Journal of Nursing Administration*, *26*, 39–47.

Sheehan, M., McCarthy, P., Barker, M. and Henderson, M. (2001) *A model for assessing the impact and costs of workplace bullying*. Paper presented at the Standing Conference on Organizational Symbolism (SCOS), Trinity College, Dublin, 30 June–4 July.

Sinclair, J., Ironside, M. and Seifert, R. (1996) Classroom struggle? Market oriented education reforms and their impact on the teacher labour process. *Work, Employment and Society*, *10*, 641–661.

Sparks, K., Cooper, C. L., Fried, Y. and Shirom, A. (1997) The effects of hours of work on health: A meta-analytical review. *Journal of Organisational and Organisational Psychology*, *70*, 391–408.

Steer, S. R. and Rhodes, R. M. (1978) Major influence on employee attendance: A process model. *Journal of Applied Psychology*, *63*, 391–407.

Tepper, B. J. (2000) Consequences of abusive supervision. *Academy of Management Journal*, *43*, 178ff.

Thyholdt, R., Eide, R. and Hellesøy, O. H. (1986) Arbeidsplass Statfjord. Helse og Sykdom på Statfjordfeltet, report no. (Workplace Statfjord: Health and Illness in the Statfjord sector, report no.). University of Bergen: Research Centre for Work, Health and Safety.

Tidwell, A. (1998) The role of workplace conflict in occupational health and safety. *Journal of Occupational Health and Safety – Australia and New Zealand*, *14*, 587–592.

UNISON (1997) *UNISON members' experience of bullying at work*. London: UNISON.

—— (2000) *Police staff bullying report* (no. 1777). London: UNISON.

Vartia, M. (1993) Psychological harassment (bullying, mobbing) at work. In K. Kauppinen-Toropainen (ed.), *OECD Panel Group on women, work and health. National report*. Finland, The Ministry of Social Affairs and Health, The Institute of Occupational Health, Publications 1993: 6, pp. 149–152.

—— (2001) Consequences of workplace bullying with respect to well-being of its targets and the observers of bullying. *Scandinavian Journal of Work, Environment and Health*, *27*, 63–69.

Voss, M., Floderus, B. and Diderichsen, F. (2001) Physical, psychosocial, and organisational factors: A study based on Sweden Post. *Journal of Occupational Environmental Medicine*, *58*, 178–184.

Zapf, D. and Gross, C. (2001) Conflict escalation and coping with workplace bullying: A replication and extension. *European Journal of Work and Organizational Psychology*, *10*, 497–522.

Zapf, D., Dormann, C. and Frese, M. (1996) Longitudinal studies in organizational stress research: A review of the literature with reference to methodological issues. *Journal of Occupational Health Psychology*, *1*, 145–169.

Zapf, D., Knortz, C. and Kulla, M. (1996) On the relationship between mobbing factors, and job content, social work environment, and health outcomes. *European Journal of Work and Organizational Psychology*, *5*, 215–237.

Wilson, C. B. (1991) US businesses suffer from workplace trauma. *Personnel Journal*, *July*, 47–50.

Part 3
Explaining the problem

8 Individual antecedents of bullying

Victims and perpetrators

Dieter Zapf and Ståle Einarsen

Introduction

The causes of bullying at work have been a 'hot issue' of debate in both the popular press and in the scientific community. While some argue that individual antecedents such as the personality of bullies and victims, indeed may be involved as causes of bullying (e.g. Coyne *et al.*, 2000), others have disregarded totally the role of individual characteristics. Heinz Leymann (1993, 1996), one of the founders of bullying research, categorically claimed that organisational factors relating to the organisation of work and the quality of leadership behaviour were the main causes of bullying. He rejected the idea that the personal characteristics of the victim are capable of playing any part in the development of bullying at work. This standpoint is also strongly advocated by some victims of bullying and their organisational networks. Other victims and their spokespersons have claimed that bullying is mainly caused by the psychopathic personality of the bully (e.g. Field, 1996).

On no account do we deny that organisational issues have to be considered in the discussion of bullying causes. However, our own standpoint is that no comprehensive model of workplace bullying would be satisfactory without also including personality and individual factors of both perpetrators and victims, and their contributing effects to the onset, escalation and consequences of the bullying process (Einarsen 2000; Hoel *et al.*, 1999; Zapf, 1999b). Research on bullying among children (e.g. Olweus, 1993) has shown that both victims and bullies portray personality characteristics that may contribute to their involvement in such situations. While in this line of research victims have been described as being cautious, sensitive, quiet, anxious and insecure, bullies have been described as self-confident, impulsive and generally aggressive. Similarly, studies of bullying at work indicate that personality characteristics of victims, such as neuroticism, do seem to be related to exposure to bullying (Mikkelsen and Einarsen, 2002; Zapf, 1999a). In addition, psychological literature presents a wide range of concepts relating to the personality of bullies, such as 'the abrasive personality', 'the authoritarian personality' and 'the petty tyrant' (see also Ashforth, 1994).

However, one has to tread carefully with respect to these issues, as one might easily be accused of 'blaming the victim' on the one hand, and 'witch-hunting' on the other. Yet, against the background of communication theory (Watzlawik *et al.*, 1969) and the psychology of interpersonal conflict (Thomas, 1992; van de Vliert, 1998), any one-sided and monocausal explanations are highly unlikely. Rather, one may have to take a broad range of potential causes of bullying into account, which may lie within the organisation, the perpetrator (the bully), the social psychology of the work group, and also the victim (Einarsen, 1999; Zapf, 1999b).

Furthermore, the personality of the victim may be highly relevant in explaining perceptions of and reactions to workplace bullying, but not necessarily as relevant in explaining the behaviour of the bully (Einarsen, 2000). It is also likely that the personalities of perpetrator and victim may be of more relevance in some cases than in others. Empirical evidence (Matthiesen and Einarsen, 2001; Zapf, 1999a, 1999b) indicates that bullying cases differ in the degree to which personality is involved as a potential cause. There is certainly nothing such as a victim personality (e.g. the 'notorious complainer') which can explain bullying in general. Rather, specific explanations may be valid for specific subgroups but not for every case of bullying. Moreover, it is likely that several antecedents together contribute to the development of bullying, although one antecedent may sometimes play a dominating role (see also Hoel and Cooper, 2001).

Bearing these precautions in mind, the aim of this chapter is to discuss individual antecedents of workplace bullying and to review the empirical evidence. We will focus on individual antecedents of bullying which can either be related to the perpetrator or to the victim, or both. Causes of bullying lying within perpetrators may, of course, overlap with factors relevant to both the social group and the organisation, for example in the case of a supervisor being the bully. Supervisors and managers are often made responsible for organisational circumstances that may contribute to the development of bullying, such as organisational culture and the organisation of work.

The chapter aims at discussing individual antecedents involved in severe cases of bullying and it does not intend to cover all aspects of interpersonal conflict and social stress at work. Applying a relatively stringent definition of bullying we will focus on those severe cases of highly escalated conflicts which usually last longer than one year, and in many cases even several years (Zapf, 1999a). The following definition of bullying or mobbing will be the basis for this chapter:

> Bullying at work means harassing, offending, socially excluding someone or negatively affecting someone's work tasks. In order for the label bullying (or mobbing) to be applied to a particular activity, interaction or process it has to occur repeatedly and regularly (e.g. weekly) and over a period of time (e.g. about six months). Bullying is an esca-

> lating process in the course of which the person confronted ends up in an inferior position and becomes the target of systematic negative social acts. A conflict cannot be called bullying if the incident is an isolated event or if two parties of approximately equal 'strength' are in conflict'.
>
> (Einarsen *et al.*, this volume. Cf. also Einarsen, 2000; Einarsen and Skogstad, 1996; Leymann, 1993; Zapf, 1999a)

Bullying research still suffers from a lack of studies which allow clear cause–effect analyses (Einarsen, 2000; Zapf, 1999b; Zapf *et al.*, 1996). Bullying is supposed to lead to health complaints on the part of the victim. Leymann (1996; Leymann and Gustafsson, 1996) even assumed that bullying can change the personality of the victim. However, one can equally assume that, for example, anxious, depressive or obsessive behaviour produces a negative reaction in a group, which leads to bullying after some time. Careful language is therefore necessary when talking about cause and effect. However, although single cases can be questioned, we do believe that the arguments in this chapter will apply at least to some cases of bullying.

When talking about individual antecedents of bullying, it should be noted that a distinction has to be made between cause and guilt or responsibility, and that, for example, in the case of the victim, a cause 'within the victim' may as much point to the social group as to the individual. An example may be a group which for some reason is not able to integrate a person which is different in one respect or another. However, being perceived as an antecedent may also mean that something is to be seen as a cause or in some cases even a responsibility. A manager who feels threatened by a subordinate may start to bully this person. A socially incompetent and narrow-minded person may, by her behaviour, create constant hassle for her colleagues, thus provoking an aggressive response. In other cases, a person, either bully or victim, may not be responsible for triggering a conflict but he or she may be the main reason behind the escalation of the conflict. Finally, personal characteristics may take the role of a moderator. That is, although neither cause nor mediator force of conflict escalation, bullying might happen to some people but not to others. There is, for example, evidence that the simple fact of being significantly different from the rest of the group, for example being the only woman among men, increases the risk of becoming an outsider and thus the victim of bullying (Leymann, 1993; Schuster, 1996; Zapf, 1999a). Given the present state of research, empirical results are often compatible with more than one of the mechanisms just described. These constraints should be kept in mind in the following.

Individual antecedents of bullying: the perpetrator

Bullying research has revealed that bullies seem to be male more often than female, and supervisors and managers more often than colleagues (see also Zapf *et al.*, this volume). Yet we do not know much about the characteristics of the perpetrators of bullying. For obvious reasons, it is difficult to collect valid information about them. Most of the available information comes from studies based on victim reports. In an interview study among thirty Irish victims of bullying, all victims blamed the difficult personality of the bully (Seigne, 1998). Some reported that bullying was related to the bully's transfer to a position of power. Zapf (1999b) found that 'They wanted to push me out of the company' was the most frequent reason reported by victims. A weak, unsure superior, competition for tasks, status or advancement, or competition for the supervisor's favour were other motives as perceived by victims in the studies of Björkqvist *et al.* (1994) and Vartia (1996). That is, at least from the victims' perspective, the cause of bullying is identified with a particular perpetrator.

Summarising empirical findings on the bully as a cause of workplace bullying, we suggest three main types of bullying related to certain perpetrator characteristics: (1) self-regulatory processes with regard to threatened self-esteem; (2) lack of social competencies; and (3) bullying as a result of micropolitical behaviour.

Bullying due to protection of self-esteem

Many theorists (e.g. Baumeister *et al.*, 1993, 1994; Baumeister *et al.*, 1996; Stahlberg *et al.*, 1985; Tesser, 1988) assume that protecting or enhancing one's self-esteem is a basic human motive which influences and controls human behaviour in many social situations. Self-esteem can be understood as having a favourable global evaluation of oneself (Baumeister *et al.*, 1996). In every social interaction, mutual recognition of one's status is a core issue. Social interactions go smooth as long as the interaction partners feel respected and recognised in their position, which means that self-evaluation and external evaluation are in agreement. Conflicts arise where this is not the case (Schulz von Thun, 1984; Stahlberg *et al.*, 1985; Watzlawik *et al.*, 1969). Baumeister *et al.* (1996) reviewed the literature on self-esteem, violence and aggression, and proposed that it is high rather than low self-esteem which is related to aggressive behaviour. People with low self-esteem are usually not aggressive because they fear losing the battle. Rather, they present depressive reactions and withdrawal. In contrast, one major cause of aggressive response is threatened egotism, that is a favourable self-appraisal that encounters an external unfavourable evaluation. When favourable views about oneself are questioned, contradicted or impugned, people may aggress. This is particularly so if unrealistically positive or inflated views of self prevail, and if these self-appraisals are uncertain, unstable or heavily dependent on external validation.

> People who regard themselves as superior beings might feel entitled to help themselves to the resources of other, seemingly lesser beings, and indeed they might even aggress against these beings without compunction, just as people kill insects or mice without remorse.
>
> (Baumeister *et al.*, 1996, p. 8)

Aggression is most commonly directed at the source of the negative evaluation, and serves to refute and prevent negative evaluations as well as to constitute a means of symbolic dominance and superiority over the other person. In terms of self-regulation, aggression is used to defend positive self-appraisals, instead of adapting to the more negative appraisal of oneself proposed by others, because the resulting decrease in self-esteem is aversive for nearly everyone. Thus, (too) high self-esteem can lead to tyrannical behaviour because it may be related to perfectionism, arrogance, and narcissism (Ashforth, 1994; Baumeister *et al.*, 1993, 1996; Kets de Vries and Miller, 1985).

In summarising their review of empirical research, Baumeister *et al.* concluded:

> In all spheres we examined, we found that violence emerged from threatened egotism, whether this was labelled as wounded pride, disrespect, verbal abuse, insults, anger manipulations, status inconsistency, or something else. For huge nationalities, medium and small groups, and lone individuals, the same pattern was found: Violence resulted most commonly from feeling that one's superiority was somehow being undermined, jeopardised, or contradicted by current circumstances.
>
> (Baumeister *et al.*, 1996, p. 26)

People with unstable high self-esteem may well become aggressive in response to even seemingly minor or trivial threats to self-esteem. Kernis *et al.* (1993) found, for example, that people with unstable self-esteem were most prone to respond defensively to unfavourable feedback. Kernis *et al.* (1989) reported that the highest levels of self-reported angry and hostile responses were associated with individuals who had high but unstable self-esteem scores. Similar results were found by Stucke and Sporer (2001) in a study on workplace bullying. In their study, employees indicated first whether or not they actively used different kinds of bullying behaviours and whether or not they were the receivers of such behaviour. Second, measures of narcissism and stability of self-esteem were measured. Narcissism implies high self-esteem along with the disregard of others. Active bullying behaviour was highest for a group high in narcissism but low in self-esteem stability. Obviously these individuals had to stabilise their high but unstable self-esteem by negatively treating other individuals. Qualitative data collected by the authors confirm these findings. For example, a typical constellation for bullying is that the department supervisor retires and a highly qualified employee of the department hopes to become his or her successor. Instead, for some reason, a person from another department is appointed whose qualification for the job is not obvious. Right from the beginning a conflict arises, and the new supervisor, who feels threatened by this person, uses all possible means to defend his or her position and starts bullying the potential rival. Accordingly, in the study by O'Moore *et al.* (1998) in Ireland, many victims claimed that the bully had just recently entered a position of power. Björkqvist *et al.* (1994) reported that unassertiveness of the bully was the second most common reason for bullying from the perspective of the victims.

Moreover, Baumeister *et al.* (1996) suggested that various negative emotions such as frustration, anger or anxiety play a mediating role between self-esteem and aggression. Envy can also play a significant role in this context. Envy arises when someone else has what the envious person wants, which can imply that oneself is less worthy and less deserving than the other (Salovey, 1991). Smith *et al.* (1994) found that envy leads to hostility only if the person retains a favourable view of self as deserving a particular positive outcome, in which case the envied person's advantage is seen as unjust and unfair.

In several studies on workplace bullying using reports of victims, envy on the part of the bullies is considered one main reason for bullying (Björkqvist *et al.*, 1994; Einarsen *et al.*, 1994a; Seigne, 1998; Vartia, 1996). In Vartia's study, this was the case in 68 per cent of all incidents. Two-thirds of the victims in the study by Seigne (1998) identified envy on the part of the bully, especially with regard to certain qualifications, as a main reason of bullying. Similar numbers were reported by Zapf (1999b).

Moreover, bullying as a kind of personal retaliation can be explained as a self-regulatory mechanism of self-esteem protection. A typical case relates to affairs between bosses and their secretaries. The unexpected termination of the relationship by the secretary leads to severely hurt feelings on the part of the manager, who, because everybody knows about the 'secret' relationship, feels that not only his or her self-esteem but also his or her reputation as a manager is at stake. Thus, he or she starts to bully the secretary with the goal of expelling him or her from the firm.

Bullying as a result of self-esteem protection may occur especially frequently if the bullies are managers because being dominant, self-assertive, having high self-esteem and protecting this self-esteem is normally expected from this group.

Bullying due to lack of social competencies

In other cases, a lack of social competencies seems to be a dominating factor. One aspect is lack of emotional control. A supervisor might vent his anger by regularly yelling at one of his subordinates. There are many published cases which correspond to such a pattern (e.g. Adams, 1992). Bullying might also be a consequence of a lack of self-reflection and perspective-taking. This implies that some bullies are not fully aware of what they are doing and how their behaviour affects the victims. In the study by Einarsen *et al.* (1994b), 'thoughtlessness' was seen as a cause of bullying in 46 per cent of all cases. On an anecdotal basis, bullies repeatedly claim that they were not aware of the consequences of their behaviour (Leymann, 1993; Krum, 1995). This is particularly so if several individuals bully someone. The personal contribution of a bully may then be small: a relatively insignificant event, such as gossiping, not greeting, not passing on information or horseplay, events which may occur, for example, every two or three weeks. If such behaviour is acted out by four or five individuals, this implies that the victim is bullied approximately twice a week. The bullies may not be aware of this situation due to little communication between perpetrators and victims, and due to the fact that perpetrators may not receive realistic feedback on their behaviour. Baumeister *et al.* (1990) analysed narrative interviews and found that victims of aggression, anger and harassment interpreted such experiences as a series of events. The victims experienced frequent and single acts as systematic and intentional behaviour directed against them. They tolerated the bullying behaviour for some time but then they tended suddenly to overreact. The perpetrators, by contrast, perceived the behaviours as isolated and tolerable events. Thus they were surprised by the reaction of the victims, finding the victims' behaviour exaggerated and difficult to understand.

Thus, to be able to reflect on the consequences of one's own behaviour, an individual must be able to reflect on the overall situation, and even take the perspective of the victim, which may show that the incidental behaviour of various bullies can be perceived by the victim as systematic and frequent harassment. In interviews with alleged bullies and observers of bullying we have noticed that if you ask them whether or not person X has been bullied at work, they will normally give a negative reply. If we rephrase the question into something like 'In hindsight and taking the whole situation of person X into account, are you able to see that this person may have perceived himself as a victim of bullying?', the answer is very often positive.

However, it is assumed that some bullies may lack this ability of perspective taking. In a study among 2,200 members of seven Norwegian labour unions, some 5 per cent of the respondents admitted that they had bullied others at work (Einarsen *et al.*, 1994b). The self-reported bullies differed from other employees in many respects, describing themselves as high on social anxiety, low on social competence, low on self-esteem, but generally high on aggressiveness. These bullies reacted with aggressiveness in a wide range of situations and towards a wide range of perceived provocations. Apart from cases where bullies admit 'light' forms of bullying (e.g.

Krum, 1995, who reported that the bullies admitted that they had spread rumours and had refused to talk to a certain colleague), bullies will normally not be prepared to report their behaviour, probably because of social desirability and even for legal reasons. In addition, many bullies will not view their own behaviour as bullying, but rather as a reasonable reaction to a difficult and tense situation. Hence, the results presented above must be interpreted with caution and must probably not be generalised to all alleged bullies. However, these characteristics may at least apply to a subgroup of bullies. Since bullying may come in many forms and shapes and evolve in a range of different situations (see Einarsen, 1999) it is highly unlikely that a single personality profile would be common to all bullies.

Bullying due to micropolitical behaviour

It has been suggested that some cases of bullying follow the logic of micropolitical behaviour in organisations (Neuberger, 1995, 1999). The concept of micropolitics is based on the premise that organisations do not consist of fully determined structures and processes. Rather, organisations require members of the organisation to assist and fill the gaps in the formal structure. That is, organisations both require individuals to take part in decision making, and offer them possibilities for influence and decision making. A second premise is that members in organisations try to protect and improve their status in the organisation which may correspond to the self-esteem protection discussed above. Thus, they normally use their possibilities for decision making not only in the interests of the organisation, i.e. to reach organisational goals, but also in their own interest, i.e. to reach personal goals in the sense of status protection and improvement. These decision processes are assumed to be political in nature. In order to be able to have influence on decisions, it may seem necessary to build coalitions, and sometimes to plot against competitors.

Such micropolitical behaviour cannot be equated with bullying. Micropolitical behaviour aims at protecting one's own interests and improving one's own position. This may of course include behaviour that negatively affects other persons. But overall the focus is on one's own interests and not on the destruction of others. This is particularly so because coalitions vary and current competitors may become allies later. Nevertheless, such micropolitical behaviour may occasionally take the form of bullying, and there may be a thin line between the acceptable use of power and bullying. Bullying due to micropolitical behaviour indicates harassment of another person in order to protect or improve one's own position in the organisation. From an outside perspective micropolitical bullying may appear as the most 'rational' behaviour of all forms of bullying because 'reasonable' motives (striving to get a position, influence, resources, etc.) can often be identified. Moreover, such bullying behaviour may sometimes only slightly transgress organisational norms and values such as being, dominant, competitive, high-achieving, etc. Some data on victim perceptions may support the concept of micropolitical bullying. In a re-analysis of interviews with victims, eleven of twenty-four cases could be classified as micropolitical behaviour (Bühler and Zapf, 1997). Micropolitical behaviour has been described as a phenomenon mainly occurring at the middle and higher hierarchical levels of an organisation. One may conclude that managers do profit more by using bullying as a form of micropolitical behaviour, which may be one of the explanations as to why supervisors and managers are so often among the bullies (Hoel *et al.*, 2001; Zapf, 1999a).

So far, we have presented various explanations for how individual antecedents of bullying could lie within the perpetrator. However, looking at data which support the view that certain perpetrators are responsible for bullying, we have to take into consideration that potential errors of attribution may play a part in this evidence: It is easier to attribute unpleasant feelings to a person than to invisible circumstances (cf. Neuberger, 1999). It is not unlikely that where the real causes are to be found in organisational circumstances or in the social system, they are instead attributed to certain persons, especially if the source of information is the victim.

Finally, Brodsky (1976) concludes that even if perpetrators may indeed have some common personality characteristics making them prone to bullying, they will not exhibit such behaviour unless they are in an organisational culture that rewards, or at least is permissive of, such behaviours. Also, research in the field of sexual harassment has shown that even though some men indeed have stronger proclivities for sexually harassing behaviours than others, they will only portray such behaviours if they perceive the social climate to encourage them (Pryor and Fitzgerald, this volume).

Individual antecedents of bullying: the victim

Among the reasons for bullying offered by Zapf (1999b), there were several items which suggest that the reasons may, at least partly, lie within the victim him or herself. It is interesting to note that these items were endorsed by few victims. Altogether, 37 per cent of the victims agreed with one of these items, whereas 63 per cent did not see any personal involvement in the emergence of the bullying episode. Notably, only 2 per cent of the victims 'admitted' that their performance was below average. Other examples of personal reasons were deficits in social skills, low performance, 'being difficult', for example being overly accurate, being aggressive or moaning.

Niedl (1995) claims that targets of bullying will only be victimised if they are unable to defend themselves for any reason or unable to escape the situation due to any dependency on their part. This dependency may either be of a social nature (hierarchical position, power relations, group membership), a physical nature (physical strength), an economic nature (labour market, private economy) or a psychological nature (self-esteem, personality, cognitive capacity). Hence, the issue of individual antecedents located within the victim relates to many issues. Primarily, some individuals may generally or in specific situations be at risk due to social, demographic or personal factors, which increases their chances of experiencing bullying. Second, their personality and behaviours may be a possible factor in eliciting aggressive behaviours in others. Third, psychological factors may be involved in the ability to defend oneself in highly escalated conflicts with peers and superiors. In a Norwegian survey (Einarsen *et al.*, 1994b), many victims felt that their lack of coping resources and self-efficacy, such as low self-esteem, shyness, and lack of conflict management skills, contributed to the problem. Fourth, individual factors may also be involved as potential moderating factors explaining why some more than others develop stress reactions and health problems after exposure to bullying (see also Einarsen, 2000). As in the case of the perpetrators, there seem to be various groups and mechanisms in which victim characteristics dominate in the development of bullying: (1) The exposed position of the victim; (2) social incompetence and self-esteem deficits; and (3) overachievement and conflict with group norms.

The salience and outsider position of the victim

Research on groups suggests that people who are outsiders in some respect and who differ from the rest of the group carry a risk of getting in trouble with others and may even be forced into the role of a scapegoat (Thylefors, 1987). Social psychologists have repeatedly demonstrated that individuals who do not belong to the group are devaluated whereas group members are much more positively evaluated (e.g. Brown, 1997). According to social identity theory (Tajfel and Turner, 1986), being different may cause others to see the persons as one of 'them' and not one of 'us', which again in certain circumstances may lead to displaced aggression towards the person seen as the outsider. Outsiders have a weaker social network and receive less social support (Cohen and Wills, 1985). According to labelling theory (Neuberger, 1999), deviant behaviour may be escalated when small peculiarities which are in itself unimportant are used to label someone (e.g. as a mischief-maker, moaner, failure, etc.), and socially exclude the person. Leymann (1993) also saw a socially exposed position of the victim as a risk factor for becoming a victim. In one of his studies, male kindergarten teachers who were a minority were more often among the victims than females. In another study (Lindroth and Leymann, 1993), 21.6 per cent of the handicapped employees in a non-profit organisation were bullied; but only 4.4 per cent of the non-handicapped. In the study by Zapf (1999a),

14 per cent of the victims claimed to be different from the other members of the work group according to visible characteristics such as age, gender or physical handicap, whereas only 8 per cent of the control group said so.

The vulnerable victim: social competence and self-esteem

Among laypeople, the most common view with regard to victim characteristics is certainly a belief that some people are more vulnerable than others, because they are low on self-assertiveness, unable to defend themselves and unable to manage the inevitable conflicts constructively. Therefore, they are seen as natural victims of bullying. A few studies have tried to find some evidence for this assumption. Together with professionals and bullying victims, Zapf and Bühler (in Zapf, 1999a) developed a list of forty-five items according to which victims of bullying may see themselves as being different from the rest of their work group, such as a lack of social skills and unassertive behaviour. An example item is 'Compared with my colleagues ... I do not recognise conflicts as quickly as others.' The results showed, that victims more often than a control group saw themselves as different from their colleagues. Most interestingly, the results showed that there were obviously very heterogeneous groups of victims. There was one group which, indeed, corresponded to what most people would expect: individuals low in social competencies, bad conflict managers, unassertive and weak personalities. Averaging the results of two samples reported in Zapf (1999a), 33 per cent of the victims saw themselves as more unassertive and as worse conflict managers than their colleagues, compared to only 16 per cent of a control group. Also, a study among 2,200 Norwegian employees in seven organisational settings showed that victims of bullying were characterised by being low on self-esteem, high on social anxiety and low on social competence. However, in terms of aggressiveness they did not differ from the average (Einarsen *et al.*, 1994b). Vartia (1996) found a relationship between bullying and neuroticism and a negative self-image in a sample of municipal employees. Using personality tests based on the five-factor model in a study of sixty Irish victims of bullying, Coyne *et al.* (2000) found that victims in comparison with a control sample were more anxious and suspicious, and reported having more problems coping with difficult situations. Moreover, victims were less assertive, competitive and outspoken than non-victims. Lindemeier (1996) carried out psychiatric analyses of eighty-seven victims of bullying. Thirty-one per cent of the patients reported a general tendency to avoid conflict, 27 per cent reported low self-esteem problems before bullying began. Moreover, 23 per cent reported that they had always been emotionally labile and had taken everything very seriously. Such descriptions largely overlap with the description of victims of bullying in school (Einarsen, 2000). In addition, in a study among all nurses and assistant nurses working in three Norwegian nursing homes, victims of bullying were shown to have little sense of humour (Einarsen, 1997); that is, they portrayed a negative attitude towards humour and the use of humour at work. On the basis of interviews with American victims of bullying, Brodsky (1976) claimed that many victims are of a humourless nature. On meeting a notorious teaser, they may feel they are being victimised and bullied by becoming the laughing-stock of the department, by the use of practical jokes, or by real or perceived excessive teasing.

Using the MMPI-2, Matthiesen and Einarsen (2001) investigated a group of eighty-five Norwegian victims of bullying. The MMPI-2 is a personality test developed for clinical purposes. When reaching a certain threshold, its scales are assumed to indicate levels of psychological problems and disturbances which require psychological treatment. The authors found elevated levels on various scales for this sample of victims. As a group, the victims were described as being oversensitive, suspicious, depressive and to have a tendency to convert psychological distress into psychosomatic symptoms. They were also seen as possibly having problems with understanding more subtle psychological explanations for their own problems. Interestingly enough, these results were identical to those of Gandolfo (1995), who used the same instrument on American victims claiming compensations from employers due to bullying. However, while the victims in Gandolfo's study were mostly younger men, the Norwegian study consisted of mainly older females.

On the application of cluster analysis, a statistical method to group individuals, thereby maximising similarity within the group and maximising differences across groups based on a given set of variables (here the MMPI-2 scales), three stable clusters could be identified: The 'seriously affected', the 'common' and the 'disappointed and depressed' (Matthiesen and Einarsen, 2001). While the first group reported a wide range of psychological and emotional problems and symptoms, the second group did not portray any particular psychological symptoms of a neurotic or psychotic nature. However, while the second group reported exposure to a wide range of specific bullying behaviours, the former reported exposure to fewer acts of bullying. The last group, being depressed and somewhat paranoid, consisted of those victims who were bullied at present. Matthiesen and Einarsen (2001) interpret these results to be indicative of a vulnerability factor in a specific group of victims.

Depue and Monroe (1986) and Dohrenwend *et al.* (1984) suggested that people high in negative affect (NA, including anxious, depressive and neurotic symptoms) through their behaviour create or enact adverse circumstances. They may, therefore, create or contribute to the development of problems and conflicts at work. For example, people high in NA might more often be involved in conflicts with others at work, be less able to manage their own workflow, and may perform less well in their job compared to people with low NA. In addition, there is a large literature showing that other people may respond negatively to depressed individuals (e.g. Sacco *et al.*, 1993). All this may increase the base rate of conflict and may thus increase the likelihood of conflict escalation. Applying this to bullying, it may be concluded that individuals low in self-esteem, self-assertion, social competencies, and high in anxiety and depression may not only be bullied because they are defenceless and 'easy targets'. Rather, due to their own behaviour they may actively produce conflicts that again may cause them to become the targets of aggression and harassment.

One may ask the question whether there is some relationship between bullying at school (see Olweus, this volume) and workplace bullying which would also point to individual antecedents of bullying. Smith *et al.* (2001) report some data based on retrospective reports on school bullying of individuals having taken part in a study on workplace bullying. They found a small correlation: victims of bullying at school were slightly more bullied in the workplace. The effect was somewhat larger for individuals having been both bullies and victims at school. Moreover, those who reported that they 'did not really cope' with school bullying as victims had an increased risk of becoming victims of workplace bullying. These results may be interpreted as further support that individuals with a general lack of good conflict management skills have a somewhat higher risk of being bullied at work.

In the study by Zapf (1999b), victims high on unassertiveness/avoidance more often claimed that their performance was below average, and that they were bullied because of their nationality or because of a physical disability. Victims high in unassertiveness/avoidance also portrayed the worst conflict management behaviours, being high in conflict avoiding and obliging, low in compromising and integrating. The clearest effect appeared for those avoiding conflict. Moreover, those victims who were low on unassertiveness/avoidance showed results comparable with those of a control group.

Individual antecedents may not only play a role in the onset of a bullying conflict. They may also play a role in the process of conflict escalation. Knorz and Zapf (1996) and Zapf and Gross (2001) compared the coping behaviour of successful victims (those whose overall situation substantially improved) and unsuccessful victims (those whose overall situation became worse and worse in spite of their coping trials). The successful victims more frequently avoided mistakes or behaviours, such as frequent absenteeism, which could be turned against them, and they were obviously better in recognising and avoiding behaviours which escalate rather than de-escalate the conflict.

Schuster (1996) related the literature on school and workplace bullying to the literature on social status and peer rejection among schoolchildren. According to Schuster, research consistently shows that children rejected by their peers are perceived as less pro-social and more aggressive by both external observers and peers. They assume more hostile intentions in others, and generate fewer and less effective solutions for social problems or conflicts. Moreover, they use less successful strategies to be accepted by a new group. If this line of research can be transferred to workplace bullying, it would underline that being socially incompetent as well as non-assertive may contribute to rejection on behalf of colleagues and superiors, and thus explain why some may easily become a target of workplace bullying.

Overachievement and clash with group norms

Various studies seem to include a further behavioural pattern related to overachievement and conscientiousness. In the study by Zapf (1999a), 69 per cent of the victims reported being more conscientious than their colleagues, whereas 40 per cent of the control group reported being so. Moreover, 62 per cent of the victims reported to be more achievement-oriented than their colleagues, compared to 41 per cent of the control group. These data fit with Brodsky's (1976) qualitative observations:

> The harassed victim generally tends to be conscientious, literal-minded, and somewhat unsophisticated. Often, he is an overachiever who tends to have an unrealistic view of himself and of the situation in which he finds himself. He may believe he is an ideal worker and that the job he is going to get will be the ideal job. As a result, he has great difficulty in adjusting not only to the imperfections of the situation but to the imperfections of his own functioning as well.
>
> (Brodsky, 1976, p. 89)

Similarly, Coyne *et al.* (2000) found that victims in comparison with a control group were generally more rule-bound, honest, punctual and accurate. Such persons may be highly annoying to others, which again may contribute to frustration and aggression outlets in their colleagues. Following the line of argument above on high but unstable self-esteem as a cause of aggression in perpetrators, feelings of being unjustly treated and belittled may of course also easily be generated among this subgroup of victims.

Victims showing these characteristics may also not be very high on empathy. In some cases, they may actually be highly qualified and experienced workers. The problem is that they clash with the norms of the work group to which they belong, because they often 'know it better', tend to be legalistic, insist on their own view, as well as having difficulties in taking the perspective of others. By being overcritical, they may be a constant threat to the self-esteem of their colleagues and superiors. Being at odds with group norms may imply that they challenge low performance standards, informal rules and privileges. In some cases, they may actually be the 'good guys' as seen from an employer point of view. However, in practice, the management is more dependent on the group than on the victim.

Therefore, and because information about the conflict situation is likely to be biased in favour of the group, the management tends to take the view of the group rather than that of the victim, thus leaving the victim in a hopeless position (cf. Leymann, 1993).

Claiming victim status

A final question is whether there can be any positive effects involved in declaring oneself a victim. When interpreting the frequencies of bullying in the various studies (cf. Zapf *et al.*, this volume), the question occurs whether bullying will be overreported or underreported. There is some evidence that victims of bullying would hesitate to label themselves so, especially if the bullying is subtle, of low intensity and consisting of indirect forms of aggression (Zapf, 1999a). Given that being a victim implies not being accepted among superiors and colleagues, and even being unable to solve the major problems of one's working life, one might assume that bullying tends to be underreported. In anecdotes it is repeatedly reported that victims tried to hide their problem as long as they could (e.g. Leymann, 1993).

However, there are also reasons to believe that labelling oneself a victim may have positive implications. In many European countries the dominant public understanding of bullying is that an innocent victim is harassed by unfair bullies or by an unfair organisation. First of all, labelling oneself as a victim of bullying may then be used to obtain personal goals, for example to receive an early pension or to win a case of unfair dismissal in court. Moreover, victim status may be used as a justification of oneself in several respects: a victim neither initiates nor escalates a conflict. A victim is fair and innocent, whereas the bullies are unfair and guilty. A victim suffers and hence earns sympathy. A victim is in a powerless position and cannot do anything against the superior strength of the alleged bully. Because the victim is innocent in every respect, he or she is not responsible for solving the problem, which is seen as the responsibility of management (cf. Neuberger, 1999). In this sense, victim status can be used for the protection of self-esteem. This is one of the central problems in some of the therapeutic approaches in rehabilitation hospitals caring for victims of bullying (Schwickerath, 2001). Therapists report that victims are unwilling to become active in therapeutic meetings because they believe that as innocent victims it is not their task to change anything.

In sum, there is the epidemiological problem that some victims of bullying tend to hide their victim status. However, there is also the practical problem that some individuals have started to use 'bullying' or 'mobbing' to achieve personal goals, which means that not every person who says so is really a victim of bullying at work.

Conclusion

In this chapter we have discussed the theoretical and empirical evidence of individual antecedents of bullying both from the perspective of the perpetrator and the victim. We have come to the conclusion that protection of self-esteem, lack of social competence and micropolitically motivated behaviour on the part of the perpetrator, and being in a salient position, being low on social competence and self-assertiveness as well as overachievement and high conscientiousness on the part of the victim are likely individual antecedents that may contribute to the occurrence of workplace bullying.

However, both the analyses of Matthiesen and Einarsen (2001) and Zapf (1999b) demonstrate that personal characteristics are *not a general explanation* of bullying. Rather, both studies identified subgroups where such an explanation is likely, while also identifying individuals with normal competence, personality and health profiles but who were at the same time targets of serious and frequent bullying at work. Thus, it is suggested that not all victims of bullying have little social competencies and a neurotic personality profile or are conscientious overachievers. Rather, it is likely that there are specific subgroups to whom these characteristics apply, and there are other victims who may not differ from respective control groups and in whose cases the main reason for bullying may lie elsewhere. However, even if personality factors do play a role in the development of workplace bullying this does not undermine and change the responsibility of managers and employers in the prevention and management of such problems.

Professionals dealing with bullying have to keep in mind that they typically meet self-selected groups of victims. Hence, psychiatrists and psychotherapists may overestimate the degree that the causes of bullying lie within the victim him or herself. Victims of bullying clearly caused by organisational factors or factors within a certain bully are often extremely happy when the situation is over. They tend to cope with their problem alone, and may even leave the company or seek a new position. They may consult their family doctor, but they will seldom approach a psychiatrist or psychotherapist. On the other hand, the overachievers tend to begin battles for their rights. They may try to get every assistance needed to demand their legal rights, even if the case continues for months and even years. They may consult trade union representatives and lawyers. Hence, personal experiences with bullying at work should not be overgeneralised.

To summarise: so far there are few 'hard facts' regarding the causes of bullying. However, taking all the existing empirical data together, there is sufficient evidence that there are many possible causes and probably often multiple causes of bullying, be it causes within the organisation, within the perpetrator, within the social system or within the victim. One-sided and simplistic discussions are in this respect usually misleading. One should consider carefully the circumstances of each bullying case, as in our experience they can be extremely different.

Bibliography

Adams, A. (1992) *Bullying at work. How to confront and overcome it.* London: Virago Press.

Ashforth, B. (1994) Petty tyranny in organizations. *Human Relations*, *47*, 755–778.

Baumeister, R. F., Heatherton, T. F. and Tice, D. M. (1993) When ego threats lead to self-regulation failure: Negative consequences of high self-esteem. *Journal of Personality and Social Psychology*, *64*, 141–156.

—— (1994) *Losing control. How and why people fail at self-regulation.* San Diego, CA: Academic Press.

Baumeister, R. F., Smart, L. and Boden, J. M. (1996) Relation of threatened egotism to violence and aggression: The dark side of high self-esteem. *Psychological Review*, *103*, 5–33.

Baumeister, R. F., Stillwell, A. and Wotman, S. R. (1990) Victim and perpetrator accounts of interpersonal conflicts. Autobiographical narratives about anger. *Journal of Personality and Social Psychology*, *59*, 994–1005.

Björkqvist, K., Österman, K. and Hjelt-Bäck, M. (1994) Aggression among university employees. *Aggressive Behavior*, *20*, 173–184.

Brodsky, C. M. (1976) *The harassed worker.* Lexington, MA: D.C. Heath and Co.

Brown, R. (1997) Beziehungen zwischen Gruppen (Relationships between groups). In W. Stroebe, M. Hewstone and G. M. Stephenson (eds), *Sozialpsychologie. Eine Einführung* (pp. 545–576). Berlin: Springer.

Bühler, K. and Zapf, D. (1997) *Stigmatisierung am Arbeitsplatz. Eine qualitative Studie mit Mobbingopfern* (Stigmatisation at work. A qualitative study with bullying victims). University of Konstanz: Social Science Faculty.

Cohen, S. and Wills, T. A. (1985) Stress, social support, and the buffering hypothesis. *Psychological Bulletin*, *98*, 310–357.

Coyne, I., Seigne, E. and Randall, P. (2000) Predicting workplace victim status from personality. *European Journal of Work and Organizational Psychology*, *9*, 335–349.

Depue, R. A. and Monroe, S. M. (1986) Conceptualization and measurement of human disorder in life stress research: The problem of chronic disturbance. *Psychological Bulletin*, *99*, 36–51.

Dohrenwend, B. S., Dohrenwend, B. P., Dodson, M. and Shrout, P. E. (1984) Symptoms, hassles, social supports, and life events: Problems of confounded measures. *Journal of Abnormal Psychology*, *93*, 222–230.

Einarsen, S. (1997) *Bullying among females and their attitudes towards humour.* Paper presented at the 8th European Congress of Work and Organizational Psychology, April, Verona.

—— (1999) The nature and causes of bullying. *International Journal of Manpower*, *20*, 16–27.

—— (2000) Harassment and bullying at work: A review of the Scandinavian approach. *Aggression and Violent Behavior*, *4*, 371–401.

Einarsen, S. and Skogstad, A. (1996) Prevalence and risk groups of bullying and harassment at work. *European Journal of Work and Organizational Psychology*, *5*, 185–202.

Einarsen, S., Raknes, B. I. and Matthiesen, S. B. (1994) Bullying and harassment at work and their relationships to work environment quality. An exploratory study. *The European Work and Organizational Psychologist*, *4*, 381–401.

Einarsen, S., Raknes, B. I., Matthiesen, S. B. and Hellesøy, O. H. (1994) *Mobbing og harde personkonflikter. Helsefarlig samspill pa arbeidsplassen* (Bullying and severe interpersonal conflicts. Unhealthy interaction at work). Soreidgrend: Sigma Forlag.

Field, T. (1996) *Bullying in sight*. Wantage: Success Unlimited.

Gandolfo, R. (1995) MMPI-2 profiles of worker's compensation claimants who present with claimant of harassment. *Journal of Clinical Psychology*, *51*, 711–715.

Hoel, H. and Cooper, C. (2001) The origins of bullying. In N. Tehrani (ed.), *Building a culture of respect. Managing bullying at work*. London: Taylor and Francis.

Hoel, H., Cooper, C. L. and Faragher, B. (2001) The experience of bullying in Great Britain: The impact of organisational status. *European Journal of Work and Organizational Psychology*, *10*, 443–465.

Hoel, H., Rayner, C. and Cooper, C. L. (1999) Workplace bullying. In C. L. Cooper and I. T. Robertson (eds), *International review of industrial and organizational psychology*, vol. 14 (pp. 195–230). Chichester: Wiley.

Kernis, M. H., Grannemann, B. D. and Barclay, L. C. (1989) Stability and level of self-esteem as predictors of anger arousal and hostility. *Journal of Personality and Social Psychology*, *56*, 1013–1022.

Kernis, M. H., Cornell, D. P., Sun, C. R., Berry, A. and Harlow, T. (1993) There is more to self-esteem than whether it is high or low: The importance of stability of self-esteem. *Journal of Personality and Social Psychology*, *65*, 1190–1204.

Kets de Vries, M. F. R. and Miller, D. (1985) Narcissism and leadership. *Human Relations*, *38*, 583–601.

Knorz, C. and Zapf, D. (1996) Mobbing – eine extreme Form sozialer Stressoren am Arbeitsplatz (Mobbing – an extreme form of social stressors at work). *Zeitschrift für Arbeits- und Organisationspsychologie*, *40*, 12–21.

Krum, H. (1995) *Mobbing – eine unethische Form der Kommunikation am Arbeitsplatz* (Mobbing – an unethical form of communication at work). Unpublished diploma thesis, Technical University of Darmstadt.

Leymann, H. (1993) *Mobbing – Psychoterror am Arbeitsplatz und wie man sich dagegen wehren kann* (Mobbing – psychoterror in the workplace and how one can defend oneself). Reinbeck: Rowohlt.

—— (1996) The content and development of mobbing at work. *European Journal of Work and Organizational Psychology*, *5*, 165–184.

Leymann, H. and Gustafsson, A. (1996) Mobbing and the development of post-traumatic stress disorders. *European Journal of Work and Organizational Psychology*, *5*, 251–276.

Lindemeier, B. (1996) Mobbing. Krankheitsbild und Intervention des Betriebsarztes (Bullying. Symptoms and intervention of the company physician) *Die Berufsgenossenschaft*, *June*, 428–431.

Lindroth S. and Leymann, H. (1993) Vuxenmobbning mot en minoritetsgrupp av män inom barnomsorgen. Om mäns jämställdhet i ett kvinnodominerat yrke (Bullying of a male minority group within child-care. On men's equality in a female-dominated occupation). Stockholm: Arbetarskyddstyrelsen.

Matthiesen, S. B. and Einarsen, S. (2001) MMPI-2-configurations among victims of bullying at work. *European Journal of Work and Organizational Psychology*, *10*, 467–484.

Mikkelsen, E. G. and Einarsen, S. (2002) Relationships between exposure to bullying at work and psychological and psychosomatic health complaints: The role of state negative affectivity and generalised self efficacy. *Scandinavian Journal of Psychology*, in press.

Neuberger, O. (1995) *Mikropolitik. Der alltägliche Aufbau und Einsatz von Macht in Organisationen* (Micropolitics. The everyday construction and use of power in organisations). Stuttgart: Enke.

—— (1999) *Mobbing. Übel mitspielen in Organisationen* (Mobbing. Bad games in organisations), 3rd edn. Munich and Mering: Rainer Hampp.

Niedl, K. (1995) *Mobbing/bullying am Arbeitsplatz. Eine empirische Analyse zum Phänomen sowie zu personalwirtschaftlich relevanten Effekten von systematischen Feindseligkeiten* (Mobbing/bullying at work. An empirical analysis of the phenomenon and of the effects of systematic harassment on human resource management). Munich: Hampp.

Olweus, D. (1993) *Bullying at schools, What we know and what we can do.* Oxford: Blackwell.

O'Moore, M., Seigne, E., McGuire, L. and Smith, M. (1998) Victims of bullying at work in Ireland. *Journal of Occupational Health and Safety – Australia and New Zealand*, *14*, 568–574.

Sacco, W. P., Dumont, C. P. and Dow, M. G. (1993) Attributional, perceptual, and affective responses to depressed and nondepressed marital partners. *Journal of Consulting and Clinical Psychology*, *61*, 1076–1082.

Salovey, P. (1991) Social comparison processes in envy and jealousy. In J. Suls and T. A. Wills (eds), *Social comparison: Contemporary theory and research* (pp. 261–285). Hillsdale, NJ: Erlbaum.

Schulz von Thun, F. (1984) Psychologische Vorgänge in der zwischenmenschlichen Kommunikation (Psychological processes in communication between humans). In B. Fittkau, H. -M. Müller-Wolf and F. Schulz von Thun (eds), *Kommunizieren lernen (und umlernen)* (Learning (and re-learning) to communicate) (pp. 9–100). Braunschweig: Agentur Pedersen.

Schuster, B. (1996) Rejection, exclusion, and harassment at work and in schools. *European Psychologist*, *1*, 293–317.

Schwickerath, J. (2001) Mobbing am Arbeitsplatz. Aktuelle Konzepte zu Theorie, Diagnostik und Verhaltenstherapie (Bullying in the workplace. Current concepts on theory, diagnostics and behaviour therapy). *Psychotherapeut*, *46*, 199–213.

Seigne, E. (1998) Bullying at work in Ireland. In C. Rayner, M. Sheehan and M. Barker (eds), *Bullying at work, 1998 research update conference: proceedings.* Stafford: Staffordshire University.

Smith, P. K., Singer, M., Hoel, H. and Cooper, C. L. (2001) *Victimisation in the school and in the workplace: Are there any links?* University of London: Goldsmiths College. Paper submitted for publication.

Smith, R. H., Parrott, W. G., Ozer, D. and Moniz, A. (1994) Subjective injustice and inferiority as predictors of hostile and depressive feelings in envy. *Personality and Social Psychology Bulletin*, *20*, 717–723.

Stahlberg, D., Osnabrügge, G. and Frey, D. (1985) Die Theorie des Selbstwertschutzes und der Selbstwerterhöhung (The theory of self-worth protection and self-worth enhancement). In D. Frey and M. Irle (eds), *Theorien der Sozialpsychologie, Band III: Motivations- und Informationsverarbeitungstheorien*

(Theories in social psychology, vol. III: Theories on motivation and information processing) (pp. 79–124). Bern: Huber.

Stucke, T. and Sporer, L. (in press) *Narzismus und Selbstkonzeptklarheit als Persönlichkeitskorrelate bei Mobbingtätern und Mobbingopfern* (Narcissism and clearness of the self concept as personality correlates of perpetrators and victims of mobbing). Zeitschrift für Arbeits- und Organisations- psychologie.

Tajfel, H. and Turner, J. (1986) The social identity theory of intergroup behavior. In S. Worchel and W. G. Austin (eds), *Psychology of intergroup relations* (pp. 7–24). Chicago: Nelson.

Tesser, A. (1988) Toward a self-evaluation maintenance model of social behaviour. In L. Berkowitz (ed.), *Advances in experimental social psychology*, vol. 21 (pp. 181–227). San Diego: Academic Press.

Thomas, K. W. (1992) Conflict and negotiation processes in organizations. In M. D. Dunnette and L. M. Hough (eds), *Handbook of industrial and organizational psychology*, vol. 3 (pp. 651–718). Palo Alto, CA: Consulting Psychologists Press.

Thylefors, I. (1987) *Syndabockar. Om utstötning och mobbning i arbetslivet* (Scapegoats. On expulsion and bullying in working life). Stockholm: Natur och Kulture.

Van de Vliert, E. (1998) Conflict and conflict management. In P. J. D. Drenth, H. Thierry and C. J. J. Wolff (eds), *Handbook of work and organizational psychology*, vol. 3: *Personnel Psychology*, 2nd edn (pp. 351–376). Hove: Psychology Press.

Vartia, M. (1996) The sources of bullying – psychological work environment and organizational climate. *European Journal of Work and Organizational Psychology*, *5*, 203–214.

Watzlawick, P., Beavin, J. H. and Jackson, D. D. (1969) *Menschliche Kommunikation* (Human communication). Bern: Huber.

Zapf, D. (1999a) Mobbing in Organisationen. Ein Überblick zum Stand der Forschung (Mobbing in organisations. A state of the art review). *Zeitschrift für Arbeits- and Organisationspsychologie*, *43*, 1–25.

—— (1999b) Organizational, work group related and personal causes of mobbing/bullying at work. *International Journal of Manpower*, *20*, 70–85.

Zapf, D. and Gross, C. (2001) Conflict escalation and coping with workplace bullying: A replication and extension. *European Journal of Work and Organizational Psychology*, *10*, 497–522.

Zapf, D., Dormann, C. and Frese, M. (1996) Longitudinal studies in organizational stress research: A review of the literature with reference to methodological issues. *Journal of Occupational Health Psychology*, *1*, 145–169.

9 Social antecedents of bullying

A social interactionist perspective

Joel H. Neuman and Robert A. Baron

As described in previous chapters, workplace bullying involves persistent patterns of behaviour in which one or more individuals engage in actions intended to harm others (e.g. Hoel *et al.*, 1999). It is our contention that bullying, although not identified as such, involves acts of interpersonal aggression – any form of behaviour directed towards the goal of harming or injuring another living being who is motivated to avoid such treatment (Baron and Richardson, 1994). While both phenomena involve actions that are intentional in nature, the persistence of aggression over time, evidenced in episodes of bullying, serves as a distinguishing characteristic. That is, while a single act of intentional harm-doing constitutes an act of aggression, it would not, by definition, constitute bullying. In short, and of central importance to the present chapter, we believe that workplace bullying involves repeated acts of interpersonal aggression directed against specific targets in work settings, or what we would refer to as *workplace aggression* – efforts by individuals to harm others with whom they work (Neuman and Baron, 1997a). Furthermore, we propose that anything that serves as an antecedent to aggression may contribute to – and increase the likelihood of – workplace bullying.

Having said all this, we do recognise that bullying in workplaces, like bullying in other contexts, represents a special or unique form of aggression in certain respects. For instance, the persons involved in bullying episodes are generally participants in ongoing, long-term relationships: they may work together for months, years or even decades. Second, since bullying often occurs openly, in front of many observers, it is clear that norms concerning such behaviour differ from the societal norms that regulate aggression generally, and – in most instances – condemn aggression, and especially repeated aggression against weak or helpless victims, as inappropriate. Thus, a key question to be addressed is: why do societal norms against aggression fail to apply, or apply only weakly, where workplace bullying is concerned? Related to this, we also ask: in what additional ways is workplace bullying different from aggression in many other contexts, and why is this so?

Building on these basic assertions, we draw on a substantial literature devoted to interpersonal aggression and examine what we believe to be important social antecedents to bullying. These *social factors* are distinct from *individual* causes of aggression, which centre on the characteristics (e.g. personality traits) of the persons who engage in workplace bullying. Social factors, in contrast, involve the words and/or deeds of individuals–actions that elicit or condone aggression. In addition, we will also focus on the social norms to which we referred above, for these, too, often exert strong effects on the nature, form and frequency of overt acts of aggression. Specifically, we will consider important social norms that serve to shape and reinforce aggression as well as the process by which norm violations (injustice perceptions) elicit retaliation or predispose individuals towards aggression and bullying. Then, employing modern theories of aggression, we discuss the mediating variables through which norm violations and injustice perceptions may lead to aggression; specifically, situations that produce frustration and stress, generate assaults on individual dignity and self-worth, and elicit negative affect, physiological arousal and hostile thoughts in the persons involved. Finally, we conclude by identifying several contemporary business practices that produce such mediating factors and create social conditions that are ripe for bullying – business practices, we might add, that are ubiquitous in today's work settings.

The social antecedents of aggression

When asked to describe situations that made them angry, most individuals refer to something another person said or did – something that caused them to become upset and view aggression against this person as justified (Harris, 1993). In short, the things that make people most angry are the words and deeds of other people – the social causes of aggression. As a point of departure for this chapter, we examine an important norm that serves to help shape social interaction.

The norm of reciprocity

Contrary to advice proffered in the 'Golden Rule', people tend to do unto others as others have actually done unto them. This behavioural norm has been found to '... exert a powerful influence upon various social behaviors ranging from altruism and assistance on the one hand through aggression and violence on the other' (Baron *et al.*, 1974, p. 374). Recognition of the importance of reciprocity in interpersonal relationships has a long history. 'There is no duty more indispensable than that of returning a kindness', says Cicero, adding, 'all men distrust one forgetful of a benefit' (as cited in Gouldner, 1960, p. 161). Cicero's admonition speaks to the obligation one owes a benefactor and the distrust that accrues to a person who fails to repay such an obligation. In a similar but opposite vein, when people feel attacked they often respond with an attack of comparable severity (Geen, 1968). Gouldner (1960) recognised these negative forms of reciprocity as 'sentiments of retaliation where the emphasis is placed not on the return of benefits but on the return of injuries' (p. 172). The act of 'exacting justice', 'evening the score', 'righting the wrong' and 'balancing the scales' is also prominently featured in the biblical injunction to exact 'an eye for an eye and a tooth for a tooth'. Several years ago, Stuckless and Goranson (1992) noted that '... the terms revenge and vengeance [were] not even designated as keywords in the American Psychological Abstracts but [were] subsumed under the classification of reciprocity' (p. 26). To summarise the importance of reciprocity in Simmel's (1950) words, 'all contacts among men rest on the schema of giving and returning the equivalence' (p. 387). This dynamic is so common to social experience that it led Becker (1956) to view the human species as 'Homo Reciprocus' (p. 1).

According to Gouldner (1960), reciprocity is accounted for as a result of the development of a beneficial cycle of mutual reinforcement between parties in social exchange. But what constitutes *benefit* as opposed to *detriment*? Clearly, individuals must recognise that they are being advantaged or disadvantaged in any given situation and this would presuppose a calculus by which justice perceptions are made. In the following sections, we examine these justice perceptions in some detail, and discuss the manner in which justice judgements are made and the process by which justice is restored.

Injustice perceptions

The relationship between perceptions of unfair treatment and the restoration of equity is the motivation underlying a significant portion of the research on social justice. As suggested by Homans (1974), 'we should be much less interested in injustice if it did not lead so often to anger and aggression' (p. 257). When we consider that fairness perceptions are manifest in almost every aspect of the employment relationship (e.g. personnel selection, performance evaluations, promotions, raises, merit pay, allocation of work assignments and office space, etc.), the importance of incorporating social justice theories in work settings seems obvious. Accordingly, there has been a burgeoning interest in the area of organisational justice – research related to people's perceptions of fairness on the job.

In reviewing the literatures related to social justice and human aggression, four classes of variables emerge as central to both research streams (Neuman and Baron, 1998). These factors include 'unjust' situations that: (1) violate norms, (2) produce frustration and stress, (3) induce negative affect (emotion) and, (4) assault individual dignity and self-worth. In the sections that follow, we examine each of these in turn.

Norm violations

Justice refers to an appropriate correspondence between a person's fate and that to which he or she is entitled – what is deserved. Rule and Ferguson (1984) discuss norm violations as an instance of 'is–ought discrepancy' and suggest that anger, blaming, and retaliation might ensue when people see their partly idiosyncratic norms of proper conduct ('oughts') violated. In a similar vein, referent cognitions theory (RCT, Folger, 1986) suggests that with respect to outcome allocation, '... resentment is maximized when people believe they *would* have obtained better outcomes if the decision maker had used other procedures that *should* have been implemented' (Cropanzano and Folger, 1989, pp. 293–294). Finally, Tedeschi and Felson (1994) have suggested that any factor that increases the likelihood that norm violations will be committed should lead to grievances and coercive interactions. But under what circumstances are norm violations likely to result in blame, anger, grievances, coercive interactions and retaliation? In attempting to answer this question, researchers have focused on issues related to distributive, procedural and interactional forms of (in)justice.

Distributive justice

The earliest research on social justice focused almost exclusively on the issue of distributive justice – the perceived fairness of the outcomes or allocations that an individual receives. In short, this research was concerned with how individuals make fairness judgements about their outcomes and how they react when they perceive inequity in the distribution of those outcomes. With respect to inequity, individuals may perceive that they have unfairly benefited or that they have been unfairly disadvantaged in a particular situation. In the language of equity theory (Adams, 1965), this would represent instances of overpayment and underpayment, respectively. With respect to the present chapter, our focus will be on underpayment inequity; i.e. perceptions of relative deprivation.

In the seminal research on distributive justice and relative deprivation, the connection between perceived injustice and aggression is clearly evident. As noted by Homans (1974), if a state of injustice exists and it is to a person's disadvantage – that is, the person experiences deprivation – he or she will display anger. Adams (1965) echoed this position in his original formulation of equity theory, when he observed that 'men do not simply become dissatisfied with conditions they perceive to be unjust. They usually do something about them' (p. 276). But what do they do? Fortunately, reactions to injustice do not always involve aggression. For example, in response to perceptions of underpayment inequity, individuals may simply put forth a greater effort – in the hope of increasing their outcomes/rewards. Unfortunately, the organisational justice literature is replete with examples of less acceptable responses, and many of these may be aggressive in nature.

From the organisation's perspective, theft would certainly constitute an unacceptable response to employee perceptions of distributive injustice. In a study designed to explore this issue, Greenberg (1990) compared theft rates within three manufacturing plants belonging to the same organisation during a period in which pay cuts were administered. In response to an ongoing financial crisis, the company imposed a pay cut of 15 per cent over a ten-week period. In a third and demographically similar plant, no pay cuts were administered (this served as the control condition). Although theft rates were traditionally low in all three plants (as measured by shrinkage), Greenberg found significantly greater theft rates in the two plants where the pay cuts were administered, in comparison to the control condition.

Similar examples of 'getting even' for perceived inequity in outcome distribution have been found in actual work settings by other researchers. In a study by Altheide *et al.* (1978), involving bread delivery drivers, stealing was generally viewed by the drivers as a way of compensating for a pay system that unfairly penalised them for making mistakes. Similarly, Zeitlin (1971) studied employees in a retail clothing store and found that theft was generally accepted as an appropriate means by which employees could 'get something additional for [their] work since [they weren't] getting paid enough' (p. 26).

The purpose of the preceding discussion is not to focus attention on theft, which is clearly not the central issue of this chapter or text; rather, it is meant to provide empirical support for the connection between distributive injustice and retaliation. Having said this, it is important to note that the motive for theft is not always greed – the desire to profit financially. It is often the case that the target of an 'injustice' merely wants to inflict harm on the source of that injustice, or, as we shall demonstrate, simply lash out against any convenient target. This is most clearly evident with respect to instances of sabotage, in which company or individual property is damaged as an act of revenge for unfair treatment, and there is ample evidence of this in work settings (cf., Analoui, 1995).

Before concluding this section, it is important to note that outcomes are not synonymous with wages, salaries and/or employee benefits. Other important outcomes relate to quality of work life, social support, and opportunities for growth and development, to name just a few. In a recent study by Neuman and Baron (1997b), employees indicated the extent to which they were satisfied with their opportunities for growth and development and the social conditions in which they worked. Dissatisfaction with these outcomes was significantly correlated with workplace aggression. Specifically, respondents who were dissatisfied with these outcomes were significantly more likely to engage in workplace aggression against the perceived source of the injustice.

As noted at the beginning of this section, reactions to perceived inequity may involve behaviours that are prosocial or antisocial in nature. Unfortunately, the research related to distributive justice is of little help in predicting the likelihood of one type of response over another. In part, the reason for this dilemma centres on the fact that distributive justice research has tended to focus on the outcomes but not the underlying causes (or perceived causes) for those outcomes. As it turns out, just as individuals are concerned with the fairness of the outcomes that they receive, they also are sensitive to the process used to determine those outcomes (Thibaut and Walker, 1975) and the nature of the interactions that characterise those transactions (Bies, 1987). Folger and Greenberg (1985) were the first researchers to consider procedural justice issues in work settings, and since their initial efforts a considerable amount of research has been conducted. To date, this line of inquiry strongly suggests that (1) fair process can mitigate against an aggressive response to unfavourable outcomes, and (2) unfair process may be more strongly linked to aggression than unfair outcomes. Now we turn to some evidence supporting our assertion that unfair treatment is an important antecedent to aggression and bullying.

The link between unfair treatment and workplace violence and aggression

There is a substantial and growing literature suggesting that perceptions of unfair (insensitive) treatment, on the part of management and/or co-workers often serve as antecedents to workplace aggression and violence. For example, the perpetrators of workplace homicide often point to what they believe was unfair treatment at the hands of a supervisor or co-worker as a justification for their actions. In Hoad's (1993) summary of the most common causes of workplace violence in the United Kingdom, feeling aggrieved – a sense of being treated unfairly (whether real or imagined) – was ranked as the most common cause of aggression. Similarly, in a study reported by Weide and Abbott (1994), over 80 per cent of the cases of workplace homicide they examined involved employees who 'wanted to get even for what they perceived as [their] organizations' unfair or unjust treatment of them' (p. 139). All of this is consistent with empirical research demonstrating a link between perceptions of unfair treatment and interpersonal conflict (Cropanzano and Baron, 1991).

In another related study (Neuman and Baron, 1997b), individuals who reported that they had been treated unfairly by their supervisors were significantly more likely than those who were satisfied with their treatment to indicate that they engaged in some form of workplace aggression. Additionally, this study provided evidence that this aggression was directed against the source of that perceived injustice. For example, when individuals expressed dissatisfaction with organisationally controlled outcomes (e.g. job security, pay and fringe benefits, social conditions), they were more likely to aggress against the entire organisation. However, when dissatisfaction was associated with a particular manager (e.g. respect, fair treatment, support and guidance that they received from their boss), they were more likely to indicate that they aggressed against that particular individual. Data from this study also suggest that aggression is directed against targets other than the source of the perceived injustice; that is, there is evidence of displaced aggression.

In a six-year, longitudinal study on workplace sabotage, Analoui (1995) found that 65 per cent of all acts of sabotage stemmed from discontent with management and its behaviour towards workers. Similarly, Crino and Leap (1989) suggest that in response to violations of the psychological employment contract, employer–employee loyalty can be severely damaged. Consequently, 'once that loyalty has been destroyed, an employee is more likely to commit an act of sabotage' (p. 32).

To summarise, recent theorising on organisational justice suggests that aggression is more likely to occur when unfavourable outcomes result from an 'unfair' process – highlighting the importance of interpersonal sensitivity in mitigating against aggression. But what is it about norm violations – distributive, procedural or interpersonal – that may elicit an aggressive response? To answer this question, we briefly turn our attention to modern theories of aggression and the role of some important mediating variables.

The General Affective Aggression Model: a modern perspective on human aggression

Early views of human aggression tended to emphasise the influence of one, or at most, a few variables. The most famous of these, perhaps, was the *frustration–aggression hypothesis*, a theoretical perspective that attached central importance to the role of frustration – thwarting of ongoing, goal-directed behaviour (Dollard *et al.*, 1939). As we will note below, although modern views of aggression include frustration as one potential cause of such behaviour, they assign far less importance to this variable than the sweeping suggestions that were part of the original frustration–aggression hypothesis. The original formulation proposed that (1) *all* aggression stems from frustration, and (2) frustration *always* produces aggression. Subsequent research demonstrated clearly that these assertions were false, in that aggression stems from many sources other than frustration and frustration does not always lead to increased aggression. In fact, aggression often generates feelings of resignation or despair rather than overt assaults against the perceived causes of such thwarting – reactions that often characterise a target's response to bullying.

Modern perspectives on aggression, in contrast, recognise that such behaviour stems from a wide range of social, situational and personal factors (e.g. Baron and Richardson, 1994). One such model that has attained widespread acceptance is the *General Affective Aggression Model* (GAAM) proposed by Anderson and his colleagues (Anderson *et al.*, 1996). According to this theory, aggression is triggered or elicited by a wide range of *input variables* – aspects of the current situation (the focus of this chapter) and/or tendencies or predispositions that individuals bring with them to a given context. Among variables included in the first category are frustration, provocation or attack from another person, exposure to aggressive models, the presence of cues or stimuli associated with aggression (e.g. guns or other weapons), and virtually anything that causes individuals to experience negative affect (e.g. harsh and unfair criticism, unpleasant environmental conditions, pain or discomfort produced by physical injuries, and so on). Variables in the second category (individual difference factors) include traits that predispose individuals towards aggression (e.g. high irritability, the Type A behaviour pattern, negative affectivity), certain attitudes and beliefs about violence (e.g. the view that it is acceptable or a demonstration of one's masculinity) and specific skills related to aggression (e.g. knowledge of how to use various weapons, skills useful in attacking others physically or verbally).

Central to the present chapter, the GAAM further suggests that these situational and individual difference variables lead to overt aggression through their impact on three basic intervening processes (which we refer to as *critical internal states*). First, they may increase physiological *arousal* or excitement. Second, they elicit *negative affect* – feelings of anger and other hostile emotions along with outward signs of these emotions (e.g. angry facial expressions). Finally, they elicit hostile *cognitions* – they induce individuals to bring hostile thoughts to mind, to remember aggression-related experiences, and so on. Depending on an individual's appraisals (interpretations) of the current situation and possible restraining factors (e.g. threat of retaliation, strict disciplinary and enforcement policies), aggression then occurs or does not occur.

The above description is necessarily brief but captures, we believe, the sophisticated nature of this and related views of human aggression. Such views are, indeed, much more complex than the suggestion that aggression stems from frustration or any other single factor. These contemporary views are also more accurate in reflecting the multifaceted nature of human aggression, and the wide array of variables and processes that influence its occurrence, form and targets. An overview of the General Affective Aggression Model is presented in Figure 9.1. As shown in this figure, the model also suggests the occurrence of interactions between arousal, affective states and aggression-related cognitions. This interaction is extremely important in that the elicitation of any single critical internal state (physiological, affective or cognitive) tends to evoke the others – initiating a potential cascade of hostile thoughts and feelings.

While the Affective Aggression Model is valuable in itself, it is especially pertinent here because it calls attention to variables and processes that may

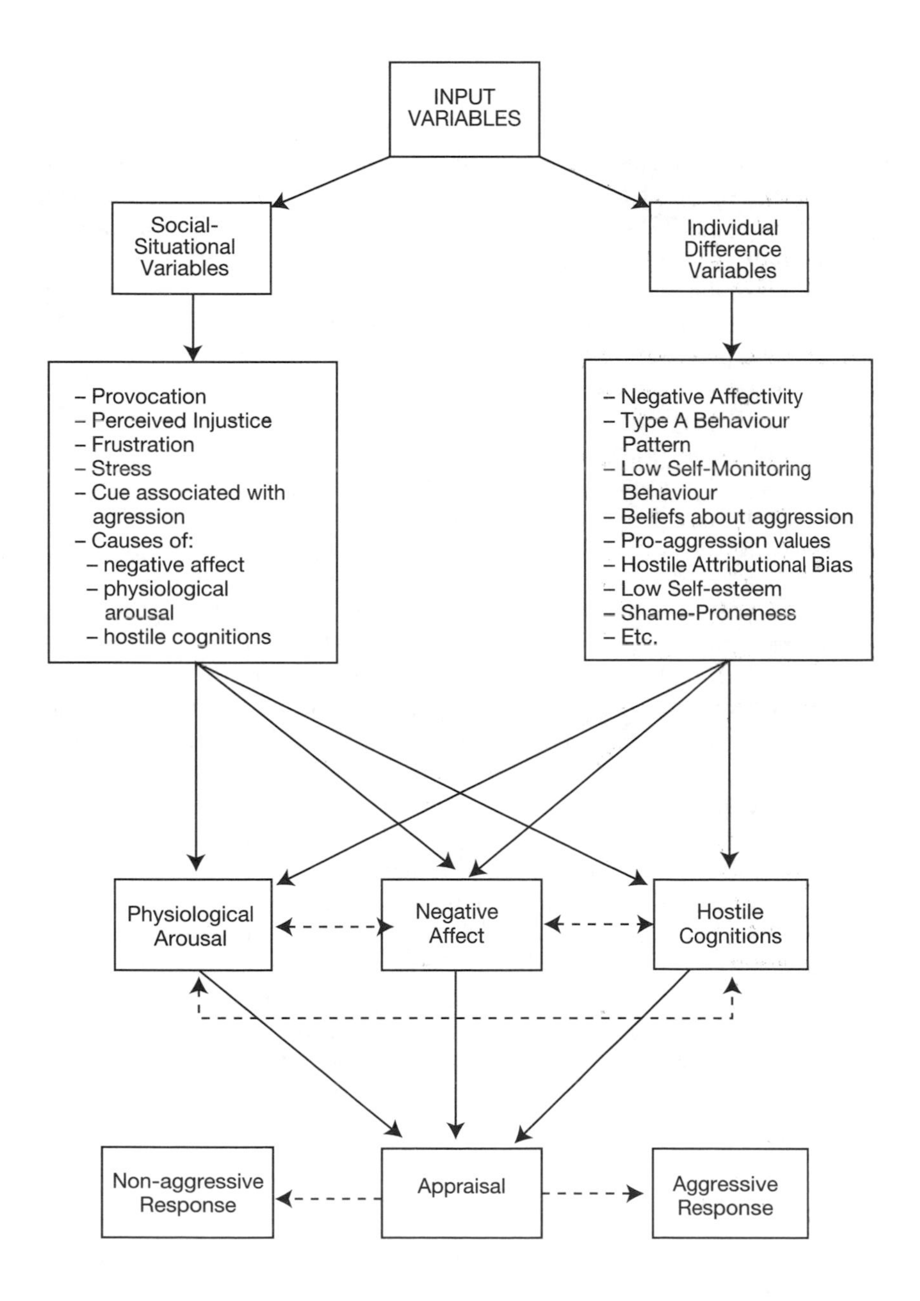

Figure 9.1 The General Affective Aggression Model

influence aggression in any social context. In the section that follows, we suggest such specific variables – ones that are ubiquitous in today's work settings.

Frustration and stress

As suggested above, norms represent entitlements – what is deserved or 'ought to be'. To the extent that norm violations block the attainment of some desired goal, a state of relative deprivation may result. This, in turn, may lead to a sense of frustration experienced as injustice. In fact, Brown and Herrnstein (1975) concluded that frustration 'may not be fundamentally different from injustice, inequity, and relative deprivation' (p. 271), and this sentiment was echoed by Crosby (1976) who observed that 'by definition, the sense of injustice is a part of relative deprivation' (p. 91). Additionally, propositions derived from cognitive dissonance theory suggest that the presence of inequity creates a state of tension (physiological and psychological), and that this tension is proportional to the magnitude of the inequity (Adams, 1965).

As noted earlier, injustice is always in the eye of the beholder, and the fact that frustration and aggression often stem from *relative* deprivation is particularly pertinent in an era in which many people feel deprived. Simply put, there is a growing sense of 'entitlement' on the part of many millions of individuals. During the past thirty years, many regions of the globe have enjoyed excellent economic growth (interrupted, of course, by various economic setbacks). One result of this long period of prosperity has been a sharp rise in expectations. Not only do many persons anticipate ever-rising personal prosperity – their belief that they are entitled to and 'deserve' such outcomes has also increased. Paradoxically, these rising expectations and beliefs (concerning entitlements) now run headlong into increasingly callous attitudes on the part of management. For instance, Hay Research conducted a survey of 750,000 middle managers over two time periods (1985–1987 and 1988–1990) and found a significant decline in managers' willingness to share information with, or listen to the problems of, their employees (Fisher, 1991). Similarly, Opinion Research Corporation of Chicago surveyed 100,000 middle managers, supervisors, professionals, and technical and clerical employees of Fortune 500 companies and found that they believed that top management was less willing to listen to them and accorded them less respect than was the case five years earlier (Farnham, 1989).

Frustration and aggression

While the assertion that frustration always precedes aggression was dismissed long ago, there is evidence to suggest that under certain limited conditions (e.g. when it is perceived as intentional and/or unjustified) frustration may produce a state of readiness or instigation to aggress (Geen, 1991). In organisational settings, frustration has been found to be positively correlated with aggression against others, interpersonal hostility, sabotage, strikes, work slowdowns, stealing and employee withdrawal (Spector, 1975).

Stress and aggression

Research, and everyday experience, suggests a relationship between stress and physical violence. With respect to work settings, Chen and Spector (1992) found a relationship between work-related stressors and interpersonal aggression, hostility, sabotage, complaints and intentions to quit. They also reported a modest relationship between work stressors and theft and employee absence. Anecdotally, workplace stressors have been cited as a causal factor in many high-profile cases of co-worker involved homicides in the United States.

With respect to the relationship between norm violations and stress, Tedeschi and Felson (1994) suggest that stress has negative effects on performance and increases violations of politeness norms. Consequently, these effects on behaviour often contribute to the development of coercive episodes and aggression. This would be very consistent with recent research demonstrating what appears to be an increase in 'workplace incivility' in the United States (Pearson, 1998).

Individual dignity and self-worth

Blows to one's self-esteem (i.e. inability to maintain a positive self-image) may, on occasion, lead to aggression. For example, Averill's (1982) examination of everyday episodes of anger among adults suggests that a common cause of anger is a loss of pride or a loss of self-esteem. The experience of shame among college students has been associated with a desire to punish others (Wicker *et al.*, 1983) as well as anger arousal, suspiciousness, resentment, irritability, a tendency to blame others for negative events and indirect expressions of hostility (Tangney *et al.*, 1992). Among fifth-grade boys, shame proneness was positively correlated with both self-reports of anger and teacher reports of aggression (Tangney *et al.*, 1991).

Performance evaluations perceived to be unfairly negative, failure to obtain the raise, promotion, choice work assignment or office to which people believe they are entitled, or being unfairly criticised by a supervisor in the presence of other employees are all blows to one's self-esteem. In fact, *all* injustices are, by their very nature, personal evaluations in which individuals have been judged to be worth less (or *worthless*) in comparison to some referent. To the extent that this results in feelings of shame and embarrassment, it may elicit defensive, retaliative anger and a tendency to project blame outward. This tendency has also been referred to as 'humiliated fury', a situation in which the acute pain of shame may elicit a rage that is directed towards the self and a real or imagined disapproving other.

When we consider the importance of a job in defining one's sense of identity, it is clear that threats to our employment relationship, hierarchical status, professional reputation, and/or chances for advancement pose significant threats to our feelings of self-worth. Unfortunately, even the most casual review of the management literature (both popular and academic) reveals a workplace environment that is fraught with ego-crushing practices. In an era of downsizing, rightsizing, layoffs, restructuring, re-engineering, outsourcing and mergers and acquisitions, the need for productivity increases is creating an environment where, according to Abraham Zaleznik, 'tough is passé. Today you're dealing with a variety of head games. That's where the cruelty is' (Dumaine, 1993, p. 39).

Instrumental forms of aggression

The preceding discussion has focused almost exclusively on affective (or reactive) aggression – where the ultimate outcome is intentionally to inflict harm on others. There are, however, forms of aggression that are more instrumental in nature – where the harm inflicted is not the ultimate aim but merely a means to a desired end. Many social antecedents of aggression fall into this category.

In a 'dog-eat-dog' business world where 'no good deed goes unpunished', a world filled with 'hostile take-overs', 'poison pills' and 'creative tension', and where you are either the 'victor or the vanquished', there appear to be a sufficient number of opportunities to learn the 'benefits' of being aggressive. Stories of emotionally abusive CEOs of highly successful companies are legion. For example, Linda Wachner of giant apparel maker Warnaco advised an executive to 'fire people if he wanted to be taken seriously' and Herbert Haft of the Dart Group fired his own wife and son because he thought they were usurping his power. The rants of Steve Jobs, of Apple Computer, and the 'Jack Attacks' of Jack Connors, founder of the Boston advertising firm Hill Holliday represent just a few visible examples of the kinds of emotional abuse (Keashly, 1998) that are rampant in today's work settings. Of course, one could easily make the case that the goal of instrumental aggression (and bullying) may be nothing more than what the Germans refer to as *Schadenfreude* – the pleasure derived from the misfortunes of others.

Regardless of the ultimate aims, hostile cultures and climates are fertile ground for the social antecedents of aggression and bullying.

From perceived injustice to workplace bullying: the role of norms, effect/danger ratio and displacement

So far, we have argued that social factors play a key role in workplace aggression, and that in this respect, violations of the norms of reciprocity and fairness are especially crucial. When individuals perceive that they have been treated unfairly, they experience stress, negative affect and other critical internal states that together set the stage for increased workplace aggression. Further, we have suggested that perceptions of unfairness are increasingly common in modern workplaces because of changes that have occurred within them in recent years. But how, precisely, does workplace bullying fit into this picture? We believe that the answer lies in three interrelated factors: (1) efforts by individuals to maximise what has been termed the *effect/danger ratio*, (Björkqvist *et al.*, 1994); (2) strong tendencies towards displaced aggression, which result from such efforts; and (3) norms that condone workplace bullying and thus assure the silence or active complicity of witnesses to such behaviour.

Effect/danger ratio

When adults aggress against others, they generally seek to maximise the harm produced, while minimising the danger to themselves through retaliation, social censure or other potential consequences (e.g. Björkqvist *et al.*, 1994). Previous research suggests that workplaces are no exception to this general rule and that, in fact, individuals often prefer disguised forms of workplace aggression for this very reason: such forms permit them to harm the victim while making it difficult for the victim to identify them as the source of such harm (Baron *et al.*, 1999).

Displaced aggression

Because of such pressures to maximise the effect/danger ratio, we believe that there are equally strong pressures towards *displacement of aggression* in the workplace. Displacement of aggression – which has recently been found to be a reliable and powerful phenomenon (e.g. Marcus-Newhall *et al.*, 2000) – refers to the tendency to aggress against someone other than the source of strong provocation because aggressing against the source of such provocation is too dangerous. Since perceived unfairness often stems from treatment by supervisors or by the entire organisation, and since aggression against such targets is dangerous and would violate the desire to maximise the effect/danger ratio, we believe that tendencies towards displaced aggression are frequently strong in many workplaces. As a result, persons 'made ready to aggress' by conditions within their workplaces often select targets who are relatively weak and defenceless, and a pattern of bullying against such individuals may then arise.

Norms supporting workplace bullying

These tendencies are further strengthened by the existence of norms emphasising 'toughness' and 'survival of the fittest' in modern organisations. This, in turn, reflects the weakening of the traditional perceived contract between organisations and their employees: no longer do organisations guarantee continued employment, even after long and faithful service. Rather, current conditions suggest that it is perfectly appropriate for organisations to close even profitable operations and dismiss loyal and hard-working employees if greater profit is to be made elsewhere. Norms of 'toughness', in turn, tend to reduce the likelihood that witnesses to workplace bullying will take action against it. On the contrary, such norms tend to increase the odds that witnesses will join in and even applaud the actions of workplace bullies. From the perspective of the effect/danger ratio, risk is reduced and the benefits are more likely to outweigh the potential costs.

In sum, from a social interactionist perspective, we suggest that many instances of workplace bullying can be understood in the following terms. Violations of the basic norm of reciprocity and/or perceived unfairness produce feelings of frustration, negative affect and stress on the part of many individuals. Since overt and/or direct aggression against the perceived sources of such feelings (i.e. the perceived causes of unfairness) is often dangerous, and therefore violates the desire to maximise the effect/danger ratio, aggression is displaced onto relatively weak and defenceless victims. Norms in modern organisations emphasising 'toughness' make it unlikely that witnesses to workplace bullying will intervene on behalf of the victims. On the contrary, such witnesses may actively support bullying or become willing participants; at the very least, they tend to ignore it or to view it as a normal feature of their workplace. As we argued at the outset, then, workplace bullying may indeed stem from social factors relating to the real or perceived treatment individuals receive from others at work.

Conclusion

In this chapter, we have provided some examples of the many ways in which social factors may serve as antecedents to aggression. We began with a discussion about the norm of reciprocity and how people are predisposed to repay 'in kind' the type of treatment they receive. Following this, we discussed aggression as a response to the violation of important social norms and perceptions of distributive, procedural, and interpersonal injustice, and focused on the specific roles of frustration and stress, negative affect, and assaults on individual dignity and self-worth. Then, drawing on contemporary theories of aggression, we demonstrated how various critical internal states might initiate a process that often concludes with aggression. Finally, we demonstrated how the effect/danger ratio, displaced aggression, and organisational norms emphasising 'toughness' serve as a bridge between perceptions of injustice and aggression and bullying.

While we believe that we have touched on several important social antecedents, and described a process by which these antecedents may lead to aggression and bullying, our discussion has been necessarily brief and far from exhaustive. For example, we have not specifically discussed the role of a victim as an antecedent to individual acts of aggression or episodes of bullying. There is literature suggesting that bullies may engage in aggression as a response to perceived provocation by a victim. While this possibility is implicit in our previous discussion about perceived injustice, we never specifically addressed this issue nor several other dynamics associated with bully–victim interaction. Also, we did not discuss the fact that people are seldom merely bullies or victims but may, on occasion, occupy both roles – another important social dynamic. While we certainly believe that these issues are important to a full understanding of bullying from a social interactionist perspective, space constraints do not permit such a discussion. Finally, with respect to the General Affective Aggression Model presented earlier, this framework is not meant to capture every aspect of social interaction but, rather, to represent the mediating process by which situational and dispositional variables may lead to individual acts of aggression. As such, it focuses heavily on affective (i.e. hostile, angry, reactive) aggression and is not meant to describe all forms of instrumental (goal-directed) aggression. Nevertheless, we believe the model describes situations present in a large proportion of cases involving both individual acts of aggression and persistent episodes of bullying.

Before concluding, it is important to describe another important connection between contextual factors and aggression and bullying. In particular, we refer to the cumulative effects of norm violations and injustices in creating a hostile work environment. As noted above, hostile cognitions represent a critical internal state that may provide an *instigation towards aggression*. It should hardly come as a surprise that people are more likely to have hostile thoughts in an environment of mistrust. As noted by Kramer (1994), individuals operating in such a climate are more likely to be subject to 'rational mistrust'. By this we mean '… a generalized expectancy or belief regarding the lack of trustworthiness of particular individuals, groups, or institutions that is predicated upon [a] specific history of interaction' (Kramer, 1994, p. 200). The development of such mistrust has been linked to patterns of exchange in which peoples' expectations of trust have been systematically and repeatedly violated. Central to the present discussion, such suspicion and mistrust predisposes one towards making sinister/hostile attributions (Dodge *et al.*, 1990; Kramer, 1994) about the behaviour of others and, as a result, towards perceptions of injustice even when no real injustice exists.

In conclusion, although it is often easy (and occasionally accurate) to attribute workplace bullying to internal, dispositional causes (e.g. he or she is just a bully), we have attempted to demonstrate here that aggression and bullying derive from a wide variety of factors – many of which are social in nature. The bad news is that these factors are all too common in everyday experience. The good news is that they are often amenable to change.

Bibliography

Adams, J. S. (1965) Inequity in social exchange. In Berkowitz (ed.), *Advances in experimental social psychology*, vol. 2 (pp. 267–299). New York: Academic Press.

Altheide, D. L., Adler, P. A., Adler, P. and Altheide, D. A. (1978) The social meanings of employee theft. In J. M. Johnson and J. D. Douglas (eds.) *Crime at the tip: Deviance in business and the professions* (pp. 90–124). Philadelphia: J. B. Lippincott.

Analoui, F. (1995) Workplace sabotage: Its styles, motives and management. *Journal of Management Development*, *14*, 48–65.

Anderson, C. A., Anderson, K. B. and Deuser, W. E. (1996) Examining an affective aggression framework: Weapon and temperature effects on aggressive thoughts, affect, and attitudes. *Personality and Social Psychology Bulletin*, *22*, 366–376.

Averill, J. R. (1982) *Anger and aggression. An essay on emotion*. New York: Springer Verlag.

Baron, R. A. and Richardson, D. R. (1994) *Human aggression*, 2nd edn. New York: Plenum.

Baron, R. A., Byrne, D. and Griffitt, W. (1974) *Social psychology: Understanding human behavior*. Boston, MA: Allyn and Bacon.

Baron, R. A., Neuman, J. H. and Geddes, D. (1999) Social and personal determinants of workplace aggression: Evidence for the impact of perceived injustice and the Type A Behavior Pattern. *Aggressive Behavior*, *25*, 281–296.

Becker, H. (1956) *Man in reciprocity*. New York: Prager.

Bies, R. J. (1987) The predicament of injustice: The management of moral outrage. In L. L. Cummings and B. M. Staw (eds) *Research in organizational behavior*, vol. 9 (pp. 289–319). Greenwich, CT: JAI Press.

Björkqvist, K., Österman, K. and Lagerspetz, K. M. J. (1994) Sex differences in covert aggression among adults. *Aggressive Behavior*, *20*, 27–33.

Brown, R. and Herrnstein, R. J. (1975) *Psychology*. Boston: Little, Brown.

Chen, P. Y. and Spector, P. E. (1992) Relationships of work stressors with aggression, withdrawal, theft and substance use: An exploratory study. *Journal of Occupational and Organizational Psychology*, *65*, 177–184.

Crino, M. D. and Leap, T. L. (1989) What HR managers must know about employee sabotage, *Personnel, May*, 31–32, 34–36, 38.

Cropanzano, R. and Baron, R. A. (1991) Injustice and organizational conflict: The moderating effect of power restoration and task type. *International Journal of Conflict Management*, *2*, 5–26.

Cropanzano, R. and Folger, R. (1989) Referent cognitions and task decision autonomy: Beyond equity theory. *Journal of Applied Psychology*, *74*, 293–299.

Crosby, F. (1976) A model of egoistical relative deprivation. *Psychological Review*, *83*, 85–113.

Dodge, K. A., Price, J. M., Bachorowski, J. A. and Newman, J. P. (1990) Hostile attributional biases in severely aggressive adolescents. *Journal of Abnormal Psychology*, *99*, 385–392.

Dollard, J., Doob, L., Miller, N., Mowrer, O. H. and Sears, R. R. (1939) *Frustration and aggression*. New Haven, CT: Yale University Press.

Dumaine, B. (1993) America's toughest bosses, *Fortune*, *18 October*, 39–42, 44, 48, 58.

Farnham, A. (1989) The trust gap. *Fortune*, *4 December*, 56–58, 62, 66, 70, 74, 78.

Fisher, A. B. (1991) Morale crisis. *Fortune*, *18 November*, 70.

Folger, R. (1986) Rethinking equity theory: A referent cognitions model. In H. W. Bierhoff, R. L. Cohen and J. Greenberg (eds), *Justice in social relations* (pp. 145–162). New York: Plenum.

Folger, R. and Greenberg, J. (1985) Procedural justice: An interpretive analysis of personnel systems. In K. Rowland and G. Ferris (eds), *Research in personnel and human resource management*, vol. 3 (pp. 141–183). Greenwich, CT: JAI Press.

Geen, R. G. (1968) Effects of frustration, attack and prior training in aggressiveness on aggressive behavior. *Journal of Personality and Social Psychology*, *9*, 316–321.

—— (1991) *Human Aggression* (Pacific Grove, CA: Brooks/Cole).

Gouldner, A. W. (1960) The norm of reciprocity. *American Sociological Review*, *25*, 161–178.

Greenberg, J. (1990) Employee theft as a reaction to underpayment inequity: The hidden cost of pay cuts. *Journal of Applied Psychology*, *75*, 561–568.

Harris, M. B. (1993) How provoking! What makes men and women angry? *Aggressive Behavior*, *19*, 199–211.

Hoad, C. D. (1993) Violence at work: Perspectives from research among 20 British employers. *Security Journal*, *4*, 64–86.

Hoel, H., Rayner, S. and Cooper, C. L. (1999) Workplace bullying. In C. L. Cooper and I. T. Robertson (eds), *International review of industrial and organizational psychology*, vol. 14 (pp. 195–229). New York: John Wiley.

Homans, G. C. (1974) *Social behavior: Its elementary forms*, rev. edn. New York: Harcourt Brace.

Keashly, L. (1998) Emotional abuse in the workplace: Conceptual and empirical issues. *Journal of Emotional Abuse*, *1*, 85–117.

Kramer, R. M. (1994) The sinister attribution error: Paranoid cognition and collective distrust in organizations. *Motivation and Emotion*, *18*, 199–230.

Marcus-Newhall, A., Pedersen, W. C. and Miller, N. (2000) Displaced aggression is alive and well: A meta-analytic review. *Journal of Personality and Social Psychology*, *78*, 670–689.

Neuman, J. H. and Baron, R. A. (1997a) Aggression in the workplace. In R. Giacalone and J. Greenberg (eds), *Antisocial behavior in organizations* (pp. 37–67). Thousand Oaks, CA: Sage.

—— (1997b) *Type A Behavior Pattern, self-monitoring, and job satisfaction as predictors of aggression in the workplace*. Paper presented at Counterproductive job performance and organizational dysfunction. Symposium conducted at the meeting of the Society for Industrial and Organizational Psychology, St Louis, Missouri.

—— (1998) *Perceived injustice as a cause of – and justification for – workplace aggression and violence.* Paper presented at Advances in organizational justice theories: The motivation to engage in dysfunctional behavior. Symposium conducted at the meeting of the Academy of Management, San Diego, CA.

Pearson, C. M. (1998) Organizations as targets and triggers of violence: Framing rational explanations for dramatic organizational deviance. In P. A. Bamberger and W. J. Sonnenstuhl (eds), *Research in the Sociology of Organizations*, vol. 15 (pp. 197–223). Stamford, CT: JAI Press.

Rule, B. G. and Ferguson, T. J. (1984) The relation among attribution, moral evaluation, anger, and aggression in children and adults. In A. Mummendey (ed.), *Social psychology of aggression. From individual behaviour to social interaction* (pp. 143–155). Berlin: Springer Verlag.

Simmel, G. (1950) *The sociology of Georg Simmel.* Glencoe, IL: Free Press.

Spector, P. E. (1975) Relationship of organizational frustration with reported behavioral reactions of employees. *Journal of Applied Psychology, 60*, 635–637.

Stuckless, N. and Goranson, R. (1992) The vengeance scale: Development of a measure of attitudes toward revenge. *Journal of Social Behavior and Personality*, 7, 25–42.

Tangney, J. P., Wagner, P. E., Burggraf, S. A., Gramzow, R. and Fletcher, C. (1991) Poster presented at the meeting of the Society for Research in Child Development, Seattle, WA.

Tangney, J. P., Wagner, P., Fletcher, C. and Gramzow, R. (1992) Shamed into anger? The relation of shame and guilt to anger and self-reported aggression. *Journal of Personality and Social Psychology*, *62*, 669–675.

Tedeschi, J. T. and Felson, R. B. (1994) *Violence, aggression, and coercive actions.* Washington, D.C: American Psychological Association.

Thibaut, J. and Walker, L. (1975) *Procedural justice: A psychological analysis.* Hillsdale, NJ: Lawrence Erlbaum Associates.

Weide, S. and Abbott, G. E. (1994) Murder at work: Managing the crisis. *Employment Relations Today*, *21*, 139–151.

Wicker, F. W., Payne, G. C. and Morgan, R. D. (1983) Participant descriptions of guilt and shame. *Motivation and Emotion*, 7, 25–39.

Zeitlin, L. R. (1971) A little larceny can do a lot for employee morale. *Psychology Today*, *June*, 22, 24, 26, 64.

10 Organisational antecedents of workplace bullying

Helge Hoel and Denise Salin

Introduction

The terms 'workplace bullying' and 'mobbing' appear to have struck a chord with large sections of the workforce even though there is little unanimity in perception or articulation of the terms among employees (see Liefooghe and Mackenzie Davey, this volume). The interest in the issue may reflect the sheer magnitude of the problem, which, depending upon the definition applied, might directly affect between 5 and 20 per cent of the working population (Hoel and Cooper, 2000; UNISON, 1997). Together with the apparent effects upon individuals and organisations (Einarsen and Mikkelsen, this volume; Hoel, Einarsen and Cooper, this volume), growing concerns about the changing context within which bullying arises can also be considered to play a part in the attention currently paid to the problem. Thus, a number of studies conclude that victims of bullying report a more negative work environment than those who were not bullied (Ashforth, 1994; Björkqvist *et al.*, 1994; Einarsen *et al.*, 1994; Vartia, 1996), and that the worst work environment is associated with those most severely bullied (Zapf *et al.*, 1996b).

There are other reasons to explain why it enjoys such a high profile. In the UK, for example, the concept has been embraced by the trade union movement, as it has enabled unions to brand a range of unfair treatment and practices as types of workplace harassment or bullying at a time when the union movement has been weakened (Lewis, 2000) The fact that bullying lends itself well to complaints from individuals, as well as to more collective strategies, also makes it an ideal campaigning issue (Lee, 2000). Moreover, by referring to the issues of violence and bullying as a health and safety issue, the unions may also succeed in airing their grievances and focusing on management practices where such opportunities may otherwise have been limited (Mullen, 1997).

Although not mutually exclusive, broadly speaking three sets of explanatory models have been advanced to account for the problem: (1) a focus on personality characteristics of perpetrator and victim; (2) inherent characteristics of human interactions within organisations; and (3) factors related to the work organisation of a contextual or environmental nature

(Einarsen, 2000; Hoel *et al.*, 1999). This chapter will focus on the last group of explanatory factors, exploring the role of situational and contextual factors in bullying scenarios.

Bullying is a complex and dynamic process, where both action and reaction should be understood within the social context in which they take place. Thus, situational factors may increase the vulnerability of targets or recipients of bullying behaviour and contribute to their response to such acts. Whilst acknowledging this fact, the present investigation will limit itself to an exploration of the influence of various organisational antecedents on the behaviour of perpetrators.

The concepts of power and control are essential to bullying. Power is understood in relative terms, expressed as an imbalance of power between the parties, where the situation of the target is identified with a perceived power deficit vis-à-vis the perpetrator (Einarsen, 1996). As such, the power imbalance may reflect formal power relationships or may refer to perceptions of powerlessness resulting from the bullying process itself (Leymann, 1996; Vartia, 1996). In the latter instances, conflicts between individuals of seemingly equal power may gradually escalate, leaving one of the parties increasingly defenceless. By contrast, where bullying is predominantly identified with managerial behaviour, the focus is on the abuse of power, arising from the power structure and associated with control over rewards and punishment (Aquino, 2000).

In the forthcoming discussion of organisational antecedents of bullying, we will explore organisational antecedents of bullying under the following headings: (1) the changing nature of work; (2) work organisation; (3) organisational culture and climate; and (4) leadership. In the final section, a model encompassing different types of antecedents, with respect to their role in the bullying process, will be introduced.

For the purpose of this investigation the following definition of bullying is used:

> Bullying emerges when one or several individuals persistently over a period of time perceive themselves to be on the receiving end of negative actions from one or several persons, in a situation where the target of bullying has difficulty in defending him or herself against these actions.
>
> (Einarsen *et al.*, 1994)

The changing nature of work

Economic globalisation has increased competition, and in order to survive in the current economic environment, organisations are restructuring and downsizing with the aim of cutting costs, with greater pressures on everyone in work as a result (Cooper, 1999). For example, in a major UK study of managers, it was found that 60 per cent had experienced large-scale restructuring within the last twelve months, with downsizing forming an integral part of the change process (Worrall and Cooper, 1999). Moreover, according to McCarthy *et al.* (1995), 60 per cent of respondents in a study of businesses undergoing restructuring reported an increase in working hours and greater pressure to bring work home. Tendering for contracts has also become increasingly common, resulting in more economic risk-taking and work-intensification in order to reduce costs (Quinlan, 1999).

With leaner organisational structures emerging from de-layering processes (Sheehan, 1999), the pressure on individual managers are likely to grow as the span of control widens. According to Sheehan (1999), such structural change processes need to be considered in connection with the greater focus on managerial accountability in general (including responsibility for performance), and the tendency to devolve to line management many functions previously undertaken by personnel departments. However, although a relationship between such structural changes and bullying is plausible, the evidence is sparse.

It is also argued that by introducing market philosophies into areas previously unaffected by such pressures, for example within the health service and the educational sector, the relationship between managers and staff has changed, with work intensification and increased managerial discretion and control as a result (Ironside and Seifert, this volume; Lee, 2000).

The potential impact of such organisational change and restructuring processes on bullying has received attention from a number of writers in the field (McCarthy *et al.*, 1995; Sheehan, 1999), suggesting that the risk of bullying may increase as managers tend to adopt more autocratic practices to bring about change.

Even if little evidence has emerged to suggest a direct relationship between work pressure and bullying, one may assume the presence of an indirect relationship, with managers, for instrumental reasons, more likely to apply strategies which could be construed as bullying in order to fulfil performance objectives or 'getting the job done' (Hoel and Cooper, 1999). Thus, the need for restructuring may encourage more authoritarian management practices with the effect of 'lowering of thresholds at which inappropriately coercive managerial behaviours [become] manifest in organisational life' (McCarthy *et al.*, 1995, p. 47). With greater job insecurity (OECD, 1999), employees may also become less resistant to managerial pressure, and more unlikely to challenge unfair and aggressive treatment on the part of managers.

Nevertheless, some researchers, notably Sjøtveit (1992), and Liefooghe and Mackenzie Davey (2001), argue that bullying may stem not so much from abusive or illegitimate use of power as from power which is considered legitimate, and tightly related to the labour process and the managerial prerogative to manage.

The compression of career structures resulting from de-layering processes also represents fewer opportunities for advancement, thereby increasing the competition between managers for promotion to a shrinking pool of jobs (Mullen, 1997; Sheehan, 1999), with growing interpersonal conflict and bullying as possible outcomes.

Kräkel (1997), who understood bullying from a 'rent-seeking' perspective, argued that the change processes referred to above, for example downsizing, de-layering, also contributed to unclear power relationships between individuals. Following such a line of argument, bullying has been found to be associated with highly politicised and competitive work environments (O'Moore *et al.*, 1998, Salin, 2001a; Vartia, 1996).

With a growing number of people seemingly working in self-directed and autonomous teams, tension and conflict between workers may increase as a result of increasing pressure. Moreover, with greater emphasis on customer satisfaction, workers may be increasingly exposed to abusive behaviour and excessive demands from clients and customers in a climate where the needs and rights of customers take precedence over the rights of employees. Thus, Hoel and Cooper (2000) in their UK-wide survey of bullying reported high levels of bullying by clients in the service sectors, for example retailing, hotel and catering and the health service.

Baron and Neuman (1996) explored the relationship between a range of organisational changes and aggression. Backed up by confirmatory statements from observers, the strongest predictors of aggression emerging from their study were found to be: use of part-time workers; change in management; pay cuts or pay freezes. As far as change of supervisor or manager is concerned, this factor has frequently been found to be associated with bullying (Hoel and Cooper, 2000; UNISON, 1997). These latter studies also identified factors such as 'a recent change in job' and 'a change in the way the organisation operated' as antecedents of bullying.

In a further study the authors identified four categories of organisational change associated with aggression: cost-cutting, organisational change, job (in)security and social change (Baron and Neuman, 1998). The last category refers to examples of change such as increasing diversity with regard to gender, ethnicity and disability status. In this respect, aggression may be a result of a lack of communication or stereotyping, and the greater the extent of change, the higher the incidence of aggression. Alternatively, aggression in such cases may be interpreted as the product of resentment (Quinlan, 1999).

As far as part-time, temporary and subcontracted workers are concerned, such types of contingent work have been associated with increasing job insecurity, whether perceived or real (OECD, 1999). Such factors may indirectly act as antecedents of bullying by influencing cost-benefit considerations (Björkqvist *et al.*, 1994) and, therefore, management behaviour, as the risk of retaliation from targets under such conditions is likely to be reduced. In an interview study with victims, Knorz and Zapf (1996) found that part-time workers were exposed to a higher degree of risk than those contracted full-time. This finding was interpreted along the lines that part-time workers had less time for conflict resolution and fewer opportunities for socialising, which tended to isolate them from their colleagues. By contrast, Kivimäki *et al.* (2000) did not find any differences in bullying rates between Finnish hospital staff, either between employees on temporary as opposed to permanent contracts, or between full- as opposed to part-time contracted employees. If anything, as far as Hoel and Cooper's (2000) study is concerned, individuals on full-time contracts were more at risk of bullying than those working part-time.

Several factors may account for this finding. With fewer opportunities to form friendships and strong ties with colleagues affecting levels of social support, contingent workers may be more wary of drawing attention to themselves, utilising avoidance strategies at times of conflict rather than complaint and confrontation. Moreover, in those cases where a perpetrator seeks personal gratification through domination and control (Randall, 1997), contingent workers may not be their preferred targets as, for such bullying scenarios to develop and escalate, they need to be nurtured by an ongoing interaction between perpetrators and recipients. Hence, the temporary element of their presence may function as a buffer against bullying and victimisation (Hoel and Cooper, 2000; Rayner *et al.*, 2002). Nevertheless, whilst the contingent status may possibly protect against victimisation, it may not shield an individual against abusive behaviour.

A recent study of the relationship between job insecurity and stress throws some further light on this relationship. According to Pearce (1998), part-time and other precarious forms of work create increasing pressure, upheaval and instability within the work group, with other employees having to update or cover for their colleagues in their absence. This may also lead to disorganisation with increasing role-conflict and role ambiguity as a result (Quinlan, 1999). With the more taxing aspects of work often transferred to permanent full-time staff, resentment may also develop (Pearce, 1998). In the same manner increased use of subcontracted personnel may also contribute to tensions between workers, as subcontracted individuals may be forced to work at a higher pace in order to secure future employment, thereby coming into conflict with core members of staff (Quinlan, 1999).

Work organisation

Bullying has frequently been associated with a negative and stressful working environment (Leymann, 1996; Einarsen *et al.*, 1994). To account for such a relationship, various work environment factors can be considered to produce or elicit occupational stress, which again may increase the risk of conflict and bullying. In some cases it can also be argued that the presence of work stressors in themselves may be perceived as harassment, particularly when they are attributed to hostile intentions (Brodsky, 1976; Einarsen *et al.*, 1994).

Faced with what they may consider unreasonable demands and poor working conditions, workers under pressure may protest or voice their concern, possibly resulting in a punitive and retaliatory response from their superior, with the effect, in some cases, of initiating and escalating the bullying process.

But a stressful environment may also elicit interpersonal conflict, with peer bullying as a possible outcome (Einarsen *et al.*, 1994). In this respect, Berkowitz (1989) argues that frustration may be mediated to cause aggression by negative affect brought about by arousal derived from external situations, i.e. stress. By contrast, the 'social interactionist' perspective (Felson, 1992; Neuman and Baron, this volume) suggests that situational or external factors may bring about aggression and bullying indirectly by giving rise to behaviours in breach of rules and norms of the group or the larger organisation. In this case it could also be argued that bullying can be seen as an intentional response to norm-violating behaviour and an instrument for social control (Hoel *et al.*, 1999).

Stress and frustration may also trigger the search for scapegoats as tension and frustration may be relieved by processes of projection (Brodsky, 1976; Thylefors, 1987). Higher levels of stress may, therefore, increase bullying down the line, as managers take their aggression out on their subordinates as well as horizontal bullying, where increased tension among staff is projected onto or taken out on colleagues.

Looking at the evidence, role-conflict and role ambiguity are features of work organisation which have been found to be linked to bullying (Einarsen *et al.*, 1994). This is particularly the case for role-conflict, which describes the extent to which employees perceive contradictory expectations, demands and values in their jobs. According to Einarsen *et al.* (1994), in a survey of Norwegian trade union members, not only were victims far more likely to report role-conflict than those reporting being victimised, but observers of bullying were also more likely to report higher levels of role-conflict. Moreover, significant correlations between bullying and role-conflict were found for all seven subsamples in Einarsen *et al.*'s study. In addition, role ambiguity seems to be associated with higher levels of bullying (Einarsen *et al.*, 1994). Thus, bullying seems to thrive where employees perceive their job situation as unpredictable and unclear. In line with this, Vartia (1996) found that victims of bullying reported less clarity of goals in their work than other employees.

Further support to a view that bullying is associated with a negative work environment is provided by Zapf (1999). Comparing victims of bullying with a control group of non-victimised individuals, victims assessed their environment more negatively than members of the control group on all features related to quality of work environment. The fact that respondents were questioned about their work environment prior to assessing their experience of bullying, appears to strengthen the view of a relationship between a bad work environment and bullying.

Attention has already been paid to work intensification and growing pressure. However, research has so far failed unequivocally to demonstrate a relationship between work pressure and pace of work, on the one hand, and bullying, on the other. So, whilst Appelberg *et al.* (1991) identified timepressure and a hectic work environment as a source of interpersonal conflict, other studies have in most cases failed to support their findings (Salin, 2001a; Vartia, 1996). Nevertheless, it has been argued that the problem of bullying comes to the fore when a high degree of pressure is present in a work environment which offers individuals little control over their own work (Einarsen *et al.*, 1994). Such an interpretation would be in line with Karasek's (1979) 'job-demand model of stress', in which strain is seen as the likely outcome of a combination of high demands and low decision latitude.

Zapf *et al.* (1996b), in a study of German 'mobbing' victims, also contradicted the view that bullying may be related to monotony and a general work-control deficit. When the study compared job complexity and task control of the mobbing victims with two other samples unconnected to bullying, one of office workers, the other of metal workers, the mobbing victims were found to have more variability in their jobs than the office workers and more work control than the metal workers. However, victims were found to have less control over time than, for example, the office workers. Zapf *et al.* interpreted these findings to the effect that pressure of time may indirectly affect bullying by undermining the opportunity to resolve conflicts. (Nevertheless, the fact that bullying affected organisational members across the organisational hierarchy, with a higher degree of complexity identified with senior positions, may have influenced the overall conclusion.) In line with this finding, they also rejected an earlier view, put forward by Thylefors (1987) and Sjøtveit (1992), that bullying may be the result of boredom.

The potential for horizontal bullying or peer bullying was also highlighted in the study by Zapf *et al.* (1996b) referred to above, with a higher degree of requirement for co-operation or team-working reported by victims than by those with whom they were compared. Enforced teamworking may, therefore, be considered a possible antecedent of bullying as a fertile ground for conflict development, particularly if linked to interteam competition for limited rewards. According to Collinson (1988), the introduction of collective bonus systems may reinforce some workers' concern to control their colleagues. In the same way bullying may be considered a way of punishing and getting rid of over- or underachieving subordinates, or colleagues, who are perceived as threats to or a burden on the group (Kräkel, 1997).

Similarly, and based on a large number of interviews with victims, Leymann (1996) concluded that bullying was closely related to poorly organised work environments where roles and command structures were unclear. As an example of such an environment, Leymann pointed to the situation of nurses in hospital settings. According to Leymann, nurses are often caught between two distinct sets of authorities, facing high and often conflicting demands from doctors, on the one hand, and from nursing managers, on the other, with increased pressure and conflicts as likely outcomes. Moreover, in times of crises, for example due to staff shortages, there is a tendency for what is referred to as spontaneous leadership to emerge. However, lacking formal authority, such 'leaders' are likely to stimulate rather than reduce conflict levels.

Whilst there is considerable evidence of bullying among nurses, not least of a horizontal nature (e.g. Farrell, 1999; Hoel and Cooper, 2000), the mechanisms that cause this are still somewhat unclear. The situation of victims is also likely to give rise to attributional processes and defensive mechanisms which will impinge upon their assessment of their environment (Hoel *et al.*, 1999; Kile, 1990). In such cases a contributory factor may be what Jones and Davies (1965) refer to as the 'fundamental attribution failure', where people tend to explain their own behaviour with reference to their environment, whilst personality is what comes to mind when explaining the behaviour of others.

Certain physical aspects of the work environment may also act as antecedents of aggressive behaviour and bullying. Hence, work undertaken under noisy, hot (or cold) circumstances or in crammed conditions has been found to be associated with increased feelings and attitudes of hostility (Anderson *et al.*, 1996; Einarsen, 1996).

Among other features of the work environment associated with increased risk of bullying are organisational size and work sector. As far as organisational size is concerned, bullying has been linked to large and bureaucratic organisations, where the threshold for bullying behaviour may be lower due to lesser chance of social condemnation (Einarsen *et al.*, 1994).

It has been argued that a higher degree of bureaucracy, and stricter rules for laying off workers in the public sector, may increase the value of using bullying as a strategy for circumventing rules and eliminating unwanted persons (Salin, 2001b). Similarly, the restructuring of the public sector in the 1990s may have created higher pressures, providing a more fertile ground for bullying. However, with regard to organisational size and work sector, the evidence is inconclusive. Thus Hoel and Cooper (2000) reported a high prevalence of bullying in small and medium-sized organisations, and Einarsen and Skogstad (1996) concluded that bullying was more prevalent in the private sector.

Organisational culture and climate

Acknowledging that any definition of organisational culture is controversial, we consider culture to be a multilevel concept based on the assumptions, beliefs, values and expectations that members take for granted and have come to share (e.g. Schein, 1985). According to Schein, at the definitional core are located those assumptions which are taken for granted, for example assumptions about human nature and the nature of truth and reality. At a somewhat less abstract level are values and norms, i.e. unwritten rules and beliefs about rights and wrongs, normal and abnormal. At the most visible surface level, culture manifests itself in different artefacts, such as rituals, routines and stories, as well as physical artefacts and symbols. However, culture also tends to manifest itself at the group level, with specific norms and rules covering the behaviour of group members, which, at times may be different from, or even contradictory to, the corporate culture (Sjøtveit, 1992).

As new members enter the organisation they will gradually adapt to the shared norms of the organisation and their work group by means of socialisation processes. Thus, in a study of bullying in the fire service, Archer (1999) explores how bullying in a 'paramilitary' setting may become institutionalised and passed on as tradition. Archer identifies the training process as a powerful source of socialisation of behaviour, in particular where every uniformed member of the organisation shares the same experience. Situational factors, such as the 'watch' culture, where the individual is allocated to the same tightly knit work team, possibly for years at a time, also suggest that there is little room for diversity. Moreover, in an autocratic leadership culture, where one's superior has been brought up within the same tradition, it is difficult to break out of the cycle and embark upon cultural change. The fact that many victims considered complaining about bullying to be an act of disloyalty further emphasises the potential strength and impact of the socialisation process.

In the above example bullying can be considered as being built into or becoming part of the culture. Organisations characterised by an extreme degree of conformity and group pressure seem to be particularly prone to bullying. Consequently, bullying seems to flourish in institution such as prisons, hospitals for the mentally ill, and the armed forces, where compliance and discipline are of overriding importance (e.g. Ashforth, 1994).

A work environment frequently identified with bullying is the luxury restaurant kitchen. During a few stressful and frenetic hours, food of top quality will be prepared and served to guests who are willing to pay high prices for the guarantee of a fully satisfactory meal. The growing interest in food preparation and cooking in the media has not only provided considerable financial benefit for some chefs, but has also meant that a number of chefs have achieved personal stardom. According to Johns and Menzel (1999), this has also given rise to the idea of the chef as an artist, whose bullying and abusive behaviour must be understood as idiosyncratic behaviour born out of their artistry and creativity. For the young apprentices, who model themselves on the artistic qualities as well as the behavioural qualities of their role models, abusive and abrasive behaviour may be considered normal and part and parcel of work as a chef. This would tend to exacerbate a problem which has riddled the hotel and catering industry for years, namely the view that quality food can only be produced in an atmosphere of blind subordination, resembling the discipline and subordination found in military environments. In this way abusive and bullying behaviour may become internalised by means of processes of learning and socialisation, handed down from one generation of chefs to the next, and affecting all kitchen staff.

Humiliating jokes, surprises and insults can also be part of the socialisation process, whereby new members are tested. However, this kind of humour can easily go sour and turn into bullying if the target, for some reason, cannot defend him or herself or does not take it as a joke (Einarsen and Raknes, 1997). In some cases, bullying, disguised as harsh humour, can be used to punish colleagues who do not conform to shared norms (Collinson, 1988). Hence, practical jokes were reported as one of the most common negative acts among male industrial workers at a Norwegian marine engineering plant (Einarsen and Raknes, 1997).

The above suggests that, in some organisations, bullying may not be an integrated part of the culture, but is still indirectly 'permitted'. If there is no policy against bullying, no monitoring policy and no punishment for those who engage in bullying, it could be interpreted that the organisation accepts the behaviour (Rayner *et al.*, 2002). Bullying is, therefore, prevalent in organisations where employees and managers feel that they have the support, or at least implicitly the blessing, of senior managers to carry on their abusive and bullying behaviour (Einarsen, 1999). This is in line with the view of Brodsky (1976), who argued that 'for harassment to occur, the harassment elements must exist within a culture that permits and rewards harassment' (p. 83). The fact that over 90 per cent of respondents in the UNISON (1997) survey of bullying identified the fact that 'bullies can get away with it' as a cause of bullying seems to confirm such a view. Furthermore, new managers will quickly come to view this form of behaviour as acceptable and 'normal' if they see others get away with it and are even rewarded for it (Rayner *et al.*, 2002). However, as has been argued by Lewis (this volume), the presence of bullying policies and procedures are no guarantee against bullying and abusive behaviours per se.

Leadership

With managers in positions of power often identified as perpetrators, a scrutiny of the impact of different leadership styles on bullying appears to be essential. However, the fact that most studies of leadership have focused on the effectiveness of leaders, where negative leadership has been seen as equivalent to ineffective leadership, little attention has traditionally been paid to more destructive aspects of leadership (Ashforth, 1994).

Together with role-conflict (reported above), low satisfaction with leadership was a second organisational feature found to be most strongly identified with bullying, in Einarsen *et al.*'s (1994) Norwegian trade union study. Furthermore, based on survey data, bullying has also been associated with lack of involvement in decision-making processes, work environments where employees are hesitant to express their views and opinions (Vartia, 1996), communication and co-operation problems, and low morale and a negative social climate (Keashly and Jagatic, 2000; Vartia, 1996).

Accordingly, an autocratic leadership and an authoritarian way of settling conflicts or dealing with disagreements have also been found to be associated with bullying (O'Moore *et al.*, 1998; Vartia, 1996). By contrast, people who had neither been bullied nor had observed bullying taking place reported that disagreements at their workplace tended to be solved by negotiation (Vartia, 1996). An authoritarian leadership or style of management may also create a climate of fear, where there is no room for criticism and where complaining may be considered futile.

Ashforth (1994) discussed what he referred to as 'petty tyranny' or a tyrannical style of leadership as a joint function of the predisposition of individuals and situational factors or facilitators, and argued that entrepreneurs may be particularly at risk of becoming 'petty tyrants' because of their strong need for independence and control, their possible distrust of others and their desire for approval.

It is important to emphasise that, whilst abuse of power has often taken centre stage, one should bear in mind that managers who perceive themselves as powerless in undertaking their tasks may resort to bullying behaviour. In such situations they may use whatever power they have to regain control (Ashforth, 1994). In such situations managers are also likely to be defensive and fearful of any voices critical of their regime.

The difference between bullying and an autocratic or 'touch' style of management has been emphasised by several commentators (e.g. Adams, 1992). However, as has been noted by Lee (2000), disagreement would remain with regard to where to draw the line as 'people with different social interests are likely to draw the boundaries between types of behaviour differently' (Lee, 2000, p. 597). Nevertheless, Hoel and Cooper (2000) report that senior managers were the group who reported the highest levels of witnessing bullying. This suggests that managers themselves in many cases have less of a problem in distinguishing bullying from so-called 'firm' management than may at first be anticipated.

So far the focus has been on abusive styles of management. However, abdication of leadership or a so-called laissez-faire style of management may also provide a fertile ground for bullying between peers or colleagues (Einarsen *et al.*, 1994; Hoel and Cooper, 2000). In addition, managers' ignorance and failure to recognise and intervene in bullying cases may indirectly contribute to bullying by conveying the message that bullying is acceptable. Similarly, dissatisfaction with the amount and quality of guidance, instructions and feedback given has been shown to be associated with higher levels of bullying (Einarsen *et al.*, 1994).

It is worth noting that the impact of individual antecedents is likely to vary between occupational settings. In a study by Einarsen *et al.* (1994) a much stronger association was found between bullying and leadership for workers in the printing or reprographic industry compared with, for example, electricians. In explaining their findings, the authors point to the fact that, whilst print workers are frequently employed in small, owner-run businesses with a strong and direct managerial presence, electricians have far less contact with their managers, frequently operating in pairs or small groups on different work sites, often at a distance from any central location. Similar discrepancies emerged when comparing groups for other organisational antecedents. Nevertheless, whilst certain features of a work environment may apply to an entire occupation or organisation, it is also likely that discrepancies between groups of workers within an occupation will exist. Moreover, different antecedents of bullying of a situational or organisational nature are also unlikely to affect different demographic groups uniformly (Hoel *et al.*, 2001).

Enabling, motivating and triggering factors

As pointed out by Zapf (1999), it is important to bear in mind that bullying is a complex and multi-causal phenomenon and can seldom be explained by one factor alone, but rather as an interaction between different factors. In accounting for the complex influence of various factors Salin (1999) introduced a model based on a framework by Boddewyn (1985), in which the antecedents of bullying were divided into three groups: enabling, motivating and triggering factors. Enabling factors describe factors which may allow bullying to occur in the first place, but which are seldom sufficient to bring about bullying on their own. This group of factors includes the presence of a perceived power imbalance between perpetrator and victim, low perceived cost to the perpetrator, and dissatisfaction with, for example, work control and the social climate. Motivating factors refer to factors which make it worthwhile for a perpetrator to engage in bullying to eliminate subordinates or colleagues perceived as 'burdens' or 'threats'. This includes, for example, high internal competition and rewards systems organised on a collective basis. Finally, triggering factors could be related to organisational changes, such as a change in manager, restructuring and downsizing. These factors thus increase insecurity, thereby making it more likely for bullying processes actually to evolve.

Whilst such a conceptual model may be useful in mapping out the role of various antecedents in bringing about scenarios of bullying, it should not be considered a straitjacket, a 'one size fits all' solution. The dynamics of individual bullying scenarios suggest that the outcomes are not predetermined, and at no stage in the process can the next stage be fully predicted.

Conclusion

In this chapter we have explored a large number of factors at the level of the organisation which, on their own or in combination, may give rise to bullying behaviour and the escalation of bullying processes. Whilst these antecedents have largely been discussed one by one, their relationship and interconnectivity have been emphasised. However, in real-life scenarios the interaction between features of the organisation and the larger environment within which it is located are likely to be far more complex than research has so far uncovered. One may also anticipate that where a number of antecedents may be present at the same time synergetic effects may occur, increasing the risk of bullying scenarios emerging. In order to reveal patterns of interaction, more sophisticated research methods need to be developed.

With a few exceptions, the findings discussed above are the results of cross-sectional studies which do not allow for robust conclusions with regard to causality (Zapf *et al.*, 1996a). This fact remains, even where findings are supported by evidence from third parties, i.e. observers or witnesses (e.g. Einarsen *et al.*, 1994; Hoel and Cooper, 2000). Thus, whilst a poor working environment may directly or indirectly give rise to bullying, alternative interpretations may be suggested. For example, anxious or depressed individuals may create tension and elicit negative reactions from colleagues and managers alike (Einarsen *et al.*, 1994). Moreover, bullying may itself have a negative effect on the work environment, for example by negatively affecting internal communication, thereby giving rise to more stress and further organisational problems (Zapf, 1999).

In the same way as organisational antecedents may combine and interact with particular characteristics of the persons involved, the effects and influence of individual antecedents are likely to vary across organisational contexts and between demographic groups. In this context one would expect that different antecedents may take on different meanings in different settings, and that factors such as gender, age and ethnicity would impinge upon such meanings. To uncover such connections and the relative impact of various antecedents in different contexts would require a greater emphasis on more qualitative research methods, and the development and application of more sophisticated research instruments.

Bibliography

Adams, A. (1992) *Bullying at work: How to confront and overcome it.* London: Virago Press.

Anderson, C. A., Anderson, K. B. and Deuser, W. E. (1996) Examining an affective aggression framework: Weapon and temperature effects on aggressive thoughts, affects and attitudes. *Personality and Social Psychology Bulletin*, 22, 366–376.

Appelberg, K., Romanov, K., Honlasalo, M. and Koskenvuo, M. (1991) Interpersonal conflicts at work and psychosocial characteristics of employees. *Social Science Medicine*, *32*, 1051–1056.

Aquino, K. (2000) Structural and individual determinants of workplace victimization: The effects of hierarchical status and conflict management style. *Journal of Management*, *26*, 171–193.

Archer, D. (1999) Exploring 'bullying' culture in the para-military organisation. *International Journal of Manpower*, *20*, 1/2, 94–105.

Ashforth, B. (1994) Petty tyranny in organizations. *Human Relations*, *47*, 7, 755–778.

Baron, R. A. and Neuman, J. H. (1996) Workplace violence and workplace aggression: Evidence on their relative frequency and potential causes. *Aggressive Behavior*, *22*, 161–173.

—— (1998) Workplace aggression – the iceberg beneath the tip of workplace violence: Evidence on its forms, frequency, and targets. *Public Administration Quarterly*, *21*, 446–464.

Berkowitz, L. (1989) The frustration-aggresion hypothesis: An examination and reformulation. *Psychological Bulletin*, *106*, 50–73.

Björkqvist, K., Österman, K. and Hjelt-bäck, M. (1994) Aggression among university employees. *Aggressive Behavior*, *20*, 173–184.

Boddewyn, J. (1985) Theories of foreign direct investment and divestment: A classificatory note. *Management International Review*, *25*, 1, 57–65.

Brodsky, C. M (1976) *The Harassed Worker*, Toronto: Lexington Books, D. C. Heath and Co.

Collinson, D. L. (1988) 'Engineering humour': Masculinity, joking and conflict in shop-floor relations. *Organization Studies*, *9*, 2, 181–199.

Cooper, C. L. (1999) The changing psychological contract at work. *European Business Journal*, *11*, 115–118.

Einarsen, S. (1996) *Bullying and harassment at work: Epidemiological and psychosocial aspects*. Doctoral dissertation, University of Bergen: Department of Psychosocial Science, Faculty of Psychology.

—— (1999) The nature and causes of bullying at work. *International Journal of Manpower*, *20*, 1/2, 16–27.

—— (2000) Harassment and bullying at work: A review of the Scandinavian approach. *Aggression and Violent Behavior*, *5*, 4, 371–401.

Einarsen, S. and Raknes, B. I. (1997) Harassment in the workplace and the victimization of men. *Violence and Victims*, *12*, 247–263.

Einarsen, S. and Skogstad, A. (1996) Bullying at work: Epidemiological findings in public and private organizations. *European Journal of Work and Organizational Psychology*, *5*, 2, 185–201.

Einarsen, S., Raknes, B. I. and Matthiesen, S. B. (1994) Bullying and harassment at work and their relationships to work environment quality: An exploratory study. *European Work and Organizational Psychologist*, *4*, 4, 381–401.

Farrell, G. A. (1999) Aggression in clinical settings: Nurses' views – a follow-up study. *Journal of Advanced Nursing*, *29*, 532–541.

Felson, R. B. (1992) 'Kick 'em when they're down': Explanations of the relationship between stress and interpersonal aggression and violence. *The Sociologist Quarterly*, *33*, 1–16.

Hoel, H. and Cooper, C. L. (1999) *The role of 'intent' in perceptions of workplace bullying*. Paper presented at the 9th European Congress on Work and Organizational Psychology: Innovations for Work, Organization and Well-Being. 12–15 May, Espoo-Helsinki, Finland.

—— (2000) *Destructive conflict and bullying at work*. Manchester School of Management, UMIST.

Hoel, H., Cooper, C. L. and Faragher, B. (2001) The experience of bullying in Great Britain: The impact of organizational status. *European Journal of Work and Organizational Psychology*, *10*, 4, 443–465.

Hoel, H., Rayner, C. and Cooper, C. L. (1999) Workplace Bullying. In C. L. Cooper and I. T. Robertson (eds), *International review of industrial and organizational psychology*, vol. 14 (pp. 195–230). Chichester: John Wiley.

Johns, N. and Menzel, P. J. (1999) 'If you can't stand the heat!' ... kitchen violence and culinary art. *Hospitality Management*, *18*, 99–109.

Jones, E. E. and Davis, K. E. (1965) From acts to dispositions: The attribution process in person perception. In I. L.Berkowitz (ed.), *Advances in experimental social psychology*, vol. 2. New York: Academic Press.

Karasek, R. A. (1979) Job demands, job decision latitude and mental strain: Implications for job redesign. *Administrative Science Quartely*, *24*, 285–308.

Keashly, L. and Jagatic, K. (2000) The nature, extent and impact of emotional abuse in the workplace: Results of a statewide survey. Paper presented at the Academy of Management Conference, Toronto, Canada.

Kile, S. M. (1990) *Helsefarleg Leiarskap: Ein Eksplorerande Studie* (Leadership with negative health implications: An exploratory study). Report from the Norwegian General Science Council, Bergen.

Kivimäki, M., Elovainio, M. and Vahtera, J. (2000) Workplace bullying and sickness absence in hospital staff. *Occupational and Environmental Medicine*, *57*, 656–660.

Knorz, C. and Zapf, D. (1996) Mobbing – eine extreme Form sozialer Stressoren am Arbeitsplatz (Mobbing: An extreme form of social stressor at work). *Zeitschrift für Arbeits- and Organisationspsychologie*, *40*, 12–21

Kräkel, M. (1997) Rent-seeking in Organisationen – eine ökonomische Analyse sozial schädlichen Verhaltens (Rent-seeking in organisations). *Schmalenbachs Zeitschrift für Betriebswirtschaftliche Forschung*, *49*, 6, 535–555.

Lee, D. (2000) An analysis of workplace bullying in the UK. *Personnel Review*, *29*, 5, 593–612.

Lewis, D. (2000) Workplace bullying – a case of moral panic? In M. Sheehan, S. Ramsay and J. Patrick (eds), *Transcending boundaries: Integrating people, processes and systems*. Conference Proceedings, Griffith University, Brisbane, Australia.

Leymann, H. (1996) The content and development of mobbing at work. *European Journal of Work and Organizational Psychology*, *5*, 2, 165–184.

Liefooghe, A. P. D. and Mackenzie Davey, K. (2001) Accounts of workplace bullying: The role of the organization. *European Journal of Work and Organizational Psychology*, *10*, 4, 375–392.

McCarthy, P., Sheehan, M. and Kearns, D. (1995) Managerial Styles and their Effect on Employees Health and Well-being in Organizations Undergoing Restructuring. Report for Worksafe Australia, Griffith University, Brisbane.

Mullen, E. (1997) Workplace violence: Cause for concern or the construction of a new category of fear. *Journal of Industrial Relations*, *39*, 21–31.

OECD (1999) *Implementing the OECD Job Strategy: Assessing performance and policy*. Paris: OECD.

O'Moore, M., Seigne, E., Mcguire, L. and Smith, M. (1998) Victims of bullying at work in Ireland. *Journal of Occupational Health and Safety – Australia and New Zealand*, *14*, 6, 569–574.

Pearce, J. L. (1998) Job insecurity is important, but not for the reasons you might think: The example of contingent workers. In Cooper and Rousseau (eds), *Trends in organisational behavior* (pp. 31–46). London: John Wiley and Sons.

Quinlan, M. (1999) The implications of labour market restructuring in industrial societies for occupational health and safety. *Economic and Industrial Democracy*, *20*, 427–460.

Randall, P. (1997) *Adult bullying: Perpetrators and victims*. London: Routledge.

Rayner, C., Hoel, H. and Cooper, C. L. (2002) *Workplace bullying: What we know, who is to blame, and what can we do?* London: Taylor and Francis

Salin, D. (1999) *Explaining workplace bullying: A review of enabling, motivating, and triggering factors in the work environment*. Working paper no. 406, Swedish School of Economics and Business Administration, Helsinki.

—— (2001a) Workplace bullying among professionals: the role of work overload and organisational politics. Presented at the 10th European Congress on Work and Organisational Psychology. 16–19 May, Prague, Czech Republic.

—— (2001b) Prevalence and forms of bullying among business professionals: A comparison of two different strategies for measuring bullying. *European Journal of Work and Organizational Psychology*, *10*, 4, 425–441.

Schein, E. (1985) *Organizational Culture and Leadership*. San Francisco: Jossey-Bass.

Sheehan, M. (1999) Workplace bullying: Responding with some emotional intelligence. *International Journal of Manpower*, *20*, 1/2, 57–69.

Sjøtveit, J. (1992) *Når veven rakner. Om samhold og mobbning på arbeidsplassen* (When the social fabric disintegrates: About solidarity and mobbing at work). Oslo: Folkets Brevskole.

Thylefors, I. (1987) *Syndabockar: om utstötning och mobbning i arbetslivet* (About expulsion and bullying in working life). Stockholm: Natur och Kultur.

UNISON (1997) *UNISON members' experience of bullying at work*. London: UNISON.

Vartia, M. (1996) The sources of bullying: Psychological work environment and organizational climate. *European Journal of Work and Organizational Psychology*, *5*, 2, 203–214.

Worrall, L. and Cooper, C. L. (1999) *The quality of working life: 1999 survey of managers' changing experiences*. London: The Institute of Management.

Zapf, D. (1999) Organisational, work group related and personal causes of mobbing/bullying at work. *International Journal of Manpower*, *20*, 1/2, 70–85.

Zapf, D., Dormann, C. and Frese, M. (1996a) Longitudinal studies in organisational stress research: A review of the literature with reference to methodological issues. *Journal of Occupational Health psychology*, *1*, 145–169

Zapf, D., Knorz, C. and Kulla, M. (1996b) On the relationship between mobbing factors, and job content, social work environment, and health outcomes. *European Journal of Work and Organizational Psychology*, *5*, 2, 215–237.

11 Explaining bullying at work

Why should we listen to employee accounts?

Andreas P. D. Liefooghe and Kate Mackenzie Davey

Introduction

Postmodernism represents a fundamental challenge to contemporary science by rejecting the taken for granted notions of rationality, order, clarity, truth and realism and the idea of intellectual progress. Instead, postmodernism draws attention to disorder, contradictory explanations and ambiguity. While modern science is concerned with the appropriate methods and procedures for establishing the truth, postmodernism uses deconstruction to reveal the strategies that are used to represent truth claims. Postmodernism is thus concerned with the use of language – as such, language becomes the unit of analysis.

On the basis of the work of Lyotard (1984), Foucault (1977, 1979, 1980) and Derrida (1974, 1978), the notion that language is a neutral tool for communicating 'facts' is undermined. Instead, language creates and imposes meaning, implying that meaning itself is unstable, depending on how it is 'read'. Postmodernism shifts attention away from the author of a text and focuses instead on the reader. Rather than asking what a text means, what the author is really trying to say or what the correct interpretation is, postmodernists are concerned with how different readers interpret the text, and what it means to them. All interpretations are regarded as equally valid, which implies that our understanding of the 'truth' will always be fragmented, selective and biased. Assuming this position allows scepticism for any approach that claims to present how things *really* are, the one best way or the truth.

In line with critical perspectives, postmodernism views truth as socially constructed – reality is a construct, a representation that is manufactured and manipulated. As such, postmodernism removes a sense of certainty and order by removing fixed reference points, by demonstrating that what we thought of as 'solid' can be seen as a socially constructed product. While modernism seeks to understand, postmodernism is continually asking whose view is being supported and whose interests are served.

Critical researchers see organisations as social historical creations, born in conditions of struggle and domination – a domination that may hide and suppress meaningful conflict (Alvesson and Deetz, 2000). Deetz (1992) describes organisations as political sites. Commercial organisations could, in theory, be positive social institutions providing forums for the articulation and resolution of important group conflicts over the use of natural resources, distribution of income, production of desirable goods and services, the development of personal qualities and the direction of society. However, in fact, various forms of power and domination have led to skewed decision making and fostered social harms and significant waste and efficiency (Alvesson and Deetz, 2000). Critical theory aims to demonstrate and critique forms of domination, asymmetry and distorted communication through showing how social constructions of reality can favour certain interests, and alternative constructions can become obscured and unrecognised. A further aim is to overcome these distortions, and to reclaim different interests, openly discuss them, leading to a resolution which is just and fair. Alvesson and Willmott (1992) argue that of special concern are forms of false consciousness, consent, systematically distorted communication, routines and normalisations which produce partial interests and keep people from genuinely understanding, expressing or acting on their own interests (also Mumby, 1988).

Critical theory research has explicit value commitments, and pays direct attention to moral and ethical issues. It is often suspicious of unconflicted accounts. The discourse also holds that people can and should act – an additional activist tone, found clearly in the work of Habermas (1975, 1987, amongst others). Thus while, like postmodernism, it seeks to interrogate unitarist analyses, unlike postmodernism, it values some accounts above others.

There are a number of themes running through postmodern/dialogic work. First, it focuses on the constructed nature of people and reality, emphasising language as a system of distinctions which are central to the construction process. In doing this it takes a stand against grand narratives and large-scale theoretical systems such as functionalism or Marxism. Second, it emphasises the power/knowledge (Foucault, 1980) connection and the role of claims of expertise in systems of domination. It emphasises the fluid and hyper-real nature of the contemporary world and the role of the mass media and information technologies in constituting that world. Finally it stresses narrative/fiction/rhetoric as central to the research process (Alvesson and Deetz, 2000).

These approaches challenge researchers to recognise the role of language in constructing the world. First, they suggest we should be suspicio[illegible] ny single account, and especially suspicious of those that claim to [illegible] credible account. Second, we should always be aware not just [illegible] eing said but of who is speaking. We should recognise that the [illegible] ikely to be heard and to be able to define meaning are those [illegible] ver. Third, we should recognise that language is a powerful [illegible] e all use it to defend our own interests and to try to make the world as we would like it to be. We should therefore be sensitive to the interests of different groups. We must be aware that those who are not being heard may tell a different story.

We formulated the following observations for bullying at work from these perspectives:

- Who talks about bullying at work is important, in particular when explanations are offered.
- Some stakeholders' voice may have been neglected and not been heard.
- The way explanations are formed may benefit those in a more dominant position, and this may lead to only part-explanations being offered, which could be biased.

This chapter is positioned in the section on 'explaining bullying'. As opposed to other contributions, we will not formulate a theoretical explanation in the strict positivist sense. Rather, we use the wider theoretical positionings of both a critical and a postmodern body of thought as a guide to chart some of the explanations that hitherto have not been heard.

Hence, in this chapter we are arguing for the importance of listening to a range of voices when explaining bullying at work. Postmodernism and critical management approaches help us to justify why it is especially important to be open to those who are seen as powerless. This approach allows us to uncover the interests that may be served by different definitions of bullying. It also allows us to challenge organisations to face up to the ways their employees are using the term 'bullying'. The chapter describes research based on these approaches in an organisation that wanted to discover why so many employees reported being bullied. We will show that while employees recognised what they called classic or school bullying, they also used the term to describe organisational practices. Employees are using the term 'bullying' to challenge many organisational systems rather than to blame individual managers or supervisors. The implications of this are that practices that many organisations see as legitimate are regarded as bullying by many employees.

Different explanations

Hence, the remainder of this chapter charts an organisational case study of bullying at work. The organisation, a UK high street bank (referred to here as Banco), approached us through their human resources department (HR). HR had just received what they described as 'worrying results' from their annual employee survey. For the first time, they had included the item 'have you been bullied at work' and 53 per cent had said 'yes'. It was our task to find out what they meant by this. Incidence rates where researches define the term and measure bullying, such as the Negative Acts Questionnaire (NAQ, Einarsen and Raknes, 1997) and Leymann's Inventory of Psychological Terrorization (LIPT, Leymann, 1990) are much lower. For a review, see Hoel *et al.* (1999). First, we will present the methodology and some of the results of our study, while second, we will consider the implications of these results for the field of research and practice related to workplace bullying.

The method

We conducted research in two departments (Technology, N = 100; Operations and Services, N = 132). The bullying incidence was so high that we treated all employees as the research population, and drew samples from this to explore explanations for the survey results. We decided to use focus groups for three reasons. First, a qualitative approach was appropriate, as the focus of our investigation was the construction of meaning. Second, we chose groups rather than one-to-one interviews to reach a larger sample of employees than would otherwise be practical. Third, we thought that the presence of others would avoid accounts that were too idiosyncratic – while we stressed to participants that we were not necessarily seeking consensus, we could nevertheless witness the discussions concerning consensus seeking amongst our participants. We decided to keep the sessions as unstructured as possible to allow maximum exploration of issues. After initial introductions and assurances that only generic information would be provided to Banco, we mentioned the survey results regarding the incidence of bullying at work, and asked them what the term 'bullying' meant to them. At no time did we offer a definition of bullying. We were interested in how employees used and defined the term.

Employee accounts of bullying

In the following section, we will focus on various elements of employee accounts. Employees did not offer a single, clear definition of bullying. To make sense of the term, they used a number of strategies. They linked bullying at work to school bullying, and contrasted their position to the powerlessness of children. They examined 'classic' bullying, which they equated with interpersonal bullying, occasionally physically violent, which is regarded as prevalent in a school environment. They also gave specific examples of incidents that they regarded as bullying at work. These tended to concern organisational practices, not interpersonal bullying. Employees recognised the 'classic' definitions, but they were using the term to describe their own experiences at work.

Ambiguity in definitions

In these employee accounts, the meaning of bullying is not at all homogeneous or clear. Participants struggle between different interpretative frameworks in order to define and ascertain what bullying is. As one participant puts it:

> F1: It can mean so many different things, I would take it as being ... being made to do something you really don't want to do, or just being told to do something rather than being asked. I mean in other people's eyes it could be something more serious, it could be physical, or emotional blackmail, depends on the situation ...

The elements she uses to define bullying are very subtle: being made to do things rather than being asked. While the literature would define this as an autocratic management style, and perhaps not view this as serious enough to be called bullying, this participant does include this as an additional use of the term 'bullying'. However, she acknowledges that this is her own definition of bullying, and that other people in other situations could use the term differently. Whether her assertions regarding what constitutes bullying *is* bullying according to the literature definitions is not important here. What we want to focus on is not on 'discovering the true nature of bullying', but on how employees use the term.

As bullying in a work situation is a relatively new concept, participants draw on the notion of bullying in a school in an attempt to define and explain what is happening in their workplace. However, the repertoire is found to be of limited use. School bullying is regarded as clearly defined, whereas work bullying depends on perception:

> F3: But surely it's ... my immediate reaction when I think of bullying is in the school situation, having kids at school, and that I think we can all relate to, it's a bigger person picking on a smaller person, but I think once you get into the working environment it gets a lot more difficult because surely it becomes a matter of perception. Because what one person takes for bullying another one is quite comfortable with that working relationship, I just think it's a very hazy area out there ...

The use of the term 'bullying' in a school environment is seen as unproblematic, in contrast to the 'hazy area out there' in the workplace. It calls for a subjective definition, rather than a definition that is clear, unambiguous and understood by all, like school bullying. Bullying offers a link between the two environments of school and work. School becomes a guiding framework participants draw upon to account for what is happening to them now in a work environment. Hence, the definition of the term 'bullying' expands:

M1: I used to think bullying is what you see in a school playground, you know. Group of kids around you and – you don't recognise it, really, in the workplace as such until, you know, until my personal experience and, on reflection I thought, 'Well actually I think I was a bit bullied.'

Yet when participants are asked to elaborate on this, and give examples of how these types of school bullying now manifest themselves in their organisation, they respond as follows:

F4: I think the old fashioned way of bullying is non-existent.
M1: Yes.
F4: Your school-ground mentality-type bullying. There's more subtle bullying going on.
M1: And it's organisational.
AL: Can you think of examples of bullying happening between two people at work? Or even if they're colleagues or whether they're a manager and one person – on an interpersonal level rather than an organisational level.
F4: Not really, no. I mean, to me – I would class that as the old-fashioned type of bullying and I don't think that happens any more. It's the more subtle – like – speaking up damages your career. That's bullying because – I mean, our Director said that speaking up won't damage your career and it has been proved that speaking up does damage your career, but it's done subtly.
AL: How ...
F4: In the appraisal process.
M3: At the appraisal process.
F4: You can't prove a lot of it now because it's subtle. It's behind the scenes, and, erm, and it – not just affects individuals, it affect – it can affect a broad base. As I say, the pay negotiations that are happening at the moment between the union and HR, it was very subtle intimidation and bullying going on there. But you can't just go up to them and say, 'Stop – will you stop bullying me', because you can't prove it. You can't prove it. They'll deny it.
F2: I find that frightening. That, to me, is somebody who's got far too much power.
M3: Well the problem is again, it goes back to the focusing on particular deliverables. HR people, to be fair, it's in their deliverables to keep the budgets down. If they can keep the budget even tighter than the target they'll get 5 on their appraisal, so they get a bit more money at the end of the year. They don't care how they do it.

In this discussion participants draw on school bullying. However, when the researcher (AL) tries to lead them into considering interpersonal bullying at work, his suggestion is emphatically rejected. Their experience of bullying at work is clearly linked to organisational systems. Two issues are considered as bullying here: lack of negotiation and speaking up, and the appraisal system and accompanying performance-related pay system, which are viewed as unfair systems. We will refer to the above quotation, and provide additional accounts to illustrate how these two issues manifest themselves in employee accounts.

Lack of negotiation and speaking up

The notion that speaking up can damage your career does not refer to one particular perpetrator. As participant F4 argues, it affects a 'broad base'. This organisational bullying is described as being perpetrated by the organisation against a group or groups of employees, rather than one individual against another. She also notes that there is a dissonance between what the official line is (the Director says speaking up won't harm you) and what happens in practice (in the appraisal process). We will turn to the link with the appraisal process later. First, we will focus on the notion of speaking up. The earlier quote included making a distinction between 'being told' or 'being asked'. Here, what is viewed as bullying is the employee response to this – are they allowed to speak up without fear of reprisals? This is not only applied to a one-to-one situation. Rather, it is extended to different groups within the organisation, such as the union and HR. In negotiations between the union and HR, F4 argues that 'very subtle intimidation and bullying' take place. So rather than placing two individuals as bully and victim, these employees use the term bullying to describe the failure of negotiations between HR and the union. A further quotation illustrates this:

F7: And at that meeting, erm, the union members felt very intimidated by the HR steering committee, Head of Industrial Relations, the Head of HR and, er, T. – what's his – don't know his second name – he's the HR – the reward package and all that. And they've put on a very intimidating act against us. They were not prepared to listen to us. They heard us but they were not prepared to listen to us, or change tactics. Erm – so, intimidation and bullying in here still goes on. And it – it goes on when it suits them, as it did today.

Rather than talking about bullying as something between two individuals, participants here construct a collective (rather than an individual) identity, and argue that it is 'them' who bully when it suits 'them'. Bullying here then means not being listened to. However, these employees do not necessarily position themselves as victims. As the following passage shows, they counter the 'them' with 'us', the 'staff', who can offer resistance:

F7: ... [A]nd as I say, I would – I would find that – if we get, erm, the pay talks are going on this afternoon with the reps and that. If we don't get an outcome we're happy with, then I think the staff will be very unhappy and morale is going to go down even more – with, like, what is being offered on the table. So I think the bank is going to have to watch this very carefully because they could see large numbers of staff walking.

M1: Well that happened before when the new terms and conditions came in we got a presentation. And one phrase I remember was that the union had been informed of the proposals. And I'd heard that the union actually had been told several hours before it was announced. And there was no negotiation. It was just, 'have been informed and are aware of what's going on'.

F57: So, yeah, so when you say, have you experienced bullying in the bank, yes we have experienced bullying and we are experiencing it now. Maybe not personally, individually, but the staff as a whole are experiencing bullying by the Banco management.

The lack of negotiation, being told what to do, forms the core of bullying here. Going through negotiations and not backing down *on anything* are regarded as bullying. While on the one hand there is a sense of not being able to do anything about this, staff also offer the threat of people leaving the company – which perhaps is the only form of resistance open to them. This is interesting, as it is in line with previous research based on single-item responses (Rayner, 1997), where the most frequent response strategy to bullying was to leave the organisation. While an individual may be able to be removed, a whole organisational practice may not appear to be that simple to tackle.

This use of the term 'bullying' as not being heard was very common in our sample. It was discussed both in traditional union–management negotiations and in individuals' daily experiences of work (Mackenzie Davey and Liefooghe, in press). We will now focus on the second aspect raised in the original quote regarding the appraisal system.

Appraisal and performance-related pay systems as being unfair

Participants F4 and M3 argued that speaking up could damage your career 'in the appraisal process'. The appraisal process in Banco is linked to reward (performance-related pay – PRP), and reward was also the topic of union and HR negotiations. They argue that this type of bullying is 'subtle' – and as the quotation below illustrates, this subtlety may be due to the fact that this does not operate at an individual level, but at an organisational one, as it is engrained in the organisational system:

F1: I think, in a way, that there's some, kind of, like, really indirect bullying goes on and it's to do with all the PRP system and stuff that we have. And it comes from right from the top, the targets and they go right down the managers. If your manager's got a target that they're really trying to achieve, I don't think – I sometimes think I feel, kind of, like the overtime one and stuff. Putting me under a lot more stress and not really wanting to understand what I'm feeling now, but just more trying to, like, cut costs.

M1: If you're overtime's cut because your manager's got to make his targets, they say, you know, 10 per cent reduction in overtime. So he's telling you there's no overtime but your customers are saying, 'I need this and need it now.' You're then getting – effectively getting bullied in two directions.

F1 and M1 discuss here how an organisational system (PRP) is not just an environmental factor facilitating interpersonal bullying, but is bullying in itself. The system itself is criticised – it is there to cut costs. Within this system, managers are put under pressure to increase staff performance, reduce overtime and cut costs to meet their targets. M1 then introduces the element of customers bullying by exerting extra pressure, leading to being bullied from both directions. Rather than identifying HR or managers as 'bullies', however, employees acknowledge that they function within the constraints of a system that requires them to act in the way they do. They have to keep their budget down. Thus, the cause of bullying is not simply attributed to individuals or to organisational groups or departments, but also includes the very organisational systems within which they operate, such as the processes by which pay and performance are negotiated, and how the appraisal system is seen to function in favour of the organisation and to the detriment of the employees. The following extract focuses more on the system itself:

M5: We're having the system imposed on us. And it still hasn't changed. We still get the system where you get given, er, you get given – targets which now, are not very, er, measurable because they've decided – you can't measure – too many of these things. They [HR] didn't want to know. They want to be able to say, we've got this bell curve, you're supposed to fit into it in each team. Now that's what I call bullying.

This participant queries the objectivity of the system through questioning the measurability. The notion of having to fit into the bell curve is used to denote that individual performance will always be relative to specific team performance. There is no reward for 'an objective good performance' as it depends where an individual falls on the imposed curve of normal distribution within their team. In other words, a whole team can never be 'excellent performers', as performance is treated as normally distributed within a specific team. Measurement, or the lack of it, becomes bullying in two senses: the lack of being able to measure, and the imposition of an artefact that is regarded as not being fair.

Summary

In sum, employees anchor their understanding of bullying in a school context. While school bullying has a relatively homogeneous meaning, workplace bullying is more ambiguous. Employees differentiate between *classic bullying*, the kind that occurs within schools amongst children and is seen to be interpersonal in its nature, and *organisational bullying*, which they argue is more relevant to their use of the term, and consists of organisational practices such as the ignoring of employee voice in pay negotiation and the appraisal system, which is seen as inherently unfair.

Explaining the findings

The original premise of this research was to explain the large incidence rate of bullying at Banco. We would argue that the additional use of the term 'bullying' to denote these organisational practices explains the higher incidence rate reported in this organisation. As the term 'bullying' is widely applied in the workplace, we need to be aware of how it is being used. Employees are clearly using the term in a very different way from researchers, employers and other interest groups.

The way in which a term is defined and used focuses attention on some aspects of the world, and leaves others unexposed (Ashforth and Humphrey, 1997). Realist approaches argue that this is necessary to ensure that meaning is shared and knowledge can progress, and defend terms from devaluation (Einarsen, 1998; Munthe, 1989 – see also Liefooghe and Olafsson, 1999, for an alternative view on this position). Alternative approaches examine the political aspects of definitions (Deetz, 1996), the socially constructed nature of reality (Gergen, 1994), the negotiated and contested nature of meaning (Weick, 1995) and the ambiguity of most communication (Mumby and Stohl, 1992). These approaches highlight the importance of examining how language is used in life outside the academy. The work we have described here focuses on the examination of the use of the term 'bullying' by groups of employees. While they acknowledge the 'classic' definition used by experts, they argue that it does not apply to their experience (consistent with the low incidence rates normally reported). However, they still use the term to describe what they perceive as the abuse of power by the organisation. This places the organisation and its systems as central to their definition of bullying at work, and in taking a systemic view changes the focus from individual bullies and victims to organisational power and control. Thus, while organisations may wish to present anti-bullying policies, stating their intolerance for bullies, employees argue that the organisation itself is responsible. The term is disputed.

While research hitherto has mainly regarded the organisation as a backdrop facilitating interpersonal bullying (e.g. Einarsen *et al.*, 1994; Zapf, 1999; amongst others), these employee accounts point to the organisation playing a much more active role, something akin to 'institutionalised bullying'. If we make the link here with findings on racism, for instance in the MacPherson (1999) Report, it becomes clear that it is not just about a few bad apples that need to be removed from the organisation, but the very practices (from recruitment to 'how things are done around here') that need to be scrutinised.

Conclusions

The additional explanation of bullying as an organisational practice rather than merely an individual or interpersonal one is important, as it allows a critical view on the day-to-day organisational exigencies facing employees. This allows the focus of explanations in the field to be extended beyond individualising, psychological explanations to more socially and organisationally based bodies of theory. Approaches such as critical theory or postmodern work can therefore be extremely useful lenses to view these wider uses of the term 'bullying'.

Bibliography

Alvesson, M. and Deetz, S. (2000) *Doing critical management research*. London: Sage.

Alvesson, M. and Willmott, H. (1992) On the idea of emancipation in management and organization studies. *Academy of Management Review*, *17*, 432–464.

Ashforth, B. E. and Humphrey, R. H. (1997) The ubiquity and potency of labelling in organizations. *Organization Science*, *8*, 1, 43–58.

Deetz, S. (1992) *Democracy in the age of corporate colonisation: Communication and the politics of everyday life*. Albany, NY: State University of New York Press.

Derrida, J. (1974) *Of grammatology*. London: Johns Hopkins University Press.

—— (1978) *Writing and difference*. London: Routledge.

Einarsen, S. (1998) Dealing with bullying at work: The Norwegian lesson. *Proceedings of the Bullying at Work Research Update Conference*, Staffordshire University Business School.

Einarsen, S. and Raknes, B. I. (1997) Harassment at work and the victimisation of men. *Violence and Victims*, *12*, 247–263.

Einarsen, S., Raknes, B. I. and Matthiesen, S. M. (1994) Bullying and harassment at work and their relationship to work environment quality. *European Work and Organizational Psychologist*, *4*, 381–401.

Deetz, S. (1996) Describing differences in approaches to organizational science: Rethinking Burrell and Morgan and their legacy. *Organization Science*, *7*, 191–207.

Foucault, M. (1977) *Discipline and punish: The birth of the prison.* Harmondsworth: Penguin.

—— (1979) Governmentality. *Ideology and Consciousness*, 6, 5–21.

—— (1980) *Power/knowledge*. New York: Pantheon.

Gergen, K. (1994) *Realities and relationships: Soundings in social construction.* Cambridge MA: Harvard University Press.

Habermas, J. (1975) *Legitimation crisis*. Boston: Beacon Press.

—— (1987) *The theory of communicative action*, vol. 2: *Lifeworld and system.* Boston: Beacon Press.

Hoel, H., Rayner, C. and Cooper, C. L. (1999) Workplace bullying. In C. L. Cooper and I. T Robertson, *International review of industrial and organizational psychology*, vol. 14 (189–230). Chichester: John Wiley.

Leymann, H. (1990) *Presentation of the LIPT Questionnaire: Construct, validation and outcome*. Stockholm: Violen inom Praktikertjänst.

Liefooghe, A. P. D. and Olafsson, R. (1999) Scientists and amateurs: Mapping the bullying domain. *International Journal of Manpower*, *20*, 1/2, 39–49.

Lyotard, J. F. (1984) *The postmodern condition: A report on knowledge*, Minneapolis, MN: University of Minnesota Press.

Mackenzie-Davey, K. and Liefooghe, A. P. D. (in press) Voice and power: Critically examining the uses of the term bullying in organizations. In A. Schorr and W. Campbell (eds), *Communication research in Europe*. Berlin: deGruyter.

Macpherson of Cluny, Sir William (1999) *The Stephen Lawrence Inquiry*. CM 4262–1, The Stationery Office.

Mumby, D. K. (1988) *Communication and power in organizations: Discourse, ideology and domination*. Norwood, NJ: Ablex.

Mumby, D. K. and Stohl, C. (1992) Power and discourse in organization studies: Absence and the dialectic of control. *Discourse and Society*, *2*, 313–332.

Munthe, E. (1989) Bullying in Scandinavia. In E. Roland and E. Munthe (eds), *Bullying: An International Perspective*. London: David Fulton Publishers.

Rayner, C. (1997) The incidence of workplace bullying. *Journal of Community and Applied Social Psychology*, 7, 3, 199–208.

Weick, K (1995) *Sense-making in organizations*. Thousand Oaks, CA: Sage.

Zapf, D. (1999) Organizational, work group related and personal causes of mobbing/bullying at work. *International Journal of Manpower*, *20*, 1/2, 70–85.

12 Bullying at work

A postmodern experience

Paul McCarthy

Introduction

What follows is an exploration of the potential of a postmodern approach to the concept of 'bullying' to extend our explanatory frameworks for this new and highly salient concept. The chapter proceeds by examining the rise of the concept as a new way of giving meaning to experiences of distress at work. In this manner, 'bullying' is treated as a new signifier that emerges from a wider discourse that attributes distress in contemporary workplaces to unacceptable behaviours. Postmodern signifiers can be recognised as concepts or names that are formed out of projections of everyday life experiences and emotions. As a new signifier, 'bullying at work' gives expression to a variety of anxieties, fears, and resentments, and it indicts perpetrators.

A key aim of the following discussion is to demonstrate that mapping the construction of 'bullying' as a signifier can provide a richer explanatory texture. The approach of treating 'bullying' as a postmodern signifier foregrounds difficulties in deciding between differing explanations of the phenomenon (Lyotard, 1988; Catley and Jones, 2001). However, while the potential of postmodernism to produce critical and innovative insights is recognised (Bauman, 1993), its multidimensionality does not fit neatly within accepted social scientific conventions.

The approach of mapping is considered a useful way of exploring the interaction of several dimensions of experience, knowledge and interest in the emergence of 'bullying' as a signifier. The process of mapping illustrates how understandings of bullying are a product of shifting alignments and tensions amongst diverse individual, organisational, professional and institutional interests (Liefooghe and Olafsson, 1999; McCarthy, 1999; Lewis, 2000). Appreciation that 'bullying at work' has emerged as a new signifier in a force-field of tensions also makes us aware that its meanings have been shaped by pre-existing meanings and by more or less powerful interests aligned to them (Foucault, 1972).

In these terms, 'bullying at work' emerges from power relations across several levels. Notably, the concept has arisen in a contest over entitlements in globalisation marked by ongoing restructuring in workplaces. The concept is empowering for employees experiencing distress in signifying a diversity of behaviours as unacceptable in respect of their effects on well-being, productivity and dignity. The emergence of the concept has been enabled by shared concerns resonating through such concepts as 'school bullying, 'stress', 'health and safety', 'harassment', 'violence', 'abuse', 'discrimination', 'equal opportunity' and 'human rights' (Liefooghe and Olafsson, 1999; McCarthy, 1999; Lewis, 2000).

Mutual recognition of the relevance of the concept 'bullying' across different community, disciplinary and institutional interests has enabled the diffusion of the concept. These interests include those concerned with: sexual harassment; equity; health and safety; employee assistance; counselling; workers' compensation and rehabilitation; industrial relations; and human resource managers, trainers and educators. The diversity of these interests, and their disciplinary alignments and terminologies, have led to differing shades of meaning being projected into the concept of bullying at work. In these respects, a contest over meanings ensues, although a mutual mobilisation of interests around the concept is evident (Liefooghe and Olafsson, 1999; McCarthy, 1999; Lewis, 2000).

Given the emotionalities projected into 'bullying', difficulties in balanced negotiation of the extremes of 'victim-blaming' and 'bully-bashing' are of ongoing concern. Bullies can be depicted as sadists, psychopaths and sociopaths at one end of a continuum, and as authoritarian, greedy, manipulative, lacking people skills, or just unaware they are bullying, towards the other. Victims can also be constructed more or less positively or negatively across a variety of positions in the discourse, for example as: being 'in the wrong place or the wrong time'; 'targets'; experiencing post-traumatic stress; displaying personality, cultural or skills characteristics that somehow render them more vulnerable to bullying; or lacking coping skills or resilience. Some psychiatrists and psychologists focus on childhood experiences in explaining why certain behaviours are seen as bullying and trigger distress. Over time, the work performance of a person experiencing bullying may decline, and that may lead others to consider them deserving of some bullying in the interests of getting the job done (Mann, 1996; Liefooghe and Olafsson, 1999; Einarsen, 2000; Zapf, 1999).

Beyond these positions, a postmodern perspective enables us to see the experience of victimisation in terms of identity-formation, as position-taking that can be more or less empowering across the spectrum of depression, coping, political activism or revenge. The 'McVeigh' and Columbine High massacres evidence that the victims experiencing bullying can turn into bullies and worse. In addition, Bauman prompts us to think more deeply of the roots of bullying, in observing that 'victims are not ethically superior to their victimisers', and it may be that their lack of power gives them 'less opportunity to commit cruelty' (1993, pp.

227–228). Starkly, 'great persecutors are recruited amongst the martyrs not quietly beheaded' (Cioran in Bauman, 1993, p. 228).

The approach to mapping the rise of 'bullying at work' that follows juxtaposes a diversity of meanings of 'bullying'. The collage so composed is proposed as a demonstration of the usefulness of a postmodern approach to the phenomenon. In these respects, it presents a topography of meaning that represents ways in which the concept has emerged in the enfoldment of prior meanings of distress and interests aligned to them. The representation is useful in illustrating the contradictions and limits of the meaning of 'bullying at work', and prompts extension of the breadth and depth of our understanding of the phenomenon. Exploring the rise of the concept as a postmodern signifier also brings us to awareness of ways the representational form and language of postmodernism inscribe themselves more or less violently in contemporary workplaces. As such, the approach seeks to open wider cultural, economic, ethical and political debate about the meanings of 'bullying' and its manifestation in the new workplaces.

A postmodern approach: prospects and difficulties

Postmodern representations are notable for the rich symbolic texture they compose about meanings of complex social phenomena. While these qualities contribute to the critical and innovative nature of postmodernism, they tend to defy categorisation in traditional social scientific terms, as Liefooghe and Olafsson (1999) observe in mapping representations of bullying. The stance taken here is that postmodernism is not necessarily at odds with the aims of critical social science, due to its potential to provide insights into the manner in which diverse threads of meaning are woven into complex social phenomena.

As a prelude to a postmodern look at 'bullying', a brief excursion into postmodern ways of seeing follows. Postmodern representations have commonly been described as multidimensional, figural, ephemeral, undecidable and ambivalent (Kearney, 1988). Architectural style perhaps realises the most visible and tangible expression of postmodernism, and is also significant since the origins of postmodern representations of social phenomena have been traced to architecture. The relevance of postmodernism to the study of a complex phenomenon such as bullying is in its potential to depict the *architecture* of the concept, in terms of the juxtaposing of meanings in its emergence from wider discourses (Rose, 1988).

A postmodern perspective helps us see parallel dynamics in architecture and the new organisational structures reflective of wider global movements. Postmodernism is writ large in the playful breaking of conventions and admixtures of old and new in the design of malls, hypermarkets, entertainment complexes and the new urban villages. An interplay of pleasure and repression can also be discerned, as access to satisfactions of symbolic goods are wrought at some human cost and are bounded by surveillance devices and gated communities, and reflect sharp divisions around wealth and status (e.g. Davis, 1990; Jameson, 1991). The emergence of equivalent reconfigurations of space in postmodern organisations, and reasons why bullying at work has arisen as a concern within them are discussed later. These understandings give insight into the economic and cultural roots of 'bullying'. They also prompt us to engage in debate about the ethical and political consequences of violence, both in postmodern ways of seeing and in the construction of signifiers such as 'bullying' (e.g. Bernstein, 1991).

While observing that the postmodernists have identified preconditions for violence in Western humanism and rationality, Bernstein (1991) also discerns roots of violence in the Nietzschean lineage in postmodernism. He raises questions about the grounding of the postmodern critique, asking 'what precisely is being affirmed ... and why?' (1991, p. 7). These observations raise concerns that postmodernism might unwittingly affirm violence, for example in the trauma of incessant change unleashed in everyday life. However, the violence in deconstruction can also be seen as positive, in the interests of critical social science. For example, Bauman points to the potential of postmodernism to tear off 'masks of illusion' and to reveal 'certain pretensions as false and certain objectives as neither attainable nor for that matter desirable' (1993, p. 3). These perspectives prompt us to think about the ethical and political consequences of 'bullying' as a postmodern signifier, and about related questions of moral responsibility.

Treating 'bullying' as a postmodern signifier does foreground critical questions about its social construction. The significance of such questions is highlighted by concerns that the mode of construction of postmodern signifiers is often left unexamined (e.g. Rose, 1988). In pursuing the construction of bullying as a postmodern sign, we can usefully draw on Foucault's (1972, 1981) innovative mapping of the interweaving of diverse disciplinary knowledges into emergent meanings of complex social phenomena and their remedies.

In taking this approach to mapping the rise of 'bullying', we become aware that interactions between diverse disciplinary, professional and institutional interests continue to shape its various shades of meaning. These interactions can be usefully depicted using Adorno's notion of 'force-field', and its 'relational interplay of attractions and aversions' (Adorno in Jay, 1973, p. 54). In this sense, the concept of 'bullying' emerges in a wider 'force-field' in which more or less powerful interests jockey over meanings.

Bullying: a new syndrome of distress?

The approach of mapping the emergence of syndromes of distress is a characteristically postmodern one, and invariably challenges accepted meanings through teasing out biases in their construction (e.g. Showalter, 1997). My purpose in representing 'bullying at work' in these terms is that it gives insights into ways in which more or less powerful disciplinary, professional and institutional interests have interacted to produce more or less dominant formations of meaning. An overarching concern driving this pursuit is to question whether commonly accepted meanings of 'bullying' have tended to accentuate an assemblage of individual, medical and therapeutic, managerial and particular legal remedies. Concerns have been raised about consequences arising from the manner in which 'bullying' is conceptualised. For example, Liefooghe and Olafsson (1999) have cautioned that, with the rise of research and media commentary, the general public might conceptualise behaviours that seem like bullying 'in ways that are less conducive to constructive action and often lead to (self-) destructive scenarios for the workers and/or the employers'. There is also the concern that the relations of power and knowledge producing more or less dominant formations of understanding and practice might tend to lock those who experience bullying into remedies of indeterminate duration, outcome and cost, at the risk of added duress and distress. These concerns are supported by evidence that, while the implementation of anti-bullying policies is less than widespread and their access fraught with difficulty, the most common remedial pathway is through anti-depressants and counselling – as if by default (McCarthy and Barker, 2000).

This is not to say that such explanations and remedies may not be useful, even absolutely necessary in cases, as Showalter (1997) recognises. Rather, their emphasis is examined to highlight limits of these meanings and to point to the need for a wider web of ethical, cultural, economic, political and legal explanations and constraints (McCarthy, 2001). The problem is that syndromes commonly construct distress as an individual, therapeutic, medical and managerial issue, and these are enlisted to common law actions, for example for negligence or breach of contract, in an increasingly litigious society. However, such remedies are limited by their cost and uncertainty. Furthermore, this emphasis can tend to marginalise proposals to extend existing health and safety, discrimination, equal opportunity and human rights laws to encompass definitions of bullying and their more accessible tribunals and obligations (e.g. Anti-Discrimination Commission, Tasmania, 2001). With that, access to less costly tribunals and to corporate policies and procedures introduced to meet related statutory obligations could also receive less consideration. In recognising such difficulties, the postmodern approach attempts to respect the uniqueness of the victim's experience. A key concern is to account for historical evidence of risks that the experience of distress can be constructed in ways that serve professional and institutional interests more so than those of the victim.

Features of the rise of 'bullying' as a new syndrome that loosely align with those modelled in Showalter's (1997) schema, include shifting alignments of meanings and interests in:

- circulation of 'horror' stories through the media and arousing of widespread public concern;
- sharing experiences of victimisation, networking of resentments and identifying of bullies;
- politicking for constraints, punishment of perpetrators and support for victims;
- posturing by 'caring' governments in promotion of self-regulation guidelines;
- ongoing struggles over meaning between interest group, researcher, professional, and institutional interests; and
- extension of explanatory frameworks to include wider ethical, cultural, political, and legal understandings and remedies, over time (McCarthy, 2001).

In this formation of meanings, 'bullying' emerges as a new signifier that is perhaps overdetermined, given the extent of the projection of new experiences of distress through older meanings of violence in its construction. For example, the concept has been invested with shades of often subtler violence hitherto recognised in school bullying, domestic violence, sexual harassment and occupational violence – as well as with meanings derived from conceptions of stress and psychological injury (Liefooghe and Olafsson, 1999; McCarthy, 1999; Lewis, 2000; Hatcher and McCarthy, 2002). In this process, the concept emerges from a *lack* of a way of expressing shared concerns about a variety of behaviours that offend and humiliate in contemporary workplaces, but fall outside existing legal constraints. Thus, 'bullying' gives sharper definition to forms of violence that mostly fall outside present legal definitions of assault, sexual harassment, discrimination, health and safety, equal opportunities and human rights (Division of Workplace Health and Safety Queensland, 1998; Mayhew, 2000).

The need for the concept and the extent of its diffusion can be located in widespread experiences of distress in workplaces undergoing restructuring to meet global market conditions. This distress is also infused with wider anxieties, including 'fears of falling' (Ehrenreich, 1990). Questions about the extent to which concerns about 'bullying' arise within a wider culture of fear and complaint, as are characteristically addressed by accommodations to risks prompted by victims' complaints, have also been raised in a study by Lewis (2000). Information supplied by colleagues in the workplace, rather than by the media or other external sources, was found to be a key factor enabling perceptions of bullying in this study.

In the naming of experiences as 'bullying at work', the term can be seen to act as a solar collector of resentments for behaviours perceived as unreasonable or inappropriate. As it arises, the concept has come to signify and mobilise concerns about a range of behaviours that inflict more subtle, conceptual, emotional, procedural or cultural 'violences' in the pressures of contemporary workplaces. The need for unions and HR professionals to respond to bullying as a risk factor in a culture driven by fear and concerns about personal safety has been discussed by Lewis (2000).

The approach to mapping taken here also reveals governmental interests in bullying. Governments have increasingly to deal with the impacts of ongoing restructuring to meet global market conditions on electorates. Resentment at loss of jobs and careers has fuelled the indictment of managerial behaviours in downsizing as heavy handed and repugnant. Widespread perceptions of these behaviours as 'bullying' and its resonance with other shades of violence have given the concept a political evocativeness. In a postmodern political scene cultures of anxiety and blaming proliferate and fuel rapid swings in political sentiments. In this scene, governments do have an interest in being seen to address social stressors such as bullying. Preventative and remedial action by individuals and organisations in accordance with government-endorsed guidelines in the third-way spirit of mutual obligation is now commonly favoured over legislative regulations. The current plethora of anti-bullying guidelines can be located within this politics (e.g. Division of Workplace Health and Safety Queensland, 1998; Office of the Employee Ombudsman, 2000; Queensland Working Women's Service, 2000; Australian Council of Trade Unions, 2000; Victorian Employers' Chamber of Commerce and Industry and Job Watch, 1999).

The rise of 'bullying' as a new signifier of distress and interest in self-regulative guidelines can also be given useful perspective by locating them within wider governmental concerns about distress in the working population. Here, a Foucauldian perspective draws our attention to ways in which self-regulative approaches to distress at work carry forward interests of pastoral care, governing at a distance, and the construction of enterprising, responsible, productive self-actualising subjects. In enterprising culture, the commitment of the state, corporations and the population to self-regulation of such problems as bullying is perhaps a more sustainable way of addressing productive concerns than command (Foucault, 1981; du Gay, 1991; Hatcher and McCarthy, 2002). To this extent, self-regulative guidelines provide a point of reference for identity formation, position taking, confession of personal and organisational skills failings, and therapeutic management.

In addition, the approach prompts consideration of the politics of resistance or resentment driving the loading of the concept 'bullying' with a plethora of meanings. In these terms, 'bullying' emerges as a new signifier of distress that challenges managerial prerogative in constructions of 'enterprise' wherein the logic of the market rules. This politics manifests in the act of naming a diversity of behaviours as 'bullying'. The projection of prior meanings, of school bullying, harassment and other violences into the concept effectively tars a wide range of behaviours that have been hitherto less questioned in the public arena with the brush of ethical repugnance. The indictment of these behaviours in terms of effects on well-being, association with skills failures, economic loss and legal risks invokes concerns that resonate across individual, managerial and institutional interests. From this perspective, the act of naming 'bullying' may be viewed as part of a wider contest over entitlements and fairness in globalisation.

This approach provides insights into the construction of the victim in 'bullying at work' conceived as a discourse. Notably, the depiction of the victim's experience in this discourse departs from earlier attributions of distress at work to such concepts as 'stress' or 'depression'. Insight is also given into the roots of these ways of conceptualising distress at work in a medical framework oriented to individual pathologies and therapeutic management. In the construction of the victim in 'bullying', a perpetrator, who is often a manager or supervisor, is more directly indicted as a cause of the distress, and efforts are made to avoid blaming the victim (Zapf, 2001).

Naming experiences as 'bullying' provides for mobilisation of the interests of victims, professionals and governmental 'caring' alike through recognition, help, guidelines and remedies. In these terms, the act of naming 'bullying' can be seen as a strategic move by outsider groups that shifts the frame of reference for distress attributed to unacceptable behaviours away from individual pathology to managerial responsibility for the quality of the work environment. In this shift, the focus turns to liabilities of the organisation and the perpetrator. Such dynamics in the emergence of new meanings have been observed by Featherstone (1988). Insights given by mapping these dynamics have been demonstrated in Fineman's (2001) exploration of the social and emotional architecture of the 'greening' of the corporation.

Bullying is produced ... and consumed

Here, similarities in the manifestation of several dimensions of violence in postmodern cities and workplaces are examined. In cities, the postmodern turn is visible in the incessant renewal of the built environment into mixed-use urban villages as conduits for service industry, new technologies and exchanges around symbolic goods. The playfulness and pleasure-seeking in these pursuits is transgressive, and an extent of violence is evident in the alignment of cities and workplaces to the consumption and production of symbolic goods. In the shadow of the pleasures of the postmodern city one finds behaviours that may easily be construed as bullying in marginalisation, exclusion, the gated community, expulsion of the poor, racism, a plethora of rages, electronic surveillance and corruption (Davis, 1990; McCarthy, 1998).

The alignment of city and work-scapes to the production and consumption of symbolic goods is most evident in the mixed-use urban villages produced by the urban renewal that shapes the postmodern city. In their rhetorical expression, both postmodern city and work-scapes are less hierarchical, flatter, more flexible and tribalised in these pursuits. The promise of liveability in postmodern architecture has its counterpart in claims of participation, quality, creativity, learning and autonomy in postmodern workplaces. The brutality at the other side of 'liveability' in city revitalisation is also paralleled in the effects of restructuring, re-engineering, downsizing, emotional management and performance surveillance in postmodern workplaces (McCarthy *et al.*, 1995). The new workplaces give entrée to consumption of symbolic goods and vicarious identification with lifestyles and qualities they signify. However, at the reverse side of these pleasures, the pressures of globalisation manifest in job insecurity, the working poor, overwork for some, underwork for others, neo-Taylorism, micro-management enabled by e-mail and computer monitoring, and ongoing loss of grounding of skills and futures. Rostered low pay, casualised service and servant work proliferates, and dirty work is relocated to third-world and developing countries.

As producers and/or consumers of symbolic goods in postmodern living, we are all unavoidably positioned as relays in the 'cruel festivity' writers such as Bernstein (1991) find pervasive in postmodernism. The most recent manifestation of postmodernism at work is marked by the rise of electronic communication networks, service industries, the mass distribution of symbolic goods, and their production in smaller, flexible, more feminised units in which labour-intensive production is diffused to the third world (Lash and Urry, 1987). For many, the fruits of postmodern style are cultivated in the pressures of contemporary workplaces, with overworking to meet performance demands common. Commitments to finance to acquire symbolic goods purveyed by the media, and to other facets of postmodern lifestyles such as sports and entertainment can also be painful. The current pandemic of victimisation can be seen to manifest from this context, and the experience of victimisation therein can be projected into a plethora of rages. Indeed, the potential of victims to act out their resentments in bullying behaviours has been identified as a key sensibility of life in postmodern culture and workplaces (McCarthy, 2000a).

To the extent that these sensibilities are driven by fears for personal security and the pursuit of increasing property values, they inscribe inequality and exclusion in public life (Caldeira, 1996). Therein, people use 'violence to make claims upon the city and use the city to make violent claims. They appropriate a space to which they then declare they belong; they violate a space which others claim' (Holston and Appardurai, 1996, p. 202). Equivalent claims can be seen in the mobilisation of interests in postmodern organisational forms in the pursuit of survival and accumulation. However, resistance manifests in signifiers such as 'bullying' that resonate widely with resentment. In this sense, concepts such as 'bullying' enable positions to be taken that challenge and reconfigure power relations, as their meanings are subsumed under guidelines, codes of conduct and legal definitions.

Here, a postmodern view is useful in illustrating that rather than the positions of the victim or bully being fixed, they may oscillate. For example, sometimes we experience victimisation and defer to bullies, and sometimes we act as bullies, or as bystanders whose fear for our own security may allow bulling to persist (Casilli, 2000; McCarthy, 2000b).

The affirmation of bullying in postmodernism

The problem that postmodern ways of seeing might themselves be a cause of bullying is considered further in this section. In discussing this question I have located 'bullying' on a 'continuum of violence' (Mayhew, 2000). In postmodernism, the play of difference in language and taking positions within it drive the ongoing progression of meaning and reconfigure power relations (Derrida, 1981; Foucault, 1972). Language games can be played more or less violently, and can occasion terror where moves are made for pleasure, power or performativity (Lyotard, 1984). This violence in language and delineation of meanings can be located in a wider lineage of writing about violence as a key driving force in modernity. This lineage has been traced through Faust, de Sade and Nietzsche, and into the postmodernists. Its emergence in the triumph of the *will to power* and attendant cruelty in postmodernism has been observed (McCarthy, 1993; Bernstein; 1991).

Understanding of this lineage offers explanations as to why bullying seems endemic in postmodern culture and organisation. Nietzsche is scornful of the celebration of the victim in the Christian ethos. In addition, de Sade offers explanations as to why greed and narcissism might underpin bullying in a postmodern world. De Sade's (1968) view is of the world as a 'matrix of maleficent molecules' (p. 400), in which, 'if reason's glacial hand waves us back, lust's fingers bear the dish towards us again, and thereafter we can no longer do without the fare' (p. 11). Therein, desire disposes itself 'toward such-and-such an object and against some other, depending on the amount of pleasure and pain I desire from these objects' (p. 34). These terms alert us to the possibility that in the molecular play of difference in language characteristic of postmodernism, the intensities of pleasure may be heightened through the violence of bullying. If there is value in de Sade's perspective, it is in prompting awareness that the act of bullying and the experience of victimisation can arouse perverse pleasures associated with sadism and masochism. In this sense there is a possibility that experiences of bullying and victimisation could be less oppositional, and act more in a mutually intensifying manner in postmodern culture and organisation.

Violence, in closure, exclusion, respatialisation and appropriation, across built environments and organisations therein, is commonly depicted as a key dynamic in the work of postmodern writers. Indeed, we learn a lot about the de facto normalisation of bullying in postmodern culture and organisation from Lyotard's observation, 'to speak is to fight, and speech acts fall within the domains of a general agonistics' (1984, p. 10). Thus, we express ourselves through taking positions in language that are more or less adversarial, as in making moves in a game.

'Postmodern' workplaces are notable for their emphasis on interactive teams as tribal organisational forms in ever-shifting terrains. It is interesting that bullying has emerged as a concern within these forms, and explanations for it can be derived from ways in which groups delineate territory in relation to external threat, and establish identities, norms and compliance (Debray, 1983). These understandings bring us to awareness that fear, closure and reaction to threatening 'Others' is at the roots of violence, including bullying, in postmodern culture and organisation. While repugnance at bullying may be overtly expressed, it may persist as a tacitly accepted strategy for securing group affiliation, identity, space and resources. Indeed, Bauman remarks, 'I shout, therefore I am – is the neotribal version of the *cogito*' (1993, pp. 235–237). Thus we find an extent of violence engraved in our identity formation and survival strategies.

Further explanation of the psychodynamics of these processes can be derived from Lacan's insight into the dynamics of the construction of self-identity in the mirror of the *Other*, as a projection out of our experience of *lack* (1977). The construction of the Other, as a reference for self-identity, cannot help but be coloured with the aura of symbolic goods in a postmodern world. However, to the extent that the Other is desired because they 'enjoy' in a certain way, and these objects of desire are beyond reach or prohibited, this self-identity may be derived at the cost of suffering (Lacan, 1977; Salecl, 1998). The violence in this formation of self-identity is that it 'splits us from ourselves and from others' (Salecl, 1998, p. 129). This 'pleasure in displeasure' can be explained in terms of Lacan's *jouissance* and provides insights into the painful dilemma at the roots of love and hate. 'We hate others,' as Salecl (1998, p. 129) observes, 'because they have enjoyment.' Thus we are alerted to the possibility that pain arising in the formation of self-identity *for Others* in postmodern culture might trigger experiences of victimisation and their acting out in bullying.

In these terms, a postmodern approach brings us to the realisation that experiencing victimisation and behaving in a bullying manner are more evenly distributed in the human condition, and at points may be mutually intensifying. In the scenario outlined above, we are brought face to face with our own implication in the production and consumption of bullying. This paradox is expressed in Rose's (1992, p. 294) lyric:

> I am abused and I abuse
> I am the victim and I am the perpetrator
> I am innocent and I am innocent
> I am guilty and I am guilty

Historically, we learn of the limits of attributions of distress heavily invested with emotions of fear and anger, as can be projected into the naming of the 'bully' as an incarnation of pure 'evil'. The events in Salem, an early European settlement in America, stand as a warning that witch-hunts with medieval overtones can manifest in modern societies. In Salem, panic about the inappropriate behaviours of some members of the community led to a rash of scapegoating of perpetrators of evil, to the extent that the basis of the very categorisation of evil eventually became untenable. If the emergence of 'bullying' as a signifier was to lead to the indiscriminate naming of bullies and their trial by public spectacle, then the social and economic fabric could be disrupted by witch-hunting (McCarthy, 1999).

Thinking through the ethical dilemmas evoked here prompts us to think of questions of moral responsibility in our construction of 'bullying', and of our political commitments and responsibilities in respect thereof (e.g. Bernstein, 1991). These insights widen our appreciation of the diversity of subject positions, meanings and interests that need to be negotiated in understanding the concept of 'bullying at work' and dealing with it in workplaces.

Conclusion

In conclusion, the approach taken demonstrates that a postmodern perspective provides novel insights into the interactions of a diversity of meanings and interests in the rise of 'bullying' as a new signifier of distress attributed to unacceptable behaviours. Beyond the bully–victim binary, a diversity of subject positions, crossovers and complicities are identified, and we are brought face to face with our own potential to be relays for bullying in the new workplaces. The approach is useful in placing difficulties in defining the concept, establishing evidence of its effects acceptable to management, tribunals and courts, and in accessing remedies, within wider explanatory frameworks. Further research into the economic, cultural, ethical and political dimensions of bullying is recommended in the interests of developing more functional, healthy interpretations, identities and possibilities for action by victims within a wider web of preventative measures and remedies.

Bibliography

Anti-Discrimination Commission, Tasmania (2001) *Recommendation 'workplace bullying, schoolyard bullying – unacceptable behaviour'*. AntiDiscrimination @justice.tas.gov.au.

Australian Council of Trade Unions (2000) *The workplace is no place for bullying*. Available at <http://www. actu.asn.au>.

Bauman, Z. (1993) *Postmodern ethics*. Oxford: Blackwell.

Bernstein, R. (1991) *The new constellation*. Cambridge: Polity Press.

Caldeira, T. (1996) Fortified enclaves: The new urban segregation. *Public culture*, *8*, 2, 303–329.

Casilli, A. (2000) *Stop mobbing*. Rome: DeriveApprodi.
Catley, B. and Jones, C. (2001) On the undecidability of violence. Paper presented at the Standing Conference on Organisational Symbolism (SCOS), Trinity College Dublin, 30 June to 4 July.
Davis, M. (1990) *City of quartz: Excavating the future in Los Angeles*. London: Vintage.
Debray, R. (1983) *Critique of political reason*. London: New Left Press.
Derrida, J. (1981) *Positions*, trans. A. Bass. Chicago: University of Chicago Press.
Division of Workplace Health and Safety (Qld) (1998) *Workplace bullying: An employer's guide*. Division of Workplace Health and Safety (Qld): Brisbane.
du Gay, P. (1991) Enterprise culture and the ideology of excellence, *New Formations*, 13: 45–61.
Einarsen, S. (2000) Harassment and bullying at work: A review of the Scandinavian approach. *Aggression and Violent Behaviour*, *5*, 4, 379–401.
Ehrenreich, B. (1990) *Fear of falling: The inner life of the middle class*. New York: Harper Perennial.
Featherstone, M. (1988) In pursuit of the postmodern: An introduction. *Theory, Culture and Society*, *5*, 2–3, 195–218.
Fineman, S. (2001) Fashioning the environment. *Organisation*, *8*, 1, 17–31.
Foucault, M. (1972) *The archaeology of knowledge*. London: Tavistock.
—— (1981). *The history of sexuality*, trans. R Hurley. Ringwood, Victoria: Penguin.
Hatcher, C. and McCarthy, P. (2002) Workplace bullying: In pursuit of truth in the bully–victim–professional practice triangle. *Australian Journal of Communication*, *29*, 3, in press.
Holston, J. and Appardurai, A. (1996) Cities and citizenship. *Public Culture*, *8*, 2, 187–204.
Jameson, F. (1991) *Postmodernism or, the cultural logic of late capitalism*. London: Verso.
Jay, M. (1973) *The dialectical imagination*. London: Heinemann.
Kearney, R. (1988) *The wake of imagination*. Minneapolis, MN: University of Minnesota Press.
Lacan, J. (1977) The four fundamental concepts of psycho-analysis, ed. J.-A. Miller, trans. A. Sheridan. New York: W. W. Norton and Co.
Lash, S. and Urry, J. (1987) *The end of organised capitalism*. Oxford: Polity Press.
Lewis, D. (2000) Workplace bullying – A case of moral panic?'. In M. Sheehan, S. Ramsay, and J. Patrick (eds), *Transcending boundaries: Integrating people, processes and systems*. Conference proceedings, 6–8 September 2000, The School of Management, Griffith University, Brisbane.
Liefooghe, A. and Olafsson, R. (1999) 'Scientists' and 'amateurs': Mapping the bullying domain. *International Journal of Manpower*, *20*, 1/2, 39–49.
Lyotard, J.-F. (1984) *The postmodern condition: A report on knowledge*. Manchester: Manchester University Press.
Lyotard, J.-F. (1988) *The differend: Phrases in dispute*. Minneapolis, MN: University of Minnesota Press.
Mann, R. (1996) Psychological abuse in the workplace. In P. McCarthy, M. Sheehan and W. Wilkie (eds), *Bullying: From backyard to boardroom*, 1st edn. Alexandria: Millennium Books.

Mayhew, C. (2000) *Preventing violence within organisations: A practical handbook*. Australian Institute of Criminology Research and Public Policy Series no 29, Canberra.
McCarthy, P. (1993) Postmodern pleasure and perversity: Scientism and sadism. In E. Amiran and J. Unsworth (eds), *Essays in postmodern culture*. New York: Oxford University Press.
—— (1998) Consulting-violence in city revitalisation. In P. McCarthy, M. Sheehan, S. Wilkie, and W. Wilkie (eds), *Bullying: Causes, costs and cures*. Brisbane: Beyond Bullying Association.
—— (1999) Strategies between managementality and victim-mentality in the pressures of continuous change. *Organisations looking ahead: Challenges and directions*. Conference proceedings, 22–23 November, School of Management, Griffith University, Brisbane.
—— (2000a) The bully-victim at work. In M. Sheehan, S. Ramsay and J. Patrick (eds), *Transcending boundaries: Integrating people, processes and systems*. Conference proceedings 6–8 September 2000, The School of Management, Griffith University, Brisbane.
—— (2000b) Prefazione. In A. Casilli, *Stop mobbing*. Rome: DeriveApprodi.
—— (2001) The bullying syndrome, Complicity and responsibility. In P. McCarthy, J. Rylance, R. Bennett and H. Zimmerman (eds), *Bullying: From backyard to boardroom*, 2nd edn. Sydney: Federation Press.
McCarthy, P. and Barker, M. (2000) Workplace bullying risk audit. *Occupational Health and Safety – Australia and New Zealand*, *16*, 5, 409–418.
McCarthy, P., Sheehan, M. and Kearns, D. (1995) *Managerial styles and their effects on employees' health and well-being in organisations undergoing restructuring*. Brisbane: School of Organisational Behaviour and Human Resource Management.
Office of the Employee Ombudsman (2000) *Bullies not wanted: Recognising and Eliminating bullying in the workplace*. Adelaide: Office of the Employee Ombudsman of South Australia.
Queensland Working Women's Service (2000) *Risky business: A useful publication for employers for preventing and resolving workplace bullying*. Brisbane: Queensland Working Women's Service.
Rose, G. (1988) Architecture to philosophy – the postmodern complicity. *Theory Culture and Society*, *5*, 2–3, 357–372.
—— (1992) *The broken middle*. Oxford: Blackwell.
Salecl, R. (1998) Spoils of freedom. In Mary Zournazi (ed.), *Foreign dialogues*. Annandale: Pluto Press.
Sade, Marquis de (1968) *Juliette*, trans. A. Warinhouse. New York: Grove.
Showalter, E. (1997) *Hystories: Hysterical epidemics and modern culture*. New York: Columbia University Press.
Victorian Employers' Chamber of Commerce and Industry and Job Watch (1999) *No bull: Say no to bullying and violence in the workplace*. Melbourne: VECCI and Job Watch.
Zapf, D. (1999) Organisational, work group related and personal causes of mobbing/bullying at work. *International Journal of Manpower*, *20*, 1/2, 70–85.
—— (2001) European research on bullying at work. In P. McCarthy, J. Rylance, R. Bennett and H. Zimmerman (eds), *Bullying: From backyard to boardroom*, 2nd edn. Sydney: Federation Press.

Part 4

Managing the problem

'Best practice'

13 Bullying policy

Development, implementation and monitoring

Jon Richards and Hope Daley

Introduction

UNISON is the largest trade union in the United Kingdom (UK) with 1.4 million members working in local government, health care, further and higher education, the water, gas, and electricity industries, transport and the voluntary sector. UNISON represents a wide range of professions and occupations, both manual and non-manual. Two-thirds of our members are women, many of whom provide caring services to the public.

We have advised on numerous bullying policies, and have written detailed general guidance on the issue for stewards and safety representatives, as well as guidance aimed at specific groups of workers. In addition, UNISON has been successful in a number of compensation cases on behalf of members bullied at work.

A policy makes a clear statement about what an organisation thinks, its relationship with staff and how it expects people to work within its culture. It also makes clear what is considered acceptable behaviour and what will not be tolerated. This is particularly important for lower-grade staff, as a large survey of UNISON members (UNISON 1997a) showed that 83 per cent of bullies were managers. Without a policy which legitimises complaints about bullying, it is difficult for staff to raise issues about their bullying manager or colleague.

Less moral reasons relate to legal, financial and organisational issues. A good employer with a sound policy should be able to demonstrate a commitment to tackling bullying. This, along with evidence that they have made reasonably practicable attempts to prevent problems, can be used to defend against litigious action. UK bodies such as the Advisory Conciliation and Arbitration Service (ACAS) recommend framing a policy to deal with potential organisational and productivity difficulties caused by bullying (ACAS, 1999).

There are a large number of bullying policies around and they are of variable quality. The most detailed in the UK appear to be found in the public sector. We have chosen to give examples from a handful which include elements of what we believe makes a good policy.

How to begin

The history of tackling a 'new' health and safety problem in the UK can be seen as one of repeated error, denial, ignorance or bewilderment as to how to face a seemingly unsolvable problem.

Policy formation should begin with data collection, both qualitative and quantitative. This can be difficult or result in patchy responses, especially in medium or large organisations, which tend to split into distinct units. Comparing data from across units can be complicated by differing methods of recording and varying perceptions of what is acceptable or unacceptable behaviour. Getting quality information from sometimes close-knit communities can be hard, especially if the units self-police. These are environments in which bullying can thrive. For example, at the Isle of Wight Healthcare Trust (Schubert and Shore, 2000) a series of complaints from patients, relatives and staff about one team was investigated but dealt with in isolation. Staff turnover and disciplinary hearings were high, but only when a serious allegation of harassment and bullying was made was the severity of the problem realised. This led to workplace interventions, which included sessions exploring the Trust's policies on harassment and bullying.

Even if a survey does not produce significant data, it gives employers the chance to show that they have begun to investigate the problem, and built up to producing a policy with resulting preventive action.

Who should draw it up, who should be involved

We recommend that a small working group is set up to draw up the initial draft of the policy. The composition of the group needs to be representative and have a status which shows the importance attached to the problem. Depending on the size of the organisation, there should be a representative of senior management to give it status, a senior member from the personnel/human resources department, representative departmental managers, plus union representatives and, if appropriate, staff from other parts of the organisation – in other words working in partnership. Obviously the group should not be so large as to be unworkable. The group should also consider the use of an 'expert' or facilitator who could provide an alternative perspective and broader overview. However, suitability of such an outsider is important, as there are many poor 'trainers' or 'consultants' around.

It is important that the policy reflects the culture of the organisation. As a result, external consultants should not be left entirely to draw up a policy, as knowledge of the organisation is crucial to the type of policy developed.

It would be surprising if we did not recommend that unions should be involved in the process of drawing up a policy from the outset. This is justified, as UNISON's survey (UNISON, 1997a) showed that 72 per cent of respondents said that if they were bullied, they would report it to their union representative.

It is crucial to its success that the policy is developed jointly for a number of reasons. Involving staff at the outset shows a commitment by employers to stamp out bullying at work; makes a clear statement to staff of the legitimacy of the issue; and implies a level of ownership to all involved. Staff involvement in the development process will also help to ensure the success of the policy.

Policy contents

Statement of commitment

The basic problems, such as the hidden nature of the problem, the reluctance to pursue a case and problems around monitoring will be the same for all organisations, and consequently the contents of most policies should have common themes.

Views differ on what a policy should cover. In the UK, policies come mainly in two sorts: either a basic bullying policy or one that combines bullying with other issues, referred to as a harassment policy, a bullying and harassment policy or a dignity at work policy. These include harassment on the grounds of sex, race, disability or sexual orientation. UNISON nationally recommends a separate bullying policy from other harassment policies, although they should cross-refer as appropriate. Similarly some organisations include bullying and harassment by non-employees such as patients, customers or other members of the public. We believe it is easier to tackle problems when they are made simple and distinct. The boundary between the actions that employers can take against a client/patient harassing a worker and a worker bullying another are different and require different actions – therefore they are easier to deal with separately. Whilst this might make a thicker policy book it means that once you've found the policy, you can deal directly with the problem without being side-tracked.

All polices should begin with a statement of commitment. This enforces the power and legitimacy of the policy (Box 13.1).

Box 13.1: Example of statement of commitment

Worthing Priority Care NHS Trust regards any form of harassment or bullying of staff as totally unacceptable and will respond seriously to any complaint by investigating thoroughly and taking any appropriate action.

(Worthing Priority Care NHS Trust)

As part of the introduction or the statement of commitment some general principles should be included such as:

- a statement that bullying will not be tolerated;
- the recognition that bullying is an organisational issue and can be tackled;
- a statement of the right of all staff to be treated with dignity and respect;
- commitment to promoting a working environment free from bullying;
- a statement that all staff will be made aware of the policy, that they are expected to comply with it and that disciplinary action will be taken against those who do not;
- acknowledgement that bullying affects health and safety, including reference to appropriate legislation;
- reference to related harassment policies.

Definitions

A definition is crucial as it enables all staff regardless of their level or grade to understand what the organisation terms 'workplace bullying'. This is particularly important in large organisations where different departments or units exist. UNISON uses the following definition:

> Workplace bullying can be defined as offensive, intimidating, malicious, insulting, or humiliating behaviour, abuse of power or authority which attempts to undermine an individual or group of employees and which may cause them to suffer stress.

Most policies include a list of examples of bullying behaviour. An example of such a list is shown in Box 13.2.

It is important that the definition focuses on bullying behaviour rather than on bullies. This approach recognises that some people may bully unintentionally. It avoids labelling people as bullies, which can escalate conflict, especially if the perpetrator is not being dismissed.

Bullying can be part of the whole culture of an organisation. To deal with it as an organisational issue, it is important for the policy to focus attention on stopping the behaviour in general. If the policy concentrates on individual acts of bullying, the underlying causes, such as a management culture of bullying, will not be tackled.

Duties of managers

The duties of managers and the consequences of failing to implement or abide by the policy should be spelt out. Line managers should be given responsibility for preventing workplace bullying and for taking appro-

Box 13.2: Examples of bullying behaviour

- severe verbal abuse
- abusive or intimidatory written communication (including via electronic mail)
- intimidating or aggressive behaviour
- the setting of impossible deadlines or intolerable workload burdens
- disparaging comments or remarks, often made in front of others
- constantly changing objectives and goals
- taking credit for others' initiatives and achievements
- constantly changing the remits and responsibilities of others
- isolating or shunning certain individuals, thus limiting consultation on important issues or deliberately excluding them from general or specific activities
- actual physical assault

(Aberdeen City Council)

priate action to eliminate it should it occur. Managers should be trained in recognising bullying, dealing with it and in delivering the policy. This can of course be complicated where it is the line manager who is being accused. Hence policies need to allow for flexible reporting procedures (see Box 13.3 for an example).

Trade union representatives

The role of union representatives in being recipients of complaints and in educating staff about bullying needs spelling out. Giving them equal status to managers in this matter enhances partnership working and

Box 13.3: Example of flexible reporting procedures

Appropriate manager

A number of cases of harassment or bullying are by the immediate line manager. In such cases the individual should contact:

- a manager with whom they already have a rapport
- the line manager's immediate manager
- a peer of the line manager

When making this contact, individuals should make it clear that they are raising an issue under the dignity at work policy.
(Severn NHS Trust)

builds confidence amongst employees. In the UK representatives have a legal right to be consulted over policies and changes, and they can be involved in representing complainants and the alleged perpetrator. Union representatives need equal training and equivalent time off to fulfil their duties.

Contact Officers

The policy will need to recognise that employees may not wish to approach employers or their union in the first instance. They may want independent advice, especially if their own manager is the bully. Alternatively, they might be too embarrassed to raise it with managers or the union, or they may find the prospect of going through formal action intimidating.

To help with this an independent Contact Officer (CO), sometimes known as dignity at work officer, should be designated to deal with complaints of bullying and to offer advice to employees who believe they have been bullied.

The role of the CO is to:

- provide sympathetic assistance to employees with complaints of bullying;
- explain to them the procedures for making complaints;
- establish the main details of the complaint;
- channel the complaint to the appropriate manager for action if the employee wishes to take the matter further;
- discuss cases in complete confidence and not divulge information to any other person without the consent of the bullied worker;
- provide sympathetic assistance to colleagues or employees who they believe are being bullied. The CO will need to decide whether to approach the person being bullied in a sensitive manner depending on the nature of the case;
- provide evidence to an investigation or disciplinary hearing.

The CO is a supplementary role and an employee should not be obliged to refer their complaints to them. Nor should employees who speak with a CO about bullying be obliged to take further action.

COs should receive training in carrying out their role, and details of how to get hold of them should be prominently and regularly advertised. It is important that the role of the CO is clearly defined, especially regarding their involvement in the disciplinary process.

Complaints procedure

Some people are unaware that their behaviour in some circumstances may be construed as bullying. Therefore it is sensible to ensure that these people are given the chance to alter their behaviour. Unfortunately some employees may find this difficult. Consequently we favour a two-pronged approach with an informal and formal stage.

Where possible employees who believe they are being bullied should ask the person to stop the behaviour, making it clear what aspect is offensive or unacceptable. A colleague, friend or union representative can carry out this action if appropriate.

If an employee is unable to adopt this approach or if the bullying is of a serious nature, employees should use the CO service. The CO will provide informal advice and explain possible means of combating the problem. Some employers have mediation services with trained mediators who offer an additional informal approach. Such a process is described in Box 13.4.

Box 13.4: Example of mediation at work

An individual who is feeling harassed may choose to address the issues with their harasser. In order to help employees do this, the Trust have trained mediators available to facilitate this process. The aim of mediation is to resolve the issues informally. If a harassed individual wishes to opt for mediation they should approach an appropriate manager or personnel advisor. Mediation will be co-ordinated through the Personnel Department. A personnel advisor will make an assessment of the circumstances and a mediator will be nominally allocated to the case. The Personnel Advisor will inform the appropriate manager that there has been a request for mediation. Specific details will remain confidential.

- The perceived harasser will be asked by the mediator whether or not they are willing to participate in this process. Continuation with the process will be subject to the harasser agreeing to do so. The alleged harasser may access a dignity at work officer.
- Continuation by the mediator will be subject to the agreement of both individuals. If necessary an alternative mediator will be allocated.
- Both parties will sign up to the process via a pro-forma letter.
- There may be a number of individual meetings before a joint meeting.
- The mediator may advise that a joint meeting would be inappropriate. Alternatively meetings may commence but subsequently break down. Here the appropriate manager will be informed that mediation was not successful.
- Individuals will jointly sign up to any agreement reached.
- Confidentiality – Issues discussed during mediation will remain confidential. An appropriate manager will be aware that mediation is taking place and a copy of the pro-forma letter will be kept on file.

(Severn NHS Trust)

If informal action does not result in the bullying behaviour ceasing, if an employee wishes to make a formal complaint, or if it is considered that the behaviour might constitute a disciplinary offence, the formal complaints procedure should be initiated. Throughout the procedure the complainant and the person complained about should have the right to be accompanied by a trade union representative, a friend or colleague, as appropriate. The complainant should also have the right to continue to seek the advice of the CO.

A formal procedure has to involve the employee or representative reporting the alleged act to an appropriate manager, making it clear that it is a formal complaint and giving details (Box 13.5).

It is important that complaints are investigated in a timely manner. Consequently the investigating officer should give an estimate of the timescale likely to complete the investigation. Explanations for extending the length of the investigation should be given to both parties. Both parties need to know the results as early as possible and timescales will stop long drawn-out investigations, which can adversely affect staff. It also means that investigators are prompted to be quick and not put off difficult hearings.

Confidentiality is a difficult balance to achieve but must be maintained so far as is consistent with progressing the investigation. An investigation is just that: it seeks to discover the truth in as open manner as possible whilst restricting allegations which could linger. Nonetheless, it is to be expected that gossip and rumour will happen, and after the completion of the procedure it may be necessary to have wider discussions with other members of staff to ensure that all are aware of the outcome and reasons behind it.

A manager senior to those involved should be appointed to run the investigation. Some policies allow for the involvement of the CO in the

Box 13.5: Examples of information to be provided when making a formal complaint

It should include as much of the following as possible:

- clear specific allegations against the named person or people;
- where possible, dates, times, and witnesses to incidents with direct quotes;
- factual description of events;
- indication of how each incident made the complainant feel;
- documentary evidence;
- details of any action the complainant or others have already taken.

(City of Salford Council)

investigation process. This has advantages and disadvantages. The advantages are that the complainant is likely to feel confident in the process and that the CO may have useful input and experience to give to the process. The downside may be that their role as advocate for the complainant affects the independent nature of the hearing and might be seen to be unfair to the person against whom the complaint was made.

Investigations must be seen to be fair and impartial, must involve direct interviews with all persons involved and with witnesses as appropriate. Investigators will normally need advice from their personnel departments, who will provide external advice, ensure procedures are kept to and that legislation is respected. This is particularly important when bullying is linked to other forms of harassment (racial, sexual, etc.) where specific legislation could be encountered.

After the investigation has been concluded a full report should be given to the complainant and the accused. It may be appropriate for the CO to be involved in this process, briefed by the investigator if necessary.

If the complaint is upheld, prompt action designed to stop the bullying should be taken, including relocation if appropriate. Unfortunately, in the past, for some organisations this automatically meant relocating the complainant, usually because they were the junior staff member. We are clear that if relocation is necessary then it should normally be the bully who is relocated. This gives them a chance to amend their behaviour in a different environment as their authority and standing may be damaged if they are left in their current position, as might be the morale of other staff involved. However in some cases the complainant themselves may wish to be relocated and this ought to be considered sympathetically – it may be possible to facilitate both moving in larger organisations. A badly handled relocation can send out the wrong message to employees and undermine the usefulness and belief in the policy.

In the UK most large and medium-sized organisations have a disciplinary (and grievance) procedure. ACAS issue a Code of Practice on disciplinary procedures (ACAS, 2000) which is admissible as evidence in the employment courts. The aim of these procedures is to put into writing the rules and processes that define management's responsibility for maintaining discipline and the setting of standards, conduct and capability of performance within an organisation. They may be invoked when an employee is alleged to have broken these standards. For employers, a written policy that is rigorously followed allows them to show they have acted fairly and reasonably, and maintain a defence against an unfair dismissal claim in the courts.

Where disciplinary action is considered, such action should be considered strictly in accordance with the disciplinary procedure. Disciplinary procedures should have time limits attached to the stages of action.

The policy should include a statement that complainants will be protected from intimidation, victimisation or discrimination for filing a complaint or assisting in an investigation. Any attempt to interfere in the process should result in disciplinary action. In the Severn NHS Trust policy the personnel department can form a witness support group.

The complainant and the alleged bully should have the right to appeal. Some policies restrict the right to appeal against the actual decision itself but allow for procedural challenges; for instance that vital witnesses were not interviewed. A time limit in which to appeal should be set. Complaints need to be in writing and should set out exactly what is being complained about.

UNISON is clear that there should be proper disciplinary procedures, which allow for a fair hearing for both those who claim to be bullied but also those against whom the claim is made. This includes unproven allegations and malicious complaints borne out of revenge or to cover incompetence (Box 13.6).

Training and information

All employees should receive basic training on bullying. This should include causes and effects, details of the policy, how to report incidents, how to get support, and how to assist any colleagues being bullied. Some employers include this training as part of a package, which also covers other forms of harassment and stress. Trade unions and the CO should be invited to speak, as should employees who have previous experience of being bullied, if they feel confident enough to do so. Training packages often miss some staff, such as night workers, shift workers, new and agency staff. Employers should also seek to involve contractors and their staff or others who might spend time on the premises. There might also be value in including compliance with the policy as part of any contract with subcontractors. Staff will often have to work alongside each other, and a complementary policy could well reduce conflict.

Box 13.6: Example of procedures for dealing with malicious complaints

Where an allegation is not upheld, it does not mean that the complaint was malicious. The complainant may still feel that they have been subjected to harassment or bullying, but the investigating officers have not been able to find the necessary evidence to substantiate the allegations.

Having conducted the investigation, the officers may conclude that the complaint was malicious. That is, that the complaint was entirely false and made with the deliberate intention of having action taken against another employee. In these circumstances the complainant should be dealt with under the disciplinary procedure.

(City of Salford Council)

Information should be provided on a number of levels. As well as publicising the policy – shorter leaflets, posters and information via the electronic mail need to be available in accessible places and accessible forms. The Severn Trust provide their staff with a leaflet 'Dignity at work: Are you harassed or bullied at work? Don't suffer in silence.' This has four short sections: a brief description of the Trust's policy, where and who to go to if the person has a problem, what will happen if they do and a list of contact names, addresses and phone numbers.

Support for bullied staff

Policies should state that bullying could affect job performance and lead to health problems. If this occurs then the complainant can self-refer or be encouraged to seek help from their occupational health department. There should be no discrimination against staff suffering ill health caused by bullying, and systems of rehabilitation should be used if the effects involve periods of sickness absence.

Some policies offer counselling to employees. If referred to in the policy then it should be made clear who can refer the employee for counselling and whether assessment by occupational health staff is appropriate before referral.

Support and counselling should also be made available to alleged bullies. This is particularly useful in rehabilitation if they are unaware that their behaviour has been unacceptable, and they wish to change.

Monitoring

This area is a clear weakness in all the policies that we have seen. Most make vague references to the need to monitor the effectiveness of the policy but give no indication as to how this will occur.

Monitoring is necessary to assess the effectiveness and intent of the policy. It shows continued commitment, demonstrates to staff that bullying is a legitimate issue and encourages any bullied worker to come forward. It needs to be carried out on a long-term basis, i.e. annually and continuously and after any investigation.

In the UK National Health Service (NHS) a partnership approach has produced a staff involvement self-assessment tool (Department of Health, 2000). One recommendation is that local employers run an annual staff attitude survey. An agreement at a local NHS Trust means that ten questions are set by the unions (Worthing Priority Care Trust, 1997). Aware of the lack of monitoring data, the unions chose to use one question to address staff awareness of the bullying policy.

The City of Salford Council policy includes a promise to publicise the number of complaints that have been dealt with and at what stage a resolution has been reached. These and the policy are subject to an annual review.

Other methods of monitoring used on an ad hoc basis by employers include analysing the results of staff appraisal systems, anonymous feedback from counsellors or whistleblowers, focus groups, exit interviews and stress-related sickness absence records. Such monitoring exercises need to be regularised and written into policies if they are to make long-term change.

Conclusion

Some employers and local union representatives have been worried about what they see as a failure of their policies, as reports of incidents appear to be on the increase. We would contend that when exposing a hidden problem there is always an increase in activity, and that this does not signal a failure.

In the UK bullying policies are still in their infancy. The first are now being revised to take in good practice from other organisations and lessons learned locally. Many policies we have seen are good on process, involvement and a desire to end the practice. However, there is much to be done to ensure that monitoring and evaluation guarantee those policies will ultimately deliver their aim.

Bibliography

Aberdeen City Council (1997) *Bullying and harassment at work policy*. Aberdeen: Aberdeen City Council.

Advisory Conciliation and Arbitration Service (ACAS) (1999) *Bullying and harassment, at work, a guide for managers and employers*. London: ACAS.

ACAS (2000) *Code of practice on disciplinary and grievance procedures*. London: ACAS.

City of Salford (2001) *Dignity at work, a policy to address harassment or bullying behaviour in the workplace*. Salford: City of Salford

Department of Health, National Health Service Executive (2000) *Working together: Staff involvement self-assessment tool*. London: Department of Health.

Schubert, D. and Shore, J. (2000) *Health Service Journal, 4 May*. London: Health Service Journal.

Severn NHS Trust (1999a) *Dignity at work policy*. Cheltenham: Severn NHS Trust.

—— (1999b) *Dignity at work policy leaflet, your guide to tackling workplace harassment and bullying*. Cheltenham: Severn NHS Trust.

UNISON (1997a) *Bullying at Work, Bullying survey report*. London: UNISON.

—— (1997b) *Bullying at work: Guidelines for UNISON branches, stewards and safety representatives*. London: UNISON.

Worthing Priority Care NHS Trust (1997) *Harassment at Work policy*. Worthing: Worthing Priority Care NHS Trust.

14 Investigating complaints of bullying

Vicki Merchant and Helge Hoel

Introduction

With growing evidence that workplace bullying represents a serious problem in many organisations (Hoel and Cooper, 2000; UNISON, 1997), the issue has rapidly moved upwards on many organisations' agendas. Faced with internal demands for action and the potential threat of litigation, policies and procedures for complaints and investigation are currently being developed, and in many cases have been implemented (Rayner *et al.*, 2002).

In response to a bullying complaint organisations have a choice between using internal resources and resorting to external expertise. The decision to opt for an external investigator may be particularly appropriate where the organisation has little or no experience of investigating bullying or harassment complaints, where the alleged perpetrator holds a senior management position, or otherwise when the authority of an external person may be helpful during the investigation as well as in the aftermath of an investigation. The fact that an external specialist is able to dedicate time exclusively to an investigation may also make the process more efficient and effective. Whilst this chapter explores the investigation process from the perspective of the external investigator, many of the issues discussed below would also apply to an internal investigator.

The lead writer of this chapter has considerable experience in investigating bullying and harassment complaints in the UK. Her experience includes being examined and cross-examined in court both as an investigator and an expert witness on workplace bullying in the UK and abroad.

The first part of the chapter describes the investigation process, highlighting the importance of making strategic decisions up front. Key issues about administering the process and seeking evidence are also examined. The second part discusses some contentious issues, such as how to ensure objectivity and fairness of the process, and guidance on protecting the investigator.

The investigation process

The investigator's role is to establish the facts and to come to a decision about whether a complaint is upheld or not upheld on the basis of those facts. The key element is fairness and this is of the essence.

The first step is to establish that the behaviour complained about would constitute bullying under the organisation's Policy, if the complaint were to be upheld. Any investigation must adhere strictly to the Policy and Procedures current at the time the behaviour is alleged to have taken place, if grounds for appeal are to be avoided, and the alleged behaviour must come within the definition used at the time.

The role of the investigator is then agreed with the employer, including the allocation of administrative tasks, such as writing to the complainant and alleged harasser and calling witnesses. There are advantages in leaving such tasks to the investigator so that s/he maintains control of the process, while the employer makes the relevant information available and arranges accommodation and backup. Decisions must also be made about who will receive the Investigation Report, and whether or not recommendations are to be included. The Policy and Procedures should state who is entitled to see the Report. Both the investigator and the employer must be clear and in agreement about their expectations and respective roles. Since an external investigator is responsible for the conduct of the investigation, including the content of the Report (but not for what happens subsequently) s/he must be satisfied that the conditions laid down by the employer support a fair and thorough investigation, and are consistent with the Policy and Procedures and current legislation. Professional indemnity insurance is worth considering!

The investigator then interviews the complainant and any witnesses named by the complainant, and that process is repeated for the alleged perpetrator. Other witnesses may be identified as well as anyone else with a potentially useful contribution to make. The investigator needs to be flexible, open-minded and proactive, and may need to use methods of investigation other than interviews.

Having gathered as much evidence as is reasonable under the circumstances, the investigator then analyses it in order to come to a conclusion based on the balance of probabilities. S/he will come to a reasonable belief and set out the reasons for the decision clearly in the Investigation Report.

The investigator may subsequently be required to present the Report to an internal Disciplinary Hearing and answer questions on it, or indeed to be exam ined in any subsequent legal action. S/he does not take any part in the implementation of the Report, or in disciplinary matters.

Establishing the ground rules

By agreeing a set of ground rules with the employer at the outset, many potential pitfalls may be avoided. Some of the issues to be considered are listed below.

Calling witnesses

It is expected that all witnesses named by both sides must be interviewed unless there are strong reasons not to do so. For example, if somebody cites forty witnesses, that may be excessive, bearing in mind that only a 'reasonable' and not an endless investigation is required in order to establish the facts. The list can be prioritised, with reasons for seeing each witness. There may be three categories of witness: essential (a direct witness who heard or saw something relevant); useful (e.g. having hearsay evidence, having been told something relevant or seen somebody in distress); and potentially relevant (e.g. having evidence of similar behaviour by the alleged perpetrator).

Timescales

It is difficult to conclude an investigation within an ideal time-scale of not more than about three weeks. Witnesses may be away or reluctant to be interviewed. Writing the Report can take much longer than envisaged. It is, therefore, essential not to promise that the investigation will be completed by a particular date, because the process can be unpredictable.

Access to the investigator

People often become very anxious when the investigation takes longer than they expect. It may drag on for weeks or even months, not because of any dilatory behaviour but simply because there is so much to cover. A complainant or alleged perpetrator may want to contact the investigator to ask what is happening. They may ask if they can call additional witnesses, submit further statements or new evidence, or be interviewed again. Allowing controlled access during the process may allay undue anxiety and also yield important new evidence.

The Investigation Report

Questions to be addressed in advance include to whom the Report will be addressed, who else will read it, the format and what is to be included. The recipient of the Report will decide what to do with it. It is all too easy to give the 'correct' information, in accordance with the Policy and Procedures, and then find that the recipient of the Report decides to do something different. Thus, the investigator will be cautious about giving information not strictly relevant to the case under investigation.

Suspension pending the outcome of an investigation

An alleged perpetrator (not the complainant) may be suspended under certain conditions, such as an allegation of threatened or actual violence (regardless of whether there appears to be a case to answer), where a complainant is unwilling to carry on working closely with an alleged perpetrator, or if there is any likelihood that witnesses may be coached or pressurised.

Administration of the investigation

Investigations are very stressful for everyone involved, however well administered, and if the process is poorly or inefficiently administered, interviewees may become obstructive.

Efficient and flexible administrative backup is needed, and a comfortable room with a round table, an additional table for papers, access to a telephone, refreshments and nearby toilet facilities. An anteroom is useful so that if people arrive early, or the investigator runs late, they are not all waiting together with the opportunity to discuss the case. An administrator should be accessible to rearrange the schedule if it becomes disrupted, or if the investigator needs the telephone numbers of interviewees. Such seemingly trivial things are in fact essential to the speed and smooth running of the process, as well as the ease and well-being of those involved.

Letters to interviewees

A fault in a letter could invalidate the investigation altogether. In particular, the alleged perpetrator must be given clearly worded allegations, with as much detail as possible. Time-scales must be stated and representation or support encouraged. Careful consideration must be given to how interviewees are called for interview, how the letters are delivered and how they are likely to be received, in order to safeguard confidentiality and avoid embarrassment.

Letters to the alleged perpetrator and complainant should be essentially the same, and each should be asked if they wish to call any witnesses and warned not to approach witnesses themselves. It is for the investigator or the administrator of the investigation to call witnesses. They should also be instructed not to discuss the investigation except with their representative or support person, whether in person, in writing or by any other means, or through a third party. Any breach of this rule may be considered a disciplinary offence. Similarly, any kind of intimidation will damage the investigation and may constitute a disciplinary offence.

Scheduling

The interview schedule needs to be realistic, allowing sufficient time for each interview, bearing in mind what each witness is likely to contribute and allowing for the unexpected. It is useful to know why witnesses have been called. If neither the investigator nor the witness knows this, it is difficult to focus questions. Interviews should normally not last more than two hours, including a break in the middle.

Recording interviews

Note taking is tiring and may lead to cutting corners. It can also lead to repetitive strain injury. It is difficult for an investigator to concentrate on what interviewees are saying and maintain eye contact when thinking about the notes. Various factors affect what is selected or omitted from the record. Bias may creep in. Emotional outbursts, long pauses, anger and tone of voice can be revealing and are difficult to record by hand. Handwritten notes also have to be agreed and signed as a correct record by the interviewee. This is time consuming and interviewees do not always agree with the record, or they may wish to change it when they realise what they have said. Any changes must be agreed and initialled.

Tape recordings, expertly transcribed for the Report, provide the only complete and accurate record. They cannot, effectively, be tampered with as written records can. The main disadvantage of tape-recorded interviews is that every word is registered and there is often repetition. It is tempting to edit out mannerisms and unfinished sentences, but transcripts should not be edited except to preserve confidentiality, if necessary.

Occasionally someone objects to being recorded on tape. Much depends on the confidence and the authority of the investigator. If the investigator explains that the interview will be tape-recorded and typed up for the Report, which may be seen by both sides, then it is clear that everything is on the record. Just as when someone refuses to speak, investigators are entitled to draw their own conclusions and inferences from any objections to tape-recording. Interviewees' fears may be allayed if they are permitted to make their own concurrent tape-recording.

Interviewee representation

The right to representation is fundamental and all interviewees can have somebody with them for support and advice, whatever their role in the investigation. This may be a friend, colleague or trade union representative, or lawyer, depending on the employer's Policy and Procedures. Being a witness can be stressful. People do not like having to take sides and possibly give evidence against a respected colleague, current manager or friend. They may fear victimisation. Someone should also be on call to support people who are unaccompanied and may become upset.

Handling difficult situations

Interviewing potentially stressed and distressed people, with all the emotions involved, requires timely breaks. The interviewer can become upset, for example, by a particularly distressing account of bullying. The interview can be postponed at any point, or reconvened after a break. Refreshments can be placed behind the investigator, providing the opportunity to turn away for a moment and give some space and privacy.

Gathering evidence

Conducting interviews

Interviewing the complainant first enables the interviewer to build as full a picture as possible of events from that perspective. Witnesses cited by the complainant can then be seen to consolidate that picture. This may reveal the need to call additional witnesses. Alternatively, it may be appropriate to move straight on to the alleged perpetrator. The most careful schedule can be disrupted by sickness, holidays, further complaints or the need to re-interview someone.

Interviewees should be offered the opportunity to come back if they wish. They may be afraid to talk in the first instance and may speak more openly once rapport and confidence in the process is established.

The most revealing interviews are those where the investigator only speaks when necessary, using minimal verbal prompts, giving time to the interviewee. The less the interviewer says the fewer mistakes are made. A slip of the tongue can be very damaging and cannot easily be undone. It can lead to allegations of unfairness, a breach of confidentiality or may even invalidate an investigation.

The purpose of the interview is to establish what has been happening. Failure to be proactive can lead to an injustice. Each piece of information can be regarded as a gift and a good investigator will accept it gracefully and follow it up.

Investigators are often asked if anybody else has been bullied by the same person. For the investigator to reveal such information would be a breach of confidentiality, but it may be useful to establish whether others are affected by exploring the interviewee's reasons for raising the issue.

If somebody, in the course of an interview, asks if they can speak in confidence or off the record, the answer is 'No.' There is no confidentiality because investigating is not a secret process. People are entitled to know everything that has been written or said about them and this may be protected by legislation. If the interviewee then refuses to speak, the investigator is entitled to draw conclusions appropriately.

Witnesses often want to talk when the tape-recorder is switched off for a break or after the interview has been formally concluded. It is tempting to allow people to speak off the record because they are more willing to reveal the truth. However, unrecorded information may be difficult or impossible to record subsequently and discussions 'off the record' are not acceptable.

Other methods of gathering evidence

Information obtained from interviews may need to be cross-referenced and checked, particularly where there are inconsistencies. This may involve visiting the site of an alleged incident and drawing a plan. In this way potentially useful witnesses not already cited can be identified.

Whilst the investigator does not normally gather forensic evidence, it is important to do everything possible to establish the facts, and to call in experts if necessary, perhaps to analyse handwriting or trace the origin of e-mails or telephone calls made or received, or to check sickness or holiday absence. The investigator should be proactive within the confines of a reasonable investigation and with due regard to the privacy of individuals.

Reaching a decision

The findings in an internal investigation are based on the investigator's reasonable belief and on the balance of probabilities. The belief is reasoned and the reasons must be transparent in the final Report. Investigators often feel they know what happened, but actually to present the evidence is another matter. When they look at the basis of their 'reasonable belief' they may find that it evaporates when they have to write it down and convince others.

It is preferable to use 'upheld' or 'not upheld' rather than 'proven' or 'not proven'. For a complaint to be upheld, the investigator must demonstrate a reasonable belief that the alleged behaviour is more likely to have happened than not. It is not necessary to show that it did actually happen if there is no direct witness or documentary or other evidence such as a recorded message from a telephone answer machine.

Bullying often takes place in private and without witnesses. Sometimes witnesses are not willing to come forward or will not tell the truth, so establishing the facts can be problematic. All that is needed is a reasonable belief, based on available evidence, including hearsay and circumstantial evidence.

A complaint may be upheld in whole or in part. Some of what is complained about may have happened but some may not. When a complaint is not upheld it is often because there is not sufficient reliable, independent evidence. Where there is simply one word against another and a reasoned judgement cannot be made in the absence of independent evidence of any kind, the complaint cannot be upheld. This does not mean that the allegations are unfounded or malicious, but simply that the investigator does not know what happened and cannot reach a decision. This must be acknowledged in the Report.

In other cases there may be independent evidence that is unreliable. It is always important to check the evidence and not simply to accept what is said at face value. Once it becomes apparent that the complainant, alleged perpetrator or a witness is not telling the truth, then the investigator is entitled not to believe that person in anything they say.

Writing the Report and making recommendations

All information and documents gathered by an investigator are included in the Report. It is useful to keep a diary from the outset and include that too. Detailed information on the process, including dates, may be needed in the event of an appeal or legal action.

The investigation has been commissioned by, and belongs to, the employer to whom it will ultimately be submitted, including tape-recordings and original, signed statements or interview notes. The employer will decide who should have access to it and how long to keep it.

When there is more than one complainant or more than one alleged perpetrator, information will emerge that some named individuals are not entitled to see in the final Report, for reasons of confidentiality and data protection. An interview with a witness may refer to a number of individuals and the confidentiality of each must be respected. Therefore separate Reports are needed dealing with each alleged perpetrator. For example, one person may complain against five others, or three people may complain against one or more alleged perpetrators. This involves painstaking and accurate editing to remove everything that is not relevant to each person who is going to see the Report.

The investigator is usually, but not always, required to make recommendations. Recommendations should not only be specific to the individuals concerned, but also general, perhaps aimed at cultural change, such as amendments to policies and their implementation, or management practice and training. Bullying does not usually occur where there are good management practices in place. Such behaviour flourishes in certain environments and the Report should focus not only on the individuals directly involved but also on the workplace culture and policies, processes and procedures.

Discussion

Above we have explored a number of issues central to a successful outcome of a bullying investigation. However, there are other issues that need to be considered.

Objectivity and fairness

Fairness is the key element in any investigation. The investigation will be judged on fairness or unfairness, both at an internal appeal and in court. The main points are: encouraging representation; interviewing all cited witnesses, as far as is possible and reasonable; allowing all sides sufficient opportunity to put their cases; and treating all sides equally.

It is all too easy to form a view straight away, but it is vital for the investigator to keep an open mind and maintain objectivity right up to the end, and to be aware of the influence of personal prejudice, which we probably all have to some degree. Like any human being, the investigator will warm to some people more than others, and so must strive to redress that and listen to everybody, including those who may have behaved in an unacceptable manner or failed to endear themselves to the investigator, for whatever reason. If the investigator allows prejudice or moral judgements to creep in, bias will inevitably result.

Investigators are likely to be committed to eliminating workplace bullying, otherwise they probably would not choose to do the job, but this could bias them towards the complainant. Some internal policies and procedures are biased towards the complainant, and the investigation must be strictly in accordance with the Policy and Procedures, good or bad. However, every element of the investigation, complainant, alleged perpetrator, witnesses, officials, policies and organisational process and procedures must be fully considered. Once the investigator starts thinking that one person has a better case than another, or that one person has priority, then the fairness of the investigation is at risk.

It is difficult for an investigator not to be emotionally affected by hearing about severe bullying that has undermined somebody, perhaps causing great distress and harm to mental and physical health. While acknowledging his or her own feelings privately and seeking appropriate support, and acting sensitively and humanely, the investigator needs to avoid showing his or her feelings. Whilst empathy is appropriate and necessary to the sensitivity of an investigation, sympathy may imply, or be interpreted, as taking sides.

Malicious complaints

In order to establish that a complaint is malicious a motive must be shown. The investigator will want to know why a person has complained, why s/he would fabricate a story, and what advantage that would give. People are unlikely to put themselves through the stress of an investigation for no reason. However, malicious complaints are occasionally made and the investigator must be alert to the possibility, especially if any allegation of malice is made.

When discoveries need to be divulged

The investigator may come across unexpected things that make the whole case take a totally different direction from that envisaged at the outset. A complainant may have presented only a fraction of the complaints or the case may be a symptom of wider or deeper problems, which may even lead on to a criminal investigation. As soon as it becomes apparent that a criminal offence may have been committed, this should be reported to the employer. The investigation should then proceed normally, unless the alleged perpetrator has been taken into custody or there is other compelling reason to suspend or terminate it.

It is therefore inappropriate to promise confidentiality. Confidentiality in an investigation means that the investigation is divulged on a 'need to know' basis only. In addition to the employer and various internal officials, outside bodies, such as professional associations and the police, may come within that category.

Caring for the investigator

Investigations are often very stressful for the investigator as well as others personally or directly involved. Events described may be emotionally disturbing, and the investigators need support and someone to share their thoughts and feelings with, in confidence. Also, the need to reach a fair decision can be burdensome, and intellectually and emotionally demanding. It is not unusual to spend sleepless nights worrying about getting it right, and the opportunity to discuss the case in confidence with an appropriate person is invaluable.

Dealing with undue pressure

The investigator may be pressured, for example by an employer or complainant, to take a particular side or support a particular viewpoint. S/he may be asked how long the investigation will take, who will see the Report, whether copies will be available and what will happen to an alleged perpetrator. It is prudent not to say anything at the time but to consider carefully and consult the employer about how much information should be divulged. It should be made clear that undue pressure is not acceptable and may be noted in the investigation Report.

Employers may also exert pressure, perhaps considering the complainant a nuisance and that discipline should not be recommended against the alleged perpetrator, particularly if s/he is a senior manager. It is important not to be persuaded to discuss the possible outcome before the investigation is completed and the Report submitted. It is helpful to raise such issues in principle with the employer when the ground rules are agreed at the outset. An apprehensive employer, perhaps under financial pressure, may even call off an investigation when it gets 'too hot', particularly if senior employees have come under scrutiny and the case is unlikely to lead to the employer's preferred outcome. This can backfire on the employer if the investigator is called to give evidence in any subsequent legal action.

In order to cope with 'lobbying' and avoid becoming overinvolved, it helps to leave a telephone answer machine on all the time, so that the investigator can decide whether to take a call or not, when to return a call and avoid being caught off guard. However lines of communication may be vital because of the high level of anxiety involved in bullying cases. Anything that can be done to alleviate that anxiety is likely to be helpful. The mere fact that the investigator returned a call, even if unable to say anything about the investigation, helps to establish trust and confidence.

Such contact does not interfere with the investigation, so long as the investigator is clear about what information can be shared.

Conclusion

By means of a properly conducted investigation, utilising the expertise of an external investigator with knowledge and experience of the process as well as the issues under investigation, complex and difficult bullying cases may efficiently be brought to a just conclusion. It must be remembered that the conclusion of any investigation can only be as good as the available evidence.

Mistakes are sometimes made during investigations, which can lead to unfairness and injustice. Highly skilled and trained lawyers and judges can get it wrong, which is why there are intricate appeal systems. All that can be expected of investigators is to do the best that they can with the information available to them and to act in good faith.

Bibliography

Hoel, H. and Cooper, C. L. (2000) *Destructive conflicts and bullying at work*. Unpublished report, UMIST.

Rayner, C., Hoel, H. and Cooper, C. L. (2002) *Workplace bullying: What we know, who is to blame, and what can we do?*. London: Taylor and Francis.

UNISON (1997) *UNISON members' experience of bullying at work*. London: UNISON.

15 Counselling and rehabilitating employees involved with bullying

Noreen Tehrani

Introduction

This chapter is written from the perspective of a practising occupational, health and counselling psychologist working as an independent consultant in a number of large organisations. My initial experience in developing organisational and employee interventions to address workplace bullying was in the British Post Office, where, as head of the employee support service I had the responsibility for providing psychosocial support for over 200,000 employees. A number of initiatives were developed in the Post Office including a peer support programme (Rains, 2001), employee assessment and support (Tehrani, 1996), and the establishment of a code of acceptable behaviour. Since leaving the Post Office I have been involved in helping organisations deal with workplace bullying through counselling bullies and the victims of bullying as well as working within the organisation to bring about the reconciliation of interpersonal conflict and bullying.

This chapter looks at the ways in which counsellors can be used to address some of the problems faced by employees affected by bullying. Where appropriate, illustrative case studies are provided. The first part of the chapter looks at the stages of helping victims of bullying, beginning with the psychological assessments, then a range of counselling interventions and finally a description of the importance of developing appropriate rehabilitation programmes. The second part of the chapter examines some of the issues surrounding the use of counselling as an intervention for addressing bullying at work. The difficulty in recognising who is the bully and who is the victim is also discussed. The chapter closes with a discussion of the main issues facing counsellors working to reduce the impact of bullying in the workplace.

Assessing the impact of bullying on psychosocial well-being

The psychological impact of bullying has been shown to be damaging to the physical and mental health (Hoel *et al.*, 1999), frequently resulting in extended periods of absence from work (Lipsedge, 2000). Employees seeking counselling following exposure to bullying at work present a number of common symptoms including high levels of anxiety and depression, physical complaints such as an upset stomach, panic attacks and difficulty in sleeping. In addition they may be experiencing difficulties in handling normal social relationships. The magnitude of the psychological and physiological symptoms found in victims of bullying give an indication of the destructive outcome that exposure to interpersonal conflict and bullying may have on health and well-being. In recent years a number of researchers have found that people exposed to bullying can develop symptoms of post-traumatic stress (Leymann and Gustafsson, 1996; Scott and Stradling, 1992). These symptoms include:

- re-experiencing the bullying incidents in flashbacks and nightmares;
- physiological and psychological arousal; and
- avoidance of people, places or things associated with the bullying.

Before commencing counselling it is important that the counsellor undertakes a full psychosocial assessment of the employee. The assessment not only gives the counsellor an indication of the nature and magnitude of the distress but also provides a baseline from which the effectiveness of the intervention can be measured. The assessment process outlined in Box 15.1 takes around two hours to complete. At the end of the assessment the employee should be provided with simple education and information. The information should include a description of the normal psychobiological responses to bullying and guidance on how to reduce the impact of their

Box 15.1: Outline of an assessment tool for assessing a victim of bullying

- A brief description of the bullying incidents
- The employee's immediate emotional and physiological reactions to the bullying incidents
- A brief physical, psychological and social history
- Descriptions of any re-experience symptoms
- Descriptions of any avoidance or numbing symptoms
- Descriptions of any hyperarousal symptoms
- Any changes in social well-being
- Any changes in work performance
- A description of any past experiences of bullying or harassment
- General Health Questionnaire 28 (Goldberg, 1985)
- Peritraumatic Dissociative Experiences (Marmer *et al.*, 1997)
- Extended Impact of Events Scale (Tehrani *et al.*, 2001)
- Lifestyle Questionnaire

symptoms. In addition there should be lifestyle guidance such as the need to take exercise, details of a range of relaxation techniques and information on the need to eat a balanced diet.

Interventions

Although many counsellors use a single theoretical counselling model in their counselling, when dealing with bullying it is helpful to be able to integrate a number of counselling models and interventions. This approach requires the counsellor to be experienced in the use of each of the counselling models and to have the skills to be able to identify which intervention would be most effective in a specific situation. Discussing with the employee the range of interventions that are available provides an opportunity for the employee to take an active part in the process, and is an acknowledgement of the need for them to take a full part in their recovery. The first counselling session, which takes place after the assessment, should begin with a discussion of the intervention plan. This plan takes account of the pattern of symptoms identified in the assessment and a description of the role of education in the sessions. In most cases the first intervention will be a psychological debriefing of the bullying incidents. The reason for using debriefing as the primary intervention is its usefulness in drawing out the main issues that will be addressed during the counselling.

Debriefing

There are a number of models of psychological debriefing (Tehrani and Westlake, 1994). Where debriefing is being used for victims of bullying it is important to identify a particular incident or a number of incidents that are related to the symptoms identified in the assessment. As with post-trauma debriefing it is important to begin the debriefing at a point in time before the bullying has occurred. Starting at a time when things were normal provides a description of the employee's working life, relationships and behaviours without the presence of bullying. The employee is then helped to give a factual account of the bullying experience with a particular emphasis on identifying the full range of sensory experiences that may have become associated with the bullying behaviour. This stage is important as sensory impressions such as a particular smell, sound or image can develop the power to trigger re-experience and arousal symptoms. During the second telling of the story the employee is encouraged to recall any thoughts that occurred at the time of the bullying experience. The counsellor will become aware of any irrational thinking present at the time of the bullying. The third telling of the story examines the emotional responses to the facts and thoughts that have been identified in the earlier telling of the story. By the end of the debriefing the counsellor should have identified the sensory triggers that are likely to cause re-experience symptoms together with any irrational thinking that may delay recovery. Irrational thinking is most effectively handled using the cognitive behavioural techniques described later in the chapter.

Narrative therapy

Narrative therapy builds upon psychological debriefing. It enables people to tell the stories of their lives so that their experiences are understood and validated. Narrative therapy is a relatively new therapy that has emerged from the work done in family therapy (Bor *et al.*, 1996). Understanding the bullying story constructed in the debriefing can help the employee to become aware of their tendencies to create self-destructive stories about their experience rather than a story that helps them grow and accept themselves. The experience of being bullied needs to be accommodated or assimilated within an existing network of stories. Narrative therapy provides the employee with the possibility of looking at all their life stories. Within the narrative network some stories may have become more dominant than others. Some of the stories will be helpful while others may have a negative influence. Where the life stories are not helpful narrative therapy provides the person with an opportunity to 'deconstruct' the story and create new stories. It is this 'reconstructing' of adaptive stories that allows the employee to learn something new about themselves and to understand the real meaning of the bullying experience (White and Epston, 1990). The use of narrative therapy is illustrated in the case presented in Box 15.2.

Cognitive behavioural therapy

Many of the problems faced by victims of bullying are caused by the way people think about themselves and what has happened to them. Cognitive behavioural therapy is based on the idea that the more rational and realistic a person's thoughts are, the less troubled they will be by the things that happen to them. During debriefing, people are often found to be irrational in their thinking. Some common irrational thoughts of victims of bullying are:

- I must have done something wrong for this to happen to me.
- Someone should have done something to stop it happening.
- Everyone hates me.
- I will never get over this.

These thoughts can be challenged using cognitive behavioural techniques. The cognitive behavioural process encourages the employee to regard each of the irrational beliefs as a hypothesis rather than a fact. This

Box 15.2: A case of narrative therapy

A female secretary worked for a female manager. The manager had developed a reputation for being extremely demanding and difficult to please. None of the manager's previous secretaries had stayed with her for more than a few months. The secretary found the relationship very testing but for most of the time she was able to handle the unpleasant situations the manager created. One day, the manager was having a particularly difficult day and began screaming at the secretary blaming her for something that she had not done. The secretary tried to explain but when the manager continued screaming at her, the secretary left the office in a distressed state.

The narrative approach identified that the secretary had lived her life based on a story she had created as a child in response to the demands of a critical father and unassertive mother. The main themes being:

- You are good at dealing with difficult people.
- You must always be considerate towards others.
- Your own needs do not matter.
- It is wrong to make a fuss.

The story had been the reason why she never felt able to challenge her manager's unreasonable behaviour. She used the story to explain why she always met her parent's constant demands on her. Her decision to work for her current manager was based on the fact that she was very good at coping with difficult managers and enjoying the challenges they presented. Yet now she had found herself in a situation where the story was not working. The secretary's initial response was to try to return to the dominant life story. She wanted to go back to work. She would apologise for the fuss she had created and then try to sort things out with the manager.

Further investigation revealed that the secretary had a number of alternative stories. One of these stories portrayed her as effective in dealing with difficult situations, illustrated her skills as a teacher and highlighted her ability to speak up when her values and beliefs were challenged. The alternative story provided a new way to deal with the manager. Following the adoption of the new story the secretary was able to become assertive in establishing agreed standards of behaviour with her manager, whom she also helped to develop skills in dealing with frustration.

allows the employee to begin to test their irrational beliefs against reality. Irrational beliefs lead to poorly adapted behaviour. Where an employee has the irrational belief that everyone hates him, his behaviour may become aggressive to other people; this in turn can bring about a situation in which people are less friendly, reinforcing the employee's original view. An important benefit of using cognitive behavioural therapy is that it encourages the employee to view the improvements in their well-being resulting from their increased knowledge and skill rather than the counsellor's endeavours.

Eye Movement Desensitisation Reprogramming (EMDR)

This therapy works very rapidly in helping employees to recover from the adverse impact of workplace bullying. Francine Shapiro (1995) developed EMDR as a way of helping victims of traumatic experiences rapidly reprocess disturbing thoughts and memories. The reprogramming process generally involves the client tracking the rapid right/left finger movement of the therapist. This activity has been found to help in the rapid reprocessing of disturbing thoughts and memories. In the example illustrated

Box 15.3: A case of Eye Movement Desensitisation Reprogramming

THERAPIST: So let's start with seeing your manager and the time that you were left with no desk or place to sit. From 0–10 how does it feel? (*The clerk imagines the situation and gives an initial rating of 9 (high) on the subjective units of distress scale (SUD).*)
THERAPIST: Concentrate on that feeling and follow my fingers with your eyes. (*Leads the clerk in a set of right to left eye movements.*) Good, what do you get now?
CLERK: I feel really bad, what I cannot understand is why the others did not do anything to help, I can see them all looking at me, but no-one is moving.
THERAPIST: Hold that. (*Another set of eye movements.*) Good, blank it out; take a deep breath. What do you get now?
CLERK: I feel really angry when I see people that I know well doing nothing to help me. They are sitting in their seats and all I can do is stand trying to operate my computer.
THERAPIST: Think of that. (*Another set of eye movements.*) Good, blank it out and take a deep breath. What do you get now?
CLERK: I feel less upset. Slowly I am beginning to see that the others are afraid of the manager. I feel that they are upset but that they don't know what to do.
THERAPIST: Think of that. (*Another set of eye movements*) Good, blank it out and take a deep breath. What do you get now?
CLERK: I can see that my friends want to help me and are looking very upset. Perhaps that is why they were unable to do anything. I don't feel bad about them anymore.

in the case presented in Box 15.3, a clerk who had previously got on well in his work was having difficulties with his new manager because he felt that the new manager was constantly picking on him. The session begins with the therapist asking the clerk to visualise a situation where he felt bullied by his manager. The level of anxiety this visualisation creates is measured in terms of subjective units of disturbance (SUD), ranged in this case by the clerk on a 0–10 scale. The clerk is then asked to track the finger movement of the therapist. At the end of a series of finger movements the clerk is asked to assess his current emotional state and describe any new information that emerged. This process is repeated until the SUD level falls. Following the EMDR the clerk would begin to feel less angry with colleagues and be able to look at other aspects of the bullying experience. After exploring a number of aspects of the manager's behaviour using EMDR, the clerk has a clear understanding of how the experience of bullying had caused him distress.

Traumatic Incident Reduction (TIR)

Traumatic Incident Reduction (TIR) uses repeat imaginal exposures to distressing or traumatic events in order to reduce negative emotions and facilitate adaptive thinking (Bisbey and Bisbey, 1998). The structure of the TIR session is clearly defined; however, the process gives the employee the freedom to explore the material that is most troubling to them with minimal interruption from the counsellor. TIR sessions generally take longer than the normal fifty-minute counselling session. The case presented in Box 15.4 illustrates how TIR helped a manager deal with the painful memory of a difficult team meeting. The manager was asked to describe the incident that was causing her problems before reviewing the incident in her mind.

Rehabilitation

Rehabilitation is central to the process of dealing with bullying in the workplace. For some employees the relationships will have broken down to such an extent that they prefer to find alternative employment rather than take the risk of becoming a victim of the bully for a second time. In these circumstances the main focus of the counselling will be to help the employees to build up self-esteem and confidence, and to prepare themselves for alternative employment. In a number of cases the rehabilitation process will include supporting the employee through formal complaints or legal processes. Legal and other forms of investigation are emotionally demanding processes. One of the most effective ways that counsellors can help is to encourage the employee to imagine the worst situations that may confront them and to help in the identification of strategies to deal with such situations.

Box 15.4: A case of Traumatic Incident Reduction Therapy

THERAPIST: Close your eyes and move through the incident in sequence with as much detail as possible. (*The manager silently reviews the incident, she begins to cry, then opens her eyes.*)

THERAPIST: OK, tell me what happened.

MANAGER: I had done a lot of work on my presentation and felt that this time there would be no complaints. As soon as I started my presentation my manager interrupted by picking holes in what I had done. I had been asked to write a high-level training strategy but he was asking for details. One of my colleagues spoke up criticising me for being too intellectual and not understanding what the business wanted. There was silence and I did not know whether to go on or give up. I tried to get to the end of the presentation but it was clear that it was not being listened to. My manager turned his back and some of the others began to talk to each other about other things. I said thank you and sat down.

THERAPIST: Got it. Go back to the beginning of the incident.

(*The manager silently reviews the incident again.*)

THERAPIST: OK, tell me what happened.

MANAGER: I'm talking to Mark, the guy who rubbished my report at the meeting. He is telling me that he intends to be a director by the time he is thirty-five. He is saying that if I want to get on I will have to play the game and that I should stop trying to impress everyone with my superior knowledge. It's funny but I had forgotten this conversation until now. I wonder if he had said something before the meeting to undermine my presentation.

(*In the next few tellings of the story the manager recalled a number of other incidents where she had shared information with other members of the team, things she had learnt at university. Whilst she had seen herself as being helpful she began to recognise that her enthusiasm for sharing information may have been viewed as undermining by others.*)

MANAGER: I have worked hard on my presentation. I notice that everyone is talking in groups and I am isolated from the others. I want to be able to talk to them but it is difficult. I can still see my manager's face when I began my presentation. We had got on well when I started working for the company but gradually he has become more distant. I have been so busy with just trying to keep up with the work; I never seem to have the time to talk to anyone anymore.

At this stage the focus changed to the time when the manager had been doing her school exams and spent all her free time studying and neglecting her friends. She described how it felt when her friends made fun of her.

> Following the TIR therapy, counselling was used to help the manager recognise that under pressure she had a tendency to isolate herself from her colleagues. She began to recognise that it was not enough to support her arguments with logic but that there was a need to present information in a way that was accessible to people who may not have the advantage of her knowledge and education. Gradually, the manager was able to re-establish good working relationships. She recognised that her colleague was ambitious and that at times he might try to gain advantage at her expense.

Where an employee decides to stay in the organisation, the process of rehabilitation should involve a number of people. It is sometimes helpful for the counsellor to facilitate a rehabilitation meeting where the employee, line manager and human resources can discuss the best way to support the return to work. Before the return to work occurs the counsellor or human resources manager should undertake an assessment of the working environment and culture. Where possible the managers and employees within the group should address the issues raised in the assessment survey. An important part of the rehabilitation process is the assurance that everyone shares the same expectations of the way people will behave towards each other. In some organisations these expectations have been formalised as an explicit statement called The 'Standards of Acceptable Behaviours' (Tehrani, 1996).

Rehabilitating the perceived bully can prove difficult without the full support of the organisation. Often the 'bully' is not provided with the same quality of counselling and support as the 'victim'. However, if the 'bully' is to be rehabilitated, this needs to be addressed. It is important that the bully is offered a programme of counselling designed to bring about a change of behaviours. Changes in behaviour will need to be clearly defined, monitored and supported in the workplace. Whilst some bullies have psychological or psychiatric problems most people who misuse their power are not psychopaths but rather people who have failed to learn how to use their personal power appropriately. The rehabilitation programmes designed to help people with behavioural problems will only work if the emphasis is on rehabilitation rather than retribution.

The problem with counselling

Most counselling models focus their attention on the individual as the object of analysis and intervention. Where the concern relates to interpersonal conflict it is clear that whilst it is possible to discuss the nature of relationships within a counselling session it may be difficult to understand the complexity of a troubled social system from the viewpoint of a single member. Counsellors are trained to take an objective outlook; however, this objectivity can be undermined when clients describe incidents in which they present their actions in the best possible light whilst, at the same time, placing the blame for initiating and maintaining the conflict on others (Björkqvist *et al.*, 1994). The problems of misattribution are illustrated in the case in which Ann, a manager in a large financial institution is having difficulties with her boss (see Box 15.5). What this case illustrates are the problems that occur when a counsellor focuses on the experience of the individual employee rather than examining the nature of the dysfunctional relationships. These problems can result from an overemphasis on the

Box 15.5: What is the full story?

Ann came to counselling complaining that her boss was bullying her. She described how he held back important information and undermined her in meetings. She described how he failed to give her clear directions on what to do, but if she did not deliver, he ridiculed her in public.

In the counselling Ann described the incidents that had occurred, and with the support of the counsellor came to the realisation that there was little that she could do to stop the bullying. Ann decided she did not want to cope with the bullying anymore and left the organisation.

The counsellor took the case to her counselling supervision where the supervisor accepted her view that the counselling had helped Ann to recognise and accept that she was the victim of bullying and that she had the right to leave this abusive relationship.

Both the counsellor and the supervisor believed that the counselling had helped Ann to deal with the bullying. However, they may have come to a different conclusion had all the relevant information been available. As it turned out later, Ann had not told her counsellor that:

- She did not want to join her new boss's team, as there had been a history of bad feelings between her old team and the new team.
- She had used customer meetings to raise issues on internal decisions and policies with which she disagreed. Her boss resented this and believed that she was abusing her position.
- Ann's boss's secretary had complained to personnel about the aggressive way in which Ann spoke to her.
- Ann was fifteen years older than her boss and believed that she should have got his job.

development of the empathetic relationship between the counsellor and the client (Egan, 1994). Rogers (1980) described empathy as 'entering the private perceptual world of the other and becoming thoroughly at home in it'. The use of an empathetic relationship helps the counsellor to become aware of their client's world. This close relationship can be very helpful. However, if the empathetic bond leads to the counsellor failing to appreciate the presence of distortions or misattributions in their client's story then the result will be detrimental to the counselling process. Experienced counsellors should be aware of the possibility of developing relationship with a client that is inappropriately close. A balance has to be formed between establishing an effective relationship with the client whilst maintaining an awareness that the client's statements may not always accurately reflect what actually occurred in the workplace.

Who are the victims of bullying?

Much of the recent research into the epidemiology of bullying has focused on the accounts of employees who perceive themselves as victims (Einarsen and Skogstad, 1996; Rayner, 1997; Hoel and Cooper, 2000). Whilst gathering the views of people who believe that they are victims of bullying is an appropriate starting point, it should be recognised that there should be more research into the perceptions of 'bullying' from the perspective of the 'bully', the 'victim' and the other groups of involved employees.

In my role as counselling psychologist, I have been called into workplaces to identify the best way to deal with a case of bullying and to provide support to the affected workers. It is only after studying the 'bullying story' from the viewpoint of all those involved that a picture of what is actually happening emerges. Whilst in some cases there is a clear bully/victim relationship, this is much less common than the literature on bullying would suggest. More commonly the accusation of bullying is triggered by the individual's responses to a series of interactions that are built up over a period of time (Cox and Leather, 1994). In the example illustrated in the case presented in Figure 15.1 it is possible to track how a relationship between a new manager and an existing employee deteriorated over a period of time. This particular organisation was going through a period of change and there was uncertainty over the long-term future of the department. Rachael had been brought in as a manager to improve the quality of service and reduce costs. One of the supervisors, Brian, became convinced that Rachael was bullying him. This case shows how an incident can escalate into a situation where an accusation of bullying is made.

In a confident and supportive team a wide range of incidents are seen as normal and have little impact on effective relationships. However, where the relationship is failing, behaviours such as not saying good morning or using an irritated tone of voice can be interpreted as negative or an aggres-

Rachael (Manager)	Brian (Supervisor)
1. Thoughts I need to make some big changes in this department if it is to survive. I have to make a difference.	
2. Feelings Excitement, challenge	
3. Behaviour → Presentation to team of supervisors. Importance of the plans for the future: 'If you don't like it get off the bus'	**4. Thoughts** Why can't Rachael see what we have been through? What does she mean by getting off the bus – the sack? resign?
	5. Feelings Fear, distrust, defensive
7. Thoughts ← Brian is not taking part. I have heard that he is difficult to manage. I will need to keep an eye on him.	**6. Behaviour** Sitting silently at presentation. Discusses what it might mean with peers.
8. Feelings Disappointed, determined	
9. Behaviour → Set up meeting with Brian to discuss his targets.	**10. Thoughts** Why is Rachael picking on me? She is not asking the others to have weekly meetings. She does not trust me.
	11. Feelings Undermined, upset
13. Thoughts ← Brian is undermining my position and the whole project. This could mean the end of the department. What does he want?	**12. Behaviour** Un-cooperative. Talking to peers about how to change the way things are being done.
14. Feelings Angry, frustrated	
15. Behaviour → Put everything in writing. Set clear deadlines. Meet with Brian's team to discuss performance.	**16. Thoughts** Rachael is bullying me. I have always done a good job and have good appraisals. How dare she arrange a meeting with my team? She wants me to leave.
	17. Feelings Distress, stress, anxiety about the future
	18. Behaviour Goes to see GP. Talks to union about bullying and considers initiating the grievance procedure.

Figure 15.1 A case of conflict escalation

sive act. The escalation process also shows how individuals involved in conflict introduce a bias into their thinking when attributing the reason for particular behaviours. Kelley (1972) found that when people are explaining their own behaviour they emphasise the effect of the situation rather than the influence of internal personal factors. For example, people are more likely to say that they got angry because they were being provoked than to admit to the fact that they have a bad temper. On the other hand when interpreting the behaviour of others, people are more likely to attribute these behaviours to internal factors rather than to the situation. In this particular case, Rachael regarded Brian as a difficult person who had chosen to undermine her authority. She did not recognise that his behaviour may have been due to the shock of change and the possibility of redundancy.

Discussion

There is an increasing body of evidence to show that bullying is damaging the health and well-being of employees. Current research has put the focus on people who regard themselves as victims of bullying. Yet when counsellors undertake counselling or conciliation within the workplace and are able to gain an understanding of the full range of interpersonal interactions taking place, it is often difficult to decide which employee is the bully and which the bullied. The Drama Triangle described by Karpman (1968) is helpful in showing how the roles of persecutor (bully), rescuer (counsellor) and victim are interchangeable. Employees may play each of the roles at one time or another. Counsellors and others involved in trying to resolve interpersonal conflict and bullying need to be aware that they can be lured and then trapped in the Drama Triangle as a rescuer, persecutor or victim. Counsellors need to understand that they can use the power and dynamics of the triangle without experiencing the negative outcomes. This is not easy and requires the counsellors to be aware of their own needs in addition to those of the other players. Experienced counsellors working with employees experiencing interpersonal conflict are helped if they understand the Drama Triangle and are able to make full use of their power without becoming persecutory, their vulnerability without becoming a victim and their responsiveness without becoming a rescuer (Proctor and Tehrani, 2001).

One of the difficulties facing counsellors and researchers working with bullying in the workplace is the lack of recognition of the bully who uses passive aggression as a means of achieving their goals at the expense of others. Whilst it is relatively easy to recognise a bully who threatens, blames, takes revenge and shouts at others, it is much more difficult to recognise passive aggressive behaviours. The passively aggressive bully uses secrecy, manipulation, obsessional and evasive behaviours as tactics to get their own way. However, if the victim of passive aggression retaliates, the passive aggressive bully will frequently move into the position of a hurt victim (Lindenfield, 1993; Elgin, 1995). Dickson (1982) describes two groups of passive aggressive behaviours. First, there are behaviours that present the individual as helpless and requiring the constant support of others. Where help is not given the behaviour becomes one of bemoaning their lot and blaming others for their misfortune. In the second group of behaviours the individual uses manipulation and deceit and flattery to get their own way. In understanding the extent of bullying in the workplace it is important to find ways to recognise and identify the extent of passive aggression.

Traditional counselling has limitations in dealing with workplace bullying. Whilst it can be helpful in the healing of wounds of the individual employee, it is not particularly effective in dealing with the organisational aspects of the bullying. There is, however, a growing group of counselling psychologists (e.g. Lane, 1990; Carroll, 1997; Tehrani, 1997) who have recognised the need for counsellors to provide counselling not only for individual employees but also for the organisation itself. It is possible that this shift of emphasis from the individual to the group will create new ways of dealing with workplace bullying.

Bibliography

Bisbey, S. and Bisbey, L. B. (1998) *Brief therapy for post-traumatic stress disorder – traumatic stress reduction and related techniques*. Chichester: Wiley.

Björkqvist, K., Österman, K. and Hjelt-Bäck, M. (1994) Aggression among university employees. *Aggressive Behaviour*, *20*, 173–184.

Bor, R., Legg, C. and Scher, I. (1996) The systems paradigm. In R. Woolfe, W. Dryden (eds), *Handbook of counselling psychology* (pp. 240–257). London: Sage.

Carroll, M. (1997) *Counselling in organisations – an overview*. In M. Carroll and M. Walton (eds), *Handbook of counselling in organisations* (pp. 8–28). London: Sage.

Cox. T. and Leather, P. J. (1994) The prevention of violence at work. In C. L. Cooper and I. T. Robertson (eds), *International review of industrial and organisational psychology*, vol. 9 (pp. 213–245). Chichester: Wiley

Dickson, A. (1982) *A woman in your own right*. London: Quartet Books.

Egan, G. (1994) *The skilled helper – a problem-management approach to helping*, 5th edn. Pacific Grove, CA: Brooks/Cole.

Einarsen, S. and Skogstad, A. (1996) Bullying at work: Epidemiological findings in public and private organisations. *European Journal of Work and Organisational Psychology*, *5*, 2, 185–201.

Elgin, S. H. (1995) *You can't say that to me!* New York: Wiley.

Goldberg, D. (1985) Identifying psychiatric illness among general medical patients. *British Medical Journal*, *291*, 161–162.

Hoel, H. and Cooper, C. L. (2000) *Destructive conflict and bullying at work*. Unpublished report, University of Manchester, Institute of Science and Technology.

Hoel, H., Rayner, C. and Cooper, C. L. (1999) Workplace bullying. In C. L. Cooper and I. T. Robertson (eds), *International review of industrial and organizational psychology*, vol. 14 (pp. 195–230). New York: John Wiley.

Kelley, H. H. (1972) Attribution in social interaction. In E. E. Jones (ed.), *Attributions: Perceiving the causes of behaviour* (pp. 151–174). Morristown, NJ: General Learning Press.

Karpman, S. (1968) Fairy tales and script drama analysis. *Transactional Bulletin*, 7, 39–44.

Lane, D. (1990), Counselling psychology in organisations. *The Psychologist, Bulletin of the British Psychological Society*, *12*, 540–544.

Leymann, H. and Gustafsson, A. (1996) Mobbing at work and the development of post-traumatic stress disorders. *European Journal of Work and Organisational Psychology*, *5*, 2, 251–175.

Lindenfield, G. (1993) *Managing anger*. London: Thorsons.

Lipsedge, M. (2000) Bullying, post traumatic stress disorder and violence in the workplace. In P. J. Baxter, P. H. Adams, T.-C. Aw, A. Cockcroft, J. M. Harrington (eds), *Hunter's diseases of occupations*, 9th edn (pp. 539–544). London: Arnold.

Marmar, C., Weiss, D. S. and Metzler, T. J. (1997) The Peritraumatic Dissociative Experiences Questionnaire. In J. P. Wilson, T. M. Keane, *Assessing psychological trauma and PTSD* (pp. 399–411). New York: Guildford Press.

Proctor, B. and Tehrani, N. (2001) Issues for counsellors and supporters. In N. Tehrani (ed.) *Managing bullying at work – building a culture of respect* (pp. 165–184). London: Taylor and Francis.

Rains, S. (2001) Building an effective response to bullying at work. In N. Tehrani (ed.) *Managing bullying at work – building a culture of respect* (pp. 155–164). London: Taylor and Francis.

Rayner, C. (1997) The incidence of workplace bullying. *Journal of Community and Applied Social Psychology*, 7, 199–208.

Rogers, C. R. (1980) *A way of being*. Boston: Houghton Mifflin.

Scott, M. J. and Stradling, S. G. (1992) *Counselling for post traumatic stress disorder*. London: Sage.

Shapiro, F. (1995) *Eye movement desensitisation and reprocessing – basic principles, protocols and procedures*. New York: Guildford Press.

Tehrani, N. (1996) The psychology of harassment. *Counselling Psychology Quarterly*, 9, 2, 101–117.

—— (1997) Internal counselling for organisations. In M. Carroll and M. Walton (eds), *Handbook of counselling in organisations* (pp. 42–56). London: Sage.

Tehrani, N. and Westlake, R. (1994) Debriefing individuals affected by violence. *Counselling Psychology Quarterly*, 7, 3, 251–259.

Tehrani, N., Cox, S. J. and Cox, T. (2001) Assessing the impact of stressful incidents in organisations: The development of an Extended Impact of Events Scale. Submitted to *Trauma*.

White, M. and Epston, D. (1990) *Narrative means to therapeutic ends*, New York: Norton.

16 Workplace bullying

The role of occupational health services

Maarit Vartia, Leena Korppoo, Sirkku Fallenius and Maj-Lis Mattila

aIntroduction

The aim of this chapter is to discuss the role of occupational health services (OHS) in supporting the individual victims of bullying as well as in preventing, handling and resolving bullying situations in the workplace. Several studies have found that bullying is related to many kinds of stress symptoms and ill health (Einarsen *et al.*, 1996; Kivimäki *et al.*, 2000) among both victims and observers (Vartia, 2001), making the issue important from an OHS perspective. In the early 1990s, only 18 per cent of the targets of bullying in the municipal sector in Finland reported that they had sought help from occupational health professionals. At present, it is our opinion that the targets of bullying increasingly ask for help from OHS and that it is more common for a conflict situation in the workplace to be labelled 'bullying'.

Since three of the writers of this chapter have a great deal of experience in occupational health psychology, with two working at present for the largest nationwide company offering occupational health services in Finland, the perspective of this chapter will be that of occupational health *psychology*. However, the approaches and methods discussed here should also be useful to other professionals working in OHS. It is necessary for all working in this field to learn to recognise bullying and to analyse such situations together with the client. More profound work, for example therapeutic treatment of bullying victims, should be carried out by a trained psychologist or counsellor. Occupational health units should also have experts with understanding of and professional skills in the area of work and organisational psychology if they plan to offer services to workplaces and organisations in as difficult and complex situations as bullying.

The nature of occupational health services in Finland

The Finnish Occupational Health Care Act obliges employers to organise and finance OHS for each worker, irrespective of the size of the company, sector or branch. The whole of Finland has about 1,100 OH units with 1,800 physicians, 1,900 nurses, 600 physiotherapists and 200 psychologists. About 20 per cent of Finnish OH units employ their own psychologists and 11 per cent have a contract with a part-time outside psychologist. The number of OH psychologists rapidly increased in the 1990s, reflecting the shift in working life problems in recent decades from the physical and health aspects to the psychological and motivational aspects of work. For instance, the proportion of premature retirements due to depression has been steadily growing in Finland, and depression is at present among the main causes of early retirement.

OHS are independent and neutral in relation to both management and employees and not an extended part of management as in some countries, for example the UK. Occupational health care personnel usually have two roles: as assessing and diagnosing specialists of the relationship between an individual and his work setting, and as health care professionals running out-patient clinics. Occupational health professionals work to maintain the health and well-being of their clients, and they always have to intervene when the loss of working ability threatens an individual or, more broadly, a workplace or an organisation. All OH professionals have to maintain strict confidentiality, which makes it easier for people to use OHS in a difficult situation, such as bullying.

Our experience with clients supports the notion that OH personnel can be of great help in cases of bullying at work. Co-operation between OHS and managers, supervisors and personnel in companies is generally good. Because of their independence and impartiality, OH professionals can play a practical role as analysers, supporters and mediators in conflicts and bullying. Besides being specialists in supporting and helping individuals, OH professionals can support the organisation and its departments in various difficult situations, including bullying.

In Finland, OH professionals widely accept the following assumptions as basic definitions of their role:

- The task of OH professionals is to support and promote the health and well-being of all employees of a client organisation.
- If the situation at a workplace threatens the health of employees, it is the duty of the OH professionals to provide support.
- The role of OH professionals is supportive, respecting the role of management and supervision of the client organisation.
- Surveys and questionnaire studies carried out by OH professionals are designed to develop the functioning of the work units together with the managers.
- OH professionals respect their clients as active parties in identifying ways of solving emerging problems.

Methods used by occupational health services in cases of bullying

OH professionals can work at three different levels: the individual, the group/team and the departmental/organisational levels. In the following, we will discuss how to work with an individual target of bullying, how to work with a work group, what the role of mediating is, and how to work with the whole organisation. From the OHS perspective, bullying is a systemic phenomenon. Hence, its causes cannot be found solely on an individual level. When dealing with a bullying situation, OH aims to analyse the situation together with the target or members of the work unit, looking at the features and procedures of the organisation that have allowed bullying to occur in that particular environment.

Analysing a client's situation

An OH unit may hear about bullying directly from the victim or from a supervisor, a safety representative, the personnel department or a member of the work unit. When the target of bullying contacts an OH unit, it is very important that the person is listened to and that the report of bullying is taken seriously. The feeling of being genuinely understood and believed by someone in the OH unit is the prerequisite of further co-operation. The situation at the workplace can be highly complex. The experiences and interpretations of the situation by workmates and supervisors can differ a lot from those of the target. In most cases, realities and their interpretations have become entangled, producing several 'truths'.

The number of sessions may vary according to the needs of the client. However, five sessions are usually enough for the professional to be able to determine what further steps are needed. Discussing with the client, the OH professional tries to picture the client's situation and clarify what the client expects from the OHS. Among the most important issues is the seriousness of the bullying, how far the process has escalated, how much the situation has affected the health and well-being of the client, and what the overall condition and life situation of the client are. The main issues discussed are:

- What is the nature of the client's experience?
- When and how did the bullying situation start?
- How has the process escalated in the work unit?
- What was the client's own contribution to the situation?
- What kinds of thoughts and feelings has bullying aroused in the client?
- What is the client's opinion of the reasons for bullying?
- Who does the client regard as the perpetrator and who else is involved?

- Is the bullying situation new to the client or has the client experienced bullying before at some other workplace. If the client has experienced bullying before, what did the client do about it and what kinds of coping mechanisms were considered helpful at that time?
- Identify other work stressors and the overall life situation of the client.
- What kind of social support and encouragement does the client get from workmates?
- How does the client think the situation could be resolved; what is the client willing to do and when?

These discussions help the client to understand what has happened, how the situation has escalated into bullying, as well as to understand his/her situation in a broader context. It is important for the client to be able to understand the factors and processes in the workplace that may have allowed bullying to occur. The discussions further help the OH professional to assess how serious a threat the situation is for the client, and his/her well-being and health. During these discussions, the OH professional together with the client can make up an action plan for this particular case.

Individual support

If the bullying has lasted for a long time, the client may have been deeply hurt and have developed many kinds of psychological and psychosomatic symptoms, such as depression, anxiety, sleeping problems, distress and concentration difficulties, as well as musculoskeletal complaints. In these cases, the client may be so distressed that it is impossible for him/her to face the supervisor and workmates and discuss the situation. Sometimes the client is afraid that contacting a supervisor will make the situation even worse. In such a situation, the possibilities of the OHS helping the client are limited to individual support and counselling. On the other hand, on these occasions individual support is indispensable.

Different means can be used to improve the client's ability to cope with the situation. In some cases, sick leave can be necessary to give the client time to recover and calm down. Sick leave is needed when the stress symptoms, for example panic or depression, loss of sleep, or feelings of distress are considerable. In these cases, the symptoms must be treated before any further action can be taken. It is, however, important to consider thoroughly the length of the sick leave needed for the victim to recover. The longer the sick leave, the more difficult it may be for the victim to return to work. In some situations, sick leave may in itself become a new reason for bullying.

An important task for the OH professional is then to examine, together with the client, the interaction between him/her and the perpetrator. How has the client's own behaviour affected the situation? Any possible interpretations of the situation must be explored as well as the potential misunderstandings that may have occurred in the situation. These efforts should, however, be made with genuine respect for the feelings and perceptions of the client.

Sometimes a perceived insult may be the starting point for a chain of events leading to bullying. The client may also see himself/herself as a person who is always unlucky. How does the victim explain this to himself/herself? What kind of generalised effects does this have on the client's life? Victims in these cases are not easy to work with and they require much consideration, support and time.

The clarity of roles and responsibilities of the employees, the organisational culture of the unit, and the ongoing transformation and change processes should all be discussed while counselling and supporting a victim of bullying. The client should always be encouraged to seek emotional and practical support and positive feedback from every possible source, be it from family members, friends or relatives. It is important to remind the victim of the many other roles that he/she has besides being a bullied member of the working community, such as being a spouse, a parent, a beloved son/daughter, or a member of associations and clubs, etc. These identities and roles are often forgotten by the victim as sources of positive feedback and satisfaction.

The clients should also be helped to explore the possibilities of protecting themselves against negative acts. Providing individual support and helping the client to analyse his/her situation sometimes helps the client to behave differently in the workplace, causing the bullying to decrease or in some cases to cease. Sometimes, it is even possible to encourage the victim of bullying to face the perpetrator and express his/her disapproval of the bully's behaviour and ask him/her to stop it immediately. Occupational health psychologists have in some cases also used short-term therapy to support and help a bullying victim, as in the following case of Mary, presented in Box 16.1.

Working with a small group: the role of a mediator

Sometimes, discussions with the work group and mediation are the most effective ways to deal with a bullying case. It must, however, be underlined that mediation is only possible when the conflict is at an early stage. Mediation is not useful when the conflict has escalated so far that the victim is unlikely to settle for anything less than full retribution, justice or revenge.

If mutual feelings of being hurt are very strong, the parties may need to talk individually about their experiences and handle their own feelings first. Only after both parties have been able to talk about their experiences individually may they regard the situation safe enough for mutual discussion.

Box 16.1: The case of Mary

Mary is a middle-aged secretary who came to the OHS because she felt a colleague, Jill, had been bullying her for years. She felt that Jill controlled her work, was nagging her, accusing her of different things and making sarcastic remarks about her. Mary felt that she was quite powerless and unable to defend herself. She was afraid that she would get a panic attack and not be able to do anything but cry as soon as she went back to work. Mary had sleeping problems and felt she was at a dead end. She did not want anyone in OHS to get in touch with her employer, because she thought it would only cause more difficulties for her. She did not want other people at the workplace to know that she could not tolerate the situation any longer.

At the OH unit the first thing was to try to help Mary. At first she got a sick note and some sleeping pills. She continued to see the OH psychologist. She received cognitively oriented short-term therapy (fifteen times) as rehabilitation for depression and burnout. In therapy the first part was counselling concerning what could be done at the workplace. When the situation improved, the main focus of the therapy was Mary's anxiety and panic symptoms in social situations. Mary blushed easily and she was afraid that other people would notice this, so she did not attend meetings at work. Her self-esteem was also low and she behaved in a passive manner in conflict situations. So one aim of the therapy was to raise her assertiveness in social contacts. At the end of the therapy Mary was more at ease with other people and could also take part in meetings. She could also handle her contacts with workmates in a more assertive way. During the sick leave Mary also discussed with her boss possibilities of changing work tasks in order to avoid seeing Jill regularly. Mary's situation at work improved, but Jill's troubles with other workmates seemed to continue. Later on Jill also approached the OH unit to seek help for her own stress problems.

The aim of mediation is to find a solution that allows everybody staying in the work unit. The participants in these discussions are usually the target person, his superior, the perpetrator and an OH professional. In some cases, a member of the work unit, a safety representative or a shop steward may be present. It is of utmost importance that all parties are impartial and fairly heard. The parties often have highly differing opinions and interpretations of the situation and of the incidents that have led to the conflict. The situations often turn out to be very complex, arising from unsolved conflicts.

In dispute-related bullying (Einarsen, 1999), both parties of the conflict may see themselves as victims. In such cases it is often impossible to say who is right and who is wrong. In cases of predatory bullying, the victims find themselves the targets of negative acts without any previous personal involvement in the situation. These cases can arise, for example, from situations in which someone is envious of a post someone else is holding and begins to nag him/her about it. The same process can emerge when two people compete for the same position in an organisation. In these situations mediation by an OH professional, usually a psychologist, may help.

In mediation, the following points are important:

- The mediator must be neutral and capable of managing discussions fairly and objectively.
- At the beginning of the meeting, the aims of the meeting should be explained. Everyone should know why he/she has been invited to the meeting.
- The rules of the meeting must be agreed upon. For example 'In this meeting, we'll speak about incidents, experiences and feelings, no-one is allowed to insult anyone else, and offending remarks are forbidden.'
- Everyone should have the feeling that they are being listened to and given the opportunity to express their feelings and interpretations of events.
- Efforts should be made to encourage the participants to make suggestions on solutions.
- The focus of the discussions should be on looking forward and seeking a solution that enables everybody to go on with their work. Focusing on the past is often useless and may create new conflicts.
- A mutual agreement should be reached and a note of the decisions must be made.
- Before ending the meeting, the parties should agree on the methods and schedule of the follow up of the situation.

Mediation is a highly demanding task. The mediator must be flexible and empathic and at the same time strictly adhere to the rules of the mediation. Bullying always involves strong emotions, and questions of pride are at stake. The discussion must be led in the direction of solutions and agreements.

In the case of Susan, presented in Box 16.2, mediation was used in order to prevent an interpersonal conflict from escalating into a more severe case of bullying.

Working with a work unit

OHS can also work with the whole work unit with the aim of helping it to stop a bullying process from continuing and spreading. If the victim wants

Box 16.2: A case of a conflict situation in the workplace

Susan, a clerk, comes to see the OH psychologist because she feels that she has been mistreated at work. She had attended occupational training and when she returned to work she was not allowed to do the tasks she was trained for and felt capable of doing. Instead, she was transferred to other tasks and to work on her own. She felt humiliated and angry.

A conflict between Susan and one of her workmates, Betty, was seen as the background for this incident. During her training period Susan had been practising at the office. One day when there was an urgent situation at work, Susan had noticed that Betty was sitting and drinking coffee instead of working. Susan had made angry remarks at her. The remarks had offended Betty, and she had told the supervisor about the incident. The supervisor had given Susan a written notice when she came back to work and transferred her to working on her own. Susan felt she had had no opportunity to explain her behaviour. Later on, she discussed the matter with the manager of the organisation and also with the lawyer of her trade union.

Both Susan and her organisation requested mediation by the OHS. Several meetings took place where Susan, Betty, their supervisor, an OH psychologist and a nurse, sometimes even a lawyer, were present. The situation was defined as a conflict situation in which different parties had different opinions about the background and reasons for the conflict. After several meetings it became clear that it was useless to try to settle the conflicting views of Susan and Betty. However, a solution was worked out. Susan was given a new job which suited her training better and which was situated in another department. Susan and Betty were asked to make an agreement about not gossiping about each other any more and that they would concentrate on their work and try to live in peace with each other.

to discuss the situation with the work unit, it is important that any advantages, disadvantages and risks of such an intervention are considered. It should be noted that discussing a problem of bullying in a large group is altogether different from talking about it in a small team or group. In a large group, fantasies far more easily replace realities. Feelings of guilt and the need to defend oneself often arise, and the members of the work unit can drift into viscious circles, accusing others and defending themselves. Hence, there is always the risk of worsening the situation if the case is discussed in a large setting.

Also, if a supervisor has requested the help of the OHS, different aspects of the situation should be thoroughly discussed before any decision to act is made. Bullying situations can be very complex and the expectations of different parties can be quite contradictory and sometimes unrealistic. Therefore, it is important for OH professionals to gather enough information before putting forward their suggestions for action. Gathering information usually means discussions with the victim, talking with several people from the work unit, and getting acquainted with the way the work unit usually handles difficult situations. In all these discussions it is important to find out how the nature and background of the situation are perceived in the workplace and how the supervisor himself/herself views the situation. In early discussions with the supervisor it is important to find out:

- What has activated the bullying problem?
- What kind of bullying is in question?
- Who are the parties involved and who are affected by the situation?
- What has been done to solve the situation so far?
- What is expected from the OH psychologist or other OH professional?
- What is the motivation of the work unit to solve the situation?
- Does the organisation have a policy dealing with bullying situations?

When OHS are asked to assist, some kind of a crisis already exists in the workplace. What is expected from OHS should be clarified in the first place. The helplessness, fear and despair that are experienced in these workplaces can also apply to the OH professional and he/she can rush to action too soon with insufficient information. Sometimes it turns out that the work unit or the victim expect 'miracles' to happen. Sometimes it is expected that the OH professional should take the role and responsibility of the supervisor and carry out administrative operations. Therefore, the real role of the OH professional must be made clear. OH professionals can help the work unit to deal with the bullying situation, but they cannot carry out any changes. Making changes is the responsibility of the members of the work unit, especially the supervisors or managers. OH professionals must definitely avoid acting as judges. The Medivire Occupational Health Services in Finland have developed a model for dealing with bullying in the work unit/work group (Box 16.3).

During this process the OH psychologist can assist the members of the work unit in expressing their views and feelings about the situation, help them see the bullying situation in a broader context and, in some cases, help them to understand bullying as a consequence of a long-lasting escalating situation. Pointing out the characteristics of a work unit that may turn it into a potential bullying arena is useful. For instance, the whole

Box 16.3: The Medivire model

1. Gathering and analysing basic information about the situation.
2. Agreement on how to deal with the situation.
3. Providing all parties with information on the forthcoming proceedings.
4. Individual interviews of the parties involved (structured interviews including questions about alternative ways to solve the situation).
5. Providing feedback of discussions and interviews, and assessing the probable ways of solving the situation together within the work unit.
6. Follow-up sessions after three, six and twelve months.

work unit may fear some imminent changes in the organisation, and their anxiety and feelings of threat may have been projected onto one member of the work unit, who then becomes the victim of bullying. By unravelling such projections and scapegoating processes, the whole group can be helped to restore their normal functioning. With different methods, the participants can be helped to externalise and share their thoughts, feelings and fantasies. This is a prerequisite for concrete changes in people's behaviour.

Sometimes people see and admit that a situation has gone too far and apologise for their own conduct in a safe environment. It is also important to make clear that almost anyone can end up breaking his/her normal moral code and find himself/herself treating a co-worker in a way that was earlier unfamiliar to him/her. The psychologist may explain to everybody the experience of being bullied and what an ongoing bullying process means for all those involved in the process. Most people experience bullying as frightening and may struggle with feelings of shame and guilt. It may be more productive to discuss the negative feelings and the negative behaviours as such and analyse underlying causes rather than to talk about bullying as such.

However, difficulties may arise if the case at hand has a long history and the unit has developed a strong prejudice towards the victim regarding him/her as an impossible case, with a shared conviction that there are no solutions to the problem. Box 16.4 presents a case which involved working with the whole work unit.

Preventive strategies

According to the Finnish Occupational Health Care Act, the OHS should focus on prevention. Typically the methods in prevention of workplace

Box 16.4: Working with a whole work unit

A human resources department (HRD) manager contacted the OHS because a member of a work unit consisting of eight persons had complained about being bullied. The victim had named three of her colleagues plus the manager of the unit as the perpetrators. The HRD manager told the OH nurse that the matter had been discussed in a shared meeting with the person feeling bullied, her manager and the HRD manager. After the initial phone call the OH nurse contacted the OH psychologist. Together they invited the manager accused of bullying and his superior to a meeting. In this session the views of both managers on the subject were discussed, as well as their views on the history of the case. Finally, an agreement was made on how to proceed.

The intervention was not launched as a bullying intervention. Instead it was presented as a development project to improve the working climate and co-operation. The objective was to avoid labelling members of the work unit as perpetrators before the situation had been more closely analysed.

The interviews were carried out by applying the so-called double technique. This technique facilitates the recognition and expression of feelings aroused by the situation. The members of the work unit described many problematic aspects in their work setting and expressed feelings of disappointment, helplessness and hurt. They also proposed several ideas for further development. It seemed that there were many kinds of misunderstandings and experiences of being hurt by others among the personnel. These experiences could have arisen from incoherencies in work tasks as well as from difficulties in understanding and approving personal conduct. Everyone was given the opportunity to be listened to and to describe the situation from their own point of view.

The intention of the subsequent sessions was to facilitate co-operation between group members. With the help of discussions and pictorial and action methods, the history of the group was exposed. The results of the interviews were then presented and discussed with the group. According to the interviews the members of the work group felt that performance objectives and division of labour within the group were unclear. Job descriptions also appeared to be vague. There were no regular weekly meetings, and when meetings took place the agenda was often unclear or missing altogether. The manager kept a lot of work to himself, spending little or no time in managing and supporting the work of group members. Some group members also felt that the climate in the group was intolerant of individual differences. As a result of the discussions, a plan was drawn up, emphasising the need for regular meetings and clear procedures for feedback. It was remarkable that during these discussions with the group the members were able to recognise their own 'blind spots' and express these to others. This seemed to enhance trust within the group. Consequently, instead of blaming one another, everyone admitted their own errors or misunderstandings that had led to overreactions or other embarrassing incidents in the past.

During monitoring of possible progress, the OH professional discovered that the work group had established a regular meeting schedule with clear procedures for decision making and production of 'memos'. The members of the work group felt that the information flow and communication within the group had improved and that reaching agreements was somewhat easier. There were still tensions in the group, but these were considered easier to live with. During the process the manager had acknowledged his own shortcomings as a leader and had been advised to look for further supervision and management training.

During and after this intervention process both the HRD manager and the OH professionals received positive feedback from the group. The well-being of the group had improved and there were no more complaints of bullying.

bullying would include training, counselling and consultation. As far as training is concerned, sessions would include the psychosocial work environment, the basic components of well-being at work and the features of a smoothly running work unit, as well as the importance of clear objectives at work, clarified work roles and a sound organisation of work processes. In training, the causes of bullying, the bullying process, the consequences of bullying and the handling of interpersonal conflicts at work are also key issues.

OH professionals may also assist the organisation in developing its own anti-bullying policy and instructions for dealing with bullying. Such policies are helpful not only for the victims, but also for the supervisor and other members of a work unit. It is particularly important to provide supervisors and managers with instructions and advice on how to proceed in the presence of a bullying incident. If no instructions are given, supervisors can react inappropriately or defensively, neglecting their responsibilities.

OHS also have the responsibility of reporting to the organisation about any health hazards and risks facing the members of staff. Where bullying seems to be endemic and where negative and uncivil behaviour appears to take hold within a unit, these matters should be reported to annual meetings. OH professionals should always maintain confidentiality, keeping the problems of any one person confidential if the person has not given permission to discuss his/her problem in public. Such a situation is sometimes difficult for an OH professional to tackle.

Support for the occupational health professionals

In severe conflicts and bullying situations, the expectations towards OH professional are often high. People may expect that the OH psychologist or nurse 'comes along and solve the problem'. Sometimes OH personnel are expected to act as 'saviours' of the work unit. As a principle, the OH professional should always gather enough information on the situation and its background before suggesting any steps to be taken. Such processes may be very time-consuming and strenuous for the professional. There is always a risk of being manipulated in one way or another. There is often a power struggle behind the scenes, and the reactions of the parties involved can be unexpected and overwhelming.

The OH professional can get caught in a web of hate, frustration, helplessness and fear. Most OH professionals have a very strong desire to help people who are distressed. The support role can be very taxing as the OH professional is prevented from transferring the employee to another unit, appoint a new supervisor or change the environment. The despair of the victim may be overwhelming and the danger of overreacting is real.

Hence, maintaining the professional role in bullying conflicts is highly demanding. In such situations, OH professionals should consider the possibility of seeking advice, supervision or support from colleagues.

OHS teams should also be able to recognise situations where the interventions of OH professionals are inadequate. In such situations, OH professionals should help the victim or the supervisor or manager to clarify the situation, and to seek help and support from other specialists, consultants or even lawyers.

Sometimes the best solution for the client is to leave the workplace, seeking employment elsewhere. In other cases, they may feel it is absolutely impossible to return to any workplace. They may not see any future for themselves at the workplace and may be looking for early retirement. This starts a new process that can be taxing for all parties involved. Sometimes the client is given a psychiatric diagnosis (e.g. depression) and ends up retiring on the grounds of ill health. In other cases, rehabilitation may help the client find a new job and new motivation for returning to work.

Bibliography

Einarsen, S. (1999) The nature and causes of bullying at work. *International Journal of Manpower*, *20*, 1–2, 16–27.

Einarsen, S., Raknes, B. I., Matthiesen, S. B. and Hellesøy, O. H. (1996) Bullying at work and its relationships with health complaints – moderating effects of social support and personality. *Nordisk Psykologi*, *48*, 2, 116–137.

Kivimäki, M., Elovainio, M. and Vahtera, J. (2000) Workplace bullying and sickness absence in hospital staff. *Occupational Environmental Medicine*, *57*, 656–660.

Vartia, M. (2001) Consequences of workplace bullying with respect to the well-being of its targets and the observers of bullying. *Scandinavian Journal of Work Environment Health*, *27*, 1, 63–69.

17 To prevent and overcome undesirable interaction

A systematic approach model

Adrienne B. Hubert

Introduction

Since 1994 employers in The Netherlands have been legally obliged through the Working Conditions Act to protect their employees from sexual harassment and (psychological) aggression at the workplace and their negative consequences. Several kinds of undesirable behaviour are covered by this legislation, for example sexual harassment, mobbing/bullying, racism, and aggression from clients, patients and the public towards employees. In 1996 I started the first scientific research on mobbing in The Netherlands at the University of Leiden. The taboo on mobbing being broken, there appeared to be a great need for information on this subject by journalists, victims of mobbing and, somewhat later, by Dutch companies and Occupational Health Services (OHS). In 1999 the Working Conditions Act was changed and now obliged employers to establish a policy concerning sexual harassment and (psychological) aggression. That year I started a company on applied research and consultancy regarding issues such as mobbing, conflicts and undesirable behaviour. On the basis of scientific research, I provide information to organisations and professionals involved with mobbing, discussing the way mobbing is being dealt with at the time.

In order to conform to the legislation, Dutch organisations took several measures. One-third of all organisations (34 per cent), for example, appointed one or more employees as confidential counsellors, where victims of sexual harassment (and evermore frequently of mobbing) could seek support and guidance. Seventeen per cent of Dutch organisations have a grievance committee and have a complaints procedure for sexual harassment established (Soethout and Sloep, 2000). In 2000, my own company, Hubert Consult, together with Research voor Beleid (a Dutch institute for applied research) was invited to participate in a study at the request of the Dutch Federation of trade unionism (FNV). The main questions in the study were how undesirable behaviour in organisations should be tackled and managed, and what the specific responsibilities of the various professional disciplines involved should be. In order to answer these questions, I led four group discussions with personnel representatives and representatives of institutions already involved with preventing and overcoming undesirable behaviour. The participants included representatives from: management (2), works councils (6), human resources departments (2), confidential counsellors (5), grievance committees (3), occupational welfare workers (2), company medical officers from the OHS (1), advisors/experts on work and organisation from the OHS (1), Labour Inspection (4), labour unions (1).

During these group discussion I focused especially on the following questions:

- Should there be separate policies for different kinds of undesirable interaction, such as bullying/mobbing, sexual harassment, and racism?
- What do you (or your institution) currently do to prevent and overcome undesirable behaviour?
- What in your opinion are effective and non-effective strategies?
- What future improvements can be made?

On the basis of these group discussions, a systematic approach for the prevention and management of undesirable behaviour was developed. This approach was further discussed during an expert meeting with representatives of umbrella organisations, including employers' organisations (1), labour unions (9), OHS (3), occupational welfare workers (3), Labour Inspection (2) and the national organisation for confidential counsellors (1). The 'Systematic Approach Model on how to prevent and overcome undesirable interaction' (Hubert and Scholten, 2000) was developed based on these group discussions and the comments made during the expert meeting. This model is presented and discussed in this chapter. The participants in the group discussion were all employed in medium to large organisations in the sectors of government and public administration, health care, and the hotel and catering industry. The Systematic Approach Model is therefore mainly suitable for these kinds of organisation.

The Systematic Approach Model

Our model distinguished between five different phases for action in efforts to prevent and overcome undesirable interaction at work:

Phase 1 prevention;
Phase 2 uncovering;
Phase 3 support;
Phase 4 intervention; and
Phase 5 after-care.

The Systematic Approach Model concentrates on the specific responsibilities of the various professional disciplines involved in each of the phases. A summary is given in Table 17.1.

Should there be separate policies for different kinds of undesirable interaction, like bullying/mobbing, sexual harassment and racism? The participants in the group discussions generally did not think so. They felt that these three undesirable ways of interacting are all about norms in organisations and about lack of respect, about feeling threatened, about power and interests. Moreover, it sometimes seems to be difficult to discriminate between bullying, racism and sexual harassment. One policy

Table 17.1 Tasks for persons /institutions concerned with the prevention and tackling of undesirable interaction at work

	Prevention *(including uncovering for making policy)*	*Uncovering* *(for support)*	*Support*	*Intervention*	*After-care*
Management	X			X	X
Works council	X				
Human Resources department	X				
Supervisor	X	X		X	X
Confidential counsellor	X	X	X	X	X
Grievance committee				X	
Occupational welfare workers /mediator				X	X
Company medical officer (OHS)		X			
Advisor /expert on work and organisation (OHS)	X				
Labour inspection	X				
Unions	X				X
Employers' organisation	X				

for all kinds of undesirable interaction in principle would clarify things for victims and management.

Regarding uncovering, intervention and after-care, another critical distinction was made in the group discussions: the distinction between *undesirable group behaviour* and *one-to-one harassment*. Undesirable group behaviour is visible to nearly everybody, examples being: pin-ups, ridiculing a person in a group, socially isolating someone or making racist jokes in a group. Sometimes undesirable group behaviour is not recognised as such, but rather as innocent horseplay or accepted behaviour which must be tolerated. 'One-to-one harassment', on the contrary, is less conspicuous. It is often in the interest of the offenders to hide their behaviour from others; the victim is often too ashamed or too afraid to tell anybody. Examples of one-to-one harassment are: sexual harassment by one person, threats by one person, one person making it impossible for another to do his/her work and personal discrimination.

Phase 1: prevention

The participants in the group discussions admitted that, at the moment, not much was being done in Dutch organisations to prevent undesirable interaction. Curative policies 'on paper', stressing complaint procedures, got most attention. Sometimes confidential counsellors showed some initiative, for example by arranging meetings and giving information about undesirable behaviour to colleagues and staff. A general complaint of the confidential counsellors was, that management was using them as an excuse to push off responsibility for a preventive policy.

Because confidential counsellors (being ordinary co-workers) sometimes lack recognition and power, they felt they usually did not have much impact and were sometimes being ridiculed themselves. From my own experience as a consultant I know that if a company pays no attention to prevention it often has an organisational climate in which someone complaining about undesirable interaction is seen as weak, humourless or a tell-tale. Intervention then becomes difficult, as undesirable behaviour is belittled as jokes one has to be able to take. Sometimes the victim is said to behave in such a way that he/she deserves to be ill-treated. A preventive policy changing the attitudes of management and employees towards undesirable interaction is thus required for successful (intervention) actions to take place in the following phases of the model.

According to the Systematic Approach Model, the establishment of a preventive policy, including the task of communication towards employees, is the responsibility of management. It is necessary for management to take undesirable interaction seriously and to demonstrate the attitude that undesirable interaction is not to be tolerated. A manager had the following experience in this regard:

> Previously we had a policy against harassment, but it disappeared in a desk-drawer. When an incident happened that reached the press, the policy was revised and communicated among personnel on a large scale. Management is taking it seriously now and presentations are given periodically.

To understand the importance of a policy to prevent undesirable interaction, the participants in the group discussions thought that management needs to be aware of:

- the legal obligations;
- the (financial) consequences; and
- the nature and prevalence of undesirable interaction in their organisation.

Legal obligations

First of all, a policy must be in accordance with national legislation as was emphasised by Labour Inspection during group discussions. Sometimes management in The Netherlands is not aware of the legal obligation to have a policy concerning undesirable interaction. The participants in the group discussion saw it as the task of a works council, experts from the OHS or unions to point out the responsibilities to the management of organisations. In The Netherlands Labour Inspection has the authority to fine an organisation that has no policy for these kinds of problems. However, the organisation always gets a warning first. Sometimes an employer also has to adapt the policy in order to conform to agreements previously reached by unions and employers' organisations through collective bargaining.

(Financial) consequences

A policy concerning undesirable interaction should not have the adherence of regulations as its only aim. By really reducing undesirable behaviour and interaction at work, negative consequences for organisations may be reduced, such as: work absence, turnover, loss of productivity and image damage. The groups felt it to be important that management is informed about these consequences and their costs in order to take appropriate measures. To inform management about the consequences of undesirable interaction was seen as a task for unions and employers' organisations, with the help of the results of scientific research if necessary.

Nature and prevalence of undesirable interaction

Even if management intends to meet legal obligations and is convinced of the (financial) consequences of undesirable interaction, the discussion groups felt that nothing would happen if management just thinks: 'We do not have this kind of behaviour in our organisation.' Hence, it was seen as important to demonstrate the prevalence and nature of undesirable behaviour occurring in the organisation. Experts from the OHS considered this a task for themselves. One instrument, for example, could be a survey among personnel. In The Netherlands, work councils also have the right to investigate the prevalence of undesirable interaction in an organisation. If there is a confidential counsellor present in the organisation, the number of formal complaints and informal mentions should be recorded.

With the information about legal obligations, (financial) consequences and the nature and prevalence of undesirable interaction, management, in dialogue with the works council, must then decide on how to establish a preventive policy.

A preventive policy should include various aspects. The participants in the group discussions mentioned many aspects previously recommended by 'De Stichting van de Arbeid' (1999). A compilation is given below:

Policy statement

A policy statement describes the expected organisational culture and covers the most important values of this desirable culture (e.g. respect, goodwill and tolerance). The statement makes clear that other behaviour will not be tolerated by the organisation and explains why it will not be tolerated. It further explains the negative (financial) consequences for victims and organisation.

Code of conduct

A code of conduct is more concrete than a policy statement. It should be made to suit the organisation and the kind of undesirable behaviour that may be displayed in this organisational setting. It should not just be an enumeration of undesirable behaviours; it could never be exhaustive anyway. Desired and undesirable behaviour should be described in general terms with clear examples of behaviour that is unacceptable. The process of formulating the code of conduct is at least as important as the finished product. By discussing desired and undesirable behaviour in the organisation and by making new standards of behaviour in co-operation with employees, the organisational attitude towards undesirable interaction may, in fact, change. The code of conduct also makes it easier to talk about and discuss undesirable behaviour.

Reporting and complaining procedure

The policy should deal with the reporting of undesirable interaction and the support possibilities (by the confidential counsellor). Also an elaborated complaints procedure and possible sanctions should be described. It is highly important that people in the organisation feel that it is safe to report undesirable interaction, to lodge a complaint or to support someone lodging a complaint.

Tasks and responsibilities

Preventive policy should state who is responsible for the actions in the prevention and the management of undesirable interaction.

Means of implementation

A problem with policies in general is that they may be forgotten. It is extremely important that repeated attention is given to the policy and to the issue of undesirable interaction at work, for example by making the policy part of another existing and frequently used policy. In the policy it should also be mentioned which means are available each year. Periodically, the issue of undesirable interaction should be the focus of thematic meetings and staff publications, as well as being an issue in appointment procedures or exit talks.

The code of conduct should be communicated to employees top-down and be maintained by management and supervisors. Supervisors need training, which may be arranged by the human resources department. Also, training should be arranged for confidential counsellors and for members of the grievance committee.

However, as it turns out, supervisors, although trained, do not always like to talk to employees about undesirable behaviour. Often they shut their eyes, laugh about it or even carry out undesirable behaviour themselves. Therefore, when evaluating supervisors, human resources department or management should not overlook the way supervisors are treating undesirable interaction in their department. Also, works councils may become active when supervisors do not take their duty seriously.

Evaluation

Measures taken should be evaluated after some time. If necessary, the policy may be adjusted.

Phase 2: uncovering

Who then is responsible for uncovering undesirable interaction? Here a distinction should be made between previously described undesirable group behaviour and one-to-one harassment. Undesirable group behaviour is visible to everybody. The group discussions concluded that as long as employees feel safe, they should address each other about undesirable behaviour. Ultimately, however, their opinion was that the supervisor is responsible for uncovering undesirable behaviour at work. He/she must recognise certain behaviour as not fitting the organisation's code of conduct, and therefore see it as undesirable and act accordingly.

One-to-one harassment is often not visible. Hence, it may be best for a victim to mention this behaviour to a confidential counsellor. Therefore the confidential counsellor must be accessible, trustworthy and known by victims. From research we know that a victim (after some time) often suffers from psychosomatic complaints (such as headaches, sleeping problems, trembling, heart palpitations, strain, perspiration, intestinal or stomach complaints: see, for example, Leymann, 1996). Also a company medical officer (Occupational Health Service) can recognise the victim of undesirable interaction. When a company medical officer is confronted with psychosomatic complaints he/she should empathetically inquire about the patient's contacts with supervisors and colleagues.

Phase 3: support

My own experience with victims of undesirable behaviour looking for support is that they are very often sent from one person to another and from institution to institution. At first everybody is willing to help, but help often only means referring the victim to someone else. This situation also puts pressure on the offender who may learn of it and hence try to justify his/her behaviour and to get support from other co-workers. The situation may then easily turn into a win-lose fight between victim and offender(s), and the problem may escalate further.

When someone else in the workplace recognises that undesirable interaction is going on, he/she should send the victim to the confidential counsellor. It is the confidential counsellor's job to support the victim. A layperson may cause a great deal of harm in such cases. There is a high risk that he/she may (unintentionally) blame the victim, leading to depression or even the victim's suicide. Inappropriate advice may also escalate rather than solve the conflict. Therefore a confidential counsellor (who is just an appointed employee) also needs training in how to meet and how to advise a victim. Again, such training may be arranged by the human resources department. Meeting the victim, the confidential counsellor should:

- listen to the victim;
- in the case of serious psychosomatic and psychic complaints, send the victim to either a medical officer or a psychologist; and
- inform and talk about possible intervention strategies.

It is also important that the confidential counsellor has a duty to secrecy and can never take any action without the permission of the victim.

Phases 4 and 5: intervention and after-care

Intervention and after-care are two different steps. However, the after-care should match the chosen intervention strategy. Therefore intervention and after-care are dealt with together here. In tackling the problem, a distinction between undesirable group behaviour and one-to-one harassment again has to be made.

Pin-ups, racial jokes and open ridicule are some examples of visible undesirable group behaviour. Intervention in these situations should aim at changing norms and values in the group. It is the responsibility of the supervisors to see and recognise such situations. They should, supported by the code of conduct, talk directly to employees whenever undesirable behaviour is being displayed, and make clear that this way of behaving is not acceptable and will not be tolerated.

However, this may be very difficult when undesirable behaviour has been prevalent in a group for a long time and is thus seen as normal by the employees. In that case, the supervisor should arrange a group intervention (possibly led by an external expert). During the group intervention contemporary group norms are discussed, and the code of conduct of the organisation is explained and discussed.

After-care in cases of undesirable group behaviour means that the supervisor remains alert to signals of undesirable behaviour. Every time he/she recognises this behaviour, he/she should directly intervene and talk to the persons who are displaying this kind of behaviour. Only in this way can the change of norms be ensured.

In cases of one-to-one harassment three interventions/situations were described by the participants in the group discussions:

- informal solution;
- formal complaints procedure; and
- no internal solution possible.

All three solutions were already being used in Dutch companies, but the informal solution seemed to be the most effective for victims and their relationships with co-workers. It is the practical experience of experts in the field that a formal complaint procedure often leads to a win-lose fight with only losers in the end. Even if victims 'win' in a complaint procedure, they almost always leave the organisation after some time, because they cannot carry on working with their co-workers and supervisors, who all know about the ins and outs of the fight. Confidential counsellors report, however, that a formal complaint procedure is often chosen as an intervention strategy before trying to reach an informal solution (sometimes this is due to a confidential counsellor without training). Labour Inspection mentioned during the group discussions that victims, feeling powerless, look for help in external organisations (e.g. unions or Labour Inspection). The organisation then sees the victim as a betrayer, and an internal (informal) solution may become difficult or even impossible. Thus, when intervening, the order of steps in the intervention is very important. The start should be as informal and as close to the offender as possible. If the strategy doesn't work, a following (more escalating) step may be necessary.

Informal solutions

Victim talks to perpetrator

Confidential counsellors mentioned during the group discussions that an offender may not always be aware of the fact that he/she is really bothering somebody. Sometimes a victim, after having seen the confidential counsellor, may feel strong enough to talk to the offender about his/her behaviour. In such cases the victim must describe the undesirable behaviour and ask the offender to stop. An advantage of this strategy, if it works, is that the problem may not escalate any further. Everything remains between victim and offender, and the offender can stop the behaviour without losing face. In the case of very severe one-to-one harassment (physical aggression or rape), an informal solution is, of course, not appropriate.

With this intervention strategy (if successful) there is no need for extensive after-care. The confidential counsellor only inquires, after some time, whether the undesirable behaviour has stopped permanently.

Mediation

Confidential counsellors have sometimes observed that a victim doesn't have the courage to talk to the offender or that the offender will not listen. If the offender doesn't deny the behaviour, then mediation can be a solution. Mediation has the advantage that it doesn't lead to sanctions against the offender or to the offender 'losing face'. When mediation succeeds the offender will realise that his/her behaviour is found undesirable, and agreements about future behaviour can be made. Mediation is not the task of the confidential counsellor because he/she should always be on the victim's side. A mediator, however, should be independent. This role could be filled by an occupational welfare worker, or an external mediator may be hired by the organisation.

The after-care in the case of mediation consists of the confidential counsellor staying in contact with the victim and giving support. The mediator should also keep an eye on whether the agreements are being followed.

Mediation/correction by the supervisor

When the offender is not open to mediation by an occupational welfare worker or an external mediator, a last informal possibility is mediation or correction by the supervisor. Personal experience suggests that getting the supervisor involved increases the risk of escalation of the situation because it is him/her who also evaluates the general work and efforts of both victim and offender. Thereby the situation (certainly, if a correction is given) may easily turn into a win-lose fight. If the victim 'wins', the offender may have feelings of rancour as well as wishes for revenge. The supervisor should be aware of this potential problem.

After-care in such cases implies that the supervisor should monitor whether the situation is improving and whether agreements between victim and offender are being kept. Also, the confidential counsellor should keep in touch with the victim and check if the situation for the victim has actually improved.

Formal complaints procedure

If an informal strategy turns out to fail or to be impossible, for example because the offender denies the behaviour, a victim may formulate a formal complaint. A formal complaints procedure may also be appropriate if the behaviour is too serious to be suitable for an informal solution. Generally, a grievance committee considers a formal complaint. In some organisations the grievance committee consists of internal members only; in others it consists of external experts, completely or in part. By hearing and re-hearing the plaintiff, the accused and the possible witnesses, the plausibility of the complaint is assessed. During the complaints procedure, the plaintiff is supported by the confidential counsellor. Except for external lawyers there is seldom anyone willing to support the accused (and there are at the moment no persons or institutions opting for this in The Netherlands). Normally, the grievance committee gives advice to the management of an organisation how the situation should be managed and solved. This advice should also include how one should arrange proper after-care. If a complaint is evaluated as plausible by the grievance committee, advice is often to transfer or to dismiss the accused. However, management can, of course, choose either to follow the advice or disregard it (with their justification).

It was felt by the members of the group discussions that in the case of a complaints procedure, the issue of after-care is very important. Besides after-care for the victim and the offender/accused, the department where the undesirable behaviour has happened should be informed. This information also works preventively. If rumours move quickly in the organisation, the information should be given to the entire organisation. When, on advice of the grievance committee, an offender is transferred, his/her new supervisor should also follow him/her up in order to prevent further undesirable behaviour.

No internal solution possible

Sometimes things turn out quite differently than has been described above. A particular difficult situation may arise if the offender has great (economic) value to the organisation. For instance, a company medical officer during group discussion claimed:

> Imagine that an IT-specialist, who is crucial for the survival of the organisation, harasses a person; will you shut the organisation down? We always assume a complaints procedure will do justice. But justice is often in conflict with organisational interests.

When great organisational interests are at stake the reporting and complaints procedure often fails to succeed. The confidential counsellor may not want to be involved too much, the supervisor may not want to mediate, and management may not want to follow the advice of the grievance committee.

It is generally known that it is most simple to blame the victim for this situation; 'he/she is just a difficult person'. The offender remains in the organisation, the victim leaves, sometimes due to illness, sometimes through dismissal. This situation even creates the risk of the offender feeling rewarded for his/her behaviour. Also, other employees of the organisation may observe that undesirable behaviour is tolerated by the organisation, or even rewarded. The law of the jungle may then begin to rule, as was recognised by the participants in the group discussions.

Although this kind of solution is not desirable, we should not close our eyes to it, because after-care is highly important in such situations. In such cases, it is necessary for the organisation to do everything to help the victim – perhaps by transferring him/her to another department or using outplacement procedures to help the victim find another job elsewhere. The labour unions must also serve the interests of the victim in these cases. Furthermore, the trade union should contact the Occupational Health Services and Labour Inspection to recheck the policy of the organisation.

For obvious reasons the organisation probably wants to drop the issue as quickly as possible, as was admitted by management during the group discussions. However, all participants in the group discussions saw the importance of management finally confronting the offender with his/her behaviour. Also, management should explain the policy to employees again. It should be made clear that organisational policy has indeed failed. Nobody should be proud of that, especially not the offender.

A systematic approach

The model that was presented here suggests the specific responsibilities of the various professional disciplines involved with undesirable interaction at work in five different phases, ranging from prevention to repression. It may give lead to organisations to develop a systematic approach to the establishment of a policy to prevent and overcome undesirable behaviour at work.

Bibliography

Hubert, A. B. and Scholten, C. M. (2000) *Ongewenste omgangsvormen op het werk. Een onderzoek naar pesten, seksuele intimidatie en racisme* (Undesirable interaction at work. A study on bullying, sexual harassment, and racism). Amsterdam: FNV.

Leymann, H. (1996) The content and development of mobbing at work. *European Journal of Work and Organisational Psychology*, 5, 2, 165–184.

Soethout, J. and Sloep, M. (2000) *Evaluatie Arbowet over seksuele intimidatie, agressie en geweld en pesten op het werk* (Evaluation of the Working Conditions Act on sexual harassment, aggression and violence and bullying at work). The Hague: Elsevier bedrijfsinformatie bv.

Stichting van de Arbeid (1999) *Met alle respect! Over bedrijfscultuur en omgangsvormen op de werkplek* (With all due respect! On business culture and conduct at work). Publication no. 7/99, 14 December. The Hague: Stichting van de Arbeid.

18 Challenging workplace bullying in a developing country

The example of South Africa

Susan Marais-Steinman

Workplace bullying in South Africa

'Africa is not for sissies – dare to succeed.' This billboard sign is prominently displayed on one of the busiest highways in South Africa. A culture of 'toughness'? Or does the billboard reflect the observation of psychologists and traumatologists locally and abroad that there is something like 'developing world resilience'? It is the slogan of a nation battered by violent crimes, job losses, the HIV/AIDS holocaust and a myriad of socio-economic problems of which workplace bullying is one.

South African societies are characterised by the exceptionally high levels of trauma at all levels. The traumatic experiences of the notorious 'Apartheid Wars' and The Struggle against the regime blew the basic tissues of social life apart, resulting in collective and individual traumas.

The prevailing high levels of crime and violence have resulted in trauma of *pandemic* proportions in South Africa transcending all racial, economic, gender, age, status, national and geographic divides. There is thus an urgent and dramatic need for traumatological intervention on an unprecedented scale. Trauma sufferers, without proper trauma intervention, are prevented from resuming and leading normal and productive lives. Despite the levels of trauma encountered in South Africa very little has been done, or is being done, to reverse the effects of trauma.

South Africa has always played a leading role in labour legislation in the world. A founder member of the International Labour Organisation (ILO), South Africa was the first country in the world to codify labour legislation. After the mineworkers' strike in 1922, the government introduced the Industrial Conciliation Act in 1924. Yet, a serious problem such as workplace bullying, like societal trauma, is not receiving its due attention from the authorities. In a developing country one is constantly challenged with the situational dynamics of society. There is always a more serious threat, there are always more urgent priorities. These realities apply when dealing with workplace bullying too. However, the argument that workplace bullying is part of the bigger, serious challenges facing the workplace, is gaining momentum.

Magnitude of the problem

Workplace bullying is reaching abnormal and unacceptable levels in South Africa. During 1998–1999 an Internet communication survey was conducted, as well as a survey which took place while training businesses in communication and interpersonal relationships. Of respondents, 78 per cent reported that they had been victimised at least once during their careers. These results had to be interpreted in terms of this high victim profile. The respondents ascribed classic symptoms of depression in communication to the profile of a victim, while verbal abuse, withholding information, sarcasm and aggressive body language fitted the profile of perpetrators of workplace bullying. South Africans appeared to be more stressed than northern hemisphere counterparts. Part of this questionnaire could be compared with a USA Gallup Poll (1992) on the annoying aspects of communication, for example interrupting while others talk, mumbling, talking in a too soft or too loud voice, fast and slow talking etc.[1] The South African response only correlated with the US poll on the identification of high and low irritators, but in general South Africans responded with 66 per cent more irritation than their US counterparts. South Africans are angry and easily irritated in interpersonal communication.

During 2001 research on workplace bullying was conducted at a large employer. The organisation was undergoing major changes and the workers were uncertain of their positions. The sample (n = 70) represented the lower levels in the hierarchy, representative of the demographics of the country. The results were astounding: 68 per cent of the respondents were very stressed, but not all due to bullying. Various stressors in the workplace were identified and tested, such as job environment, job structure and description, work ethics, change and company policies. Significantly:

- 25 per cent affirmed 'often' and 'way too much' dreading to go to work because of interpersonal relationships with their peers and supervisors being less than pleasant;
- 30 per cent of respondents experienced bullying 'not too often';
- 10 per cent of respondents experienced physical threats on a daily basis;
- 2.5 per cent are physically attacked on a daily basis;
- 60 per cent affirmed that they were negative towards the changes the organisation was implementing and felt that it would not improve the organisation; and
- 60 per cent affirmed being angry with their employer.

This was indeed an explosive situation and one that is symptomatic of many South African organisations undergoing major change. Although a much more comprehensive and would-be conclusive survey is under way, these indications are undesirable and one can only interpret this within the socio-economic realities.

Unique and specific to South Africa

Specific and unique socio-economic factors impact significantly on the level of workplace violence and these factors make it difficult to deal with workplace bullying in South Africa. Such factors are, for instance, the high stress levels reported in the workplace due to a variety of socio-economic factors, crime and violence being most prominent. Workers living with HIV/AIDS find themselves challenged, and despite the openness, sufferers would rather keep quiet. Affirmative Action also poses new challenges and problems unique to this society, and there are no easy answers. A rise in crimes and misdemeanours such as fraud, corruption and nepotism could be attributed to poverty, the high unemployment figures and general moral decay. These issues are dealt with in more detail elsewhere in this chapter.

Section 10 of the South African constitution states: 'Everyone has inherent dignity and a right to have their dignity respected and protected.' When dealing with workplace bullying, the constitution is considered together with the labour laws of South Africa and particularly the Employment Equity Act (EEA) (Sections 5–11), as well as the Basic Conditions of Employment Act (Section 78–81) and the Labour Relations Act (Schedule 7(2)). These acts explicitly prohibit direct or indirect unfair discrimination against an employee on any grounds including race, gender, sex, pregnancy, marital status, family responsibility, ethnic or social origin, colour, sexual orientation, age, disability, religion, HIV status, conscience, belief, political opinion, culture, language and birth. 'Harassment of an employee is a form of unfair discrimination and is prohibited on any one, or a combination of grounds of unfair discrimination listed in subsection (1)' (of Section 6 of the EEA) states the position clearly. While the labour laws implicitly deal with workplace violence referred to as discrimination, harassment and physical attacks, these laws need to be augmented by a code of conduct dealing with workplace violence, as the mere 37 per cent success rate (individual commissioners of the Council for Conciliation, Mediation and Arbitration (CCMA) put it at a 10 per cent success rate in the case of workplace bullying) of employees claiming constructive dismissal reported by the CCMA emphasises the need for an effective code of conduct in labour legislation.

The implications of fighting workplace bullying in a developing country where the HIV/AIDS holocaust is taking its toll, where poverty and employment are still the people's enemies are that the privileged class, those who are employed, should – or, so it is argued, should not – complain about bullying.

Some individual trade unions are paying attention to the problem, requesting workshops and information. There is close liaison between the Foundation for the Study of Work Trauma (FOSWOT) and the labour movement, and it can be expected that, once the full force of South Africa's vibrant labour movement addresses this problem, it will become a major issue and earn its rightful place on the agenda of the federations.

South Africans are particularly aware of human rights. There is a concerted effort and commitment to wipe out all forms of discrimination, as is evident in the Basic Conditions of Employment Act (BCEA),[2] the Labour Relations Act (LRA)[3] and the Employment Equity Act.

Difficulties in dealing with workplace bullying

To understand and combat the occurrence of workplace bullying requires a systemic approach and an insight into the socio-economic challenges facing a developing country, and post-apartheid South Africa in particular.

The popular arguments for not attending to workplace violence, namely that there is a great deal of violence in society in general and that it is simply not economically viable to fight workplace bullying, has been refuted by Dr Vittorio Di Martino of the International Labour Organisation in Geneva. He pointed out that in a situation of generalised violence it is necessary to 'concentrate efforts on one initial, well-selected target, where the possibilities of success are quite high, rather than wasting resources in an attempt to solve all problems at the same time'.[4] The workplace is a suitable forum to address the problem of violence because the unique conditions and expertise within the workplace provide a very solid basis for combating violence efficiently: tackle violence in society by starting in the workplace. While the cost of violence represents a serious, sometimes lethal threat to the efficiency and success of organisations, violence-free organisations show the benefits and would pave the way to new initiatives in this area.

While recognising that the problem exists, some trade union federations feel that workplace bullying should take a backseat for the time being, while the efforts of the Foundation for the Study of Work Trauma are encouraged. The problems making it particularly difficult to deal with workplace bullying in South Africa are:

1 A significant number of employees have been traumatised by violent crimes, and this affects performance and behaviour in the workplace. An article on the front page of *Sunday Times*, the largest Sunday paper in the country, announced in June 2000 that 44 per cent of South Africans lived in fear of crime. Society is traumatised. Employees arrive at the workplace, traumatised by an attempted or successful hijacking on their way to work, or by having been threatened at knife-point for a watch. Traumatologists in the workplace responding to trauma outside or at the workplace (armed robberies) have become a necessity, and certainly a bigger priority than attending to workplace bullying, in the present corporate mindset. High stress levels seem to fuel workplace bullying.
2 The growing unemployment figure (37 per cent)[5] has increased poverty levels. The majority of the unemployed are African, and they

are either illiterate or poorly educated. 'This means that blacks occupy the unskilled labour sector while whites and a small minority of blacks are regarded as skilled workers.'[6] However, it was announced in local newspapers that the total income of the 'black' population had exceeded that of the 'white' population in 2000 for the first time in the history of South Africa.

3 While affirmative action programmes are compulsory and vigorously introduced by all major corporations in South Africa, there is still a lack of commitment on the part of some companies to train such appointees and introduce meaningful empowerment.

4 Affirmative Action plays a significant role in the bullying scenario. Surprisingly, the black workers lower down in the hierarchy openly stated in focus group discussions that they had no racial preferences for 'bosses', but did not approve of their black supervisors mainly because they don't see them as being entitled to be 'bosses'. During the apartheid struggle these black supervisors were their equals – 'brothers and sisters' – and this is a significant problem in black empowerment facing the black manager: to exercise authority without being accused of bullying by his/her 'own'. Training to change the 'brother/sister culture' is of utmost importance. This culture could also influence perceptions of bullying and bullying statistics.

5 Racial discrimination is still a problem in the workplace. According to COSATU's[7] input to the South African Human Rights Commission Public Hearings in the KwaZulu-Natal Region on 26 June 2000, 'typical examples of racism in the workplace are where a company retrenches some of its employees and all those facing retrenchment happen to be African'. Racism is more subtle or hidden. 'There are instances, during job applications, where whites receive first preference. Or white employees get better promotions and the best skills development programmes. In these cases their African colleagues get few promotions and meaningless skills development programmes.' Because Affirmative Action is not based solely on merit, but very necessary to address past injustices, very competent and accomplished individuals benefiting from this system are viewed with suspicion and are being bullied or find themselves in some instances in positions with the status, but given insignificant tasks and too little decision-making power.

6 The criteria for workplace bullying and racial discrimination seem to overlap. Incidents had been reported to the Foundation for the Study of Work Trauma of white employers being racist, while on investigation it had been found that they were bullies and harassed their black and white employees equally. When a white employee reported a black supervisor harassing her, the reaction was that she was a racist, despite the fact that she had a bona fide case of sexual harassment, with her black female colleagues supporting her claims. Racial discrimination

should be divorced from workplace bullying because of the sensitivities and misconceptions, as well as the dangerous possibility of scapegoating the victim. This is an unhealthy mindset and poses a real danger in addressing bullying. Racism is real, but the notion that only whites could be guilty of this form of bullying is harmful and discriminatory in essence. On the positive side the Foundation for the Study of Work Trauma found that, in the majority of cases where racial discrimination had been reported, the reporting parties were of the same colour as the perpetrator.

7 There is resistance from the traditionally 'white' trade unions, in particular the Mineworkers' Union (MWU), who argues that white workers, in particular white men, are discriminated against and disadvantaged in the job market. The MWU feels that their white male members' human rights are being infringed through Affirmative Action, and that white workers in general are being traumatised through job losses and early retirements.

8 The World Health Organisation (WHO) reports that 13.5 per cent of the South African workforce is HIV positive and there are 1,500 new infections daily. A large majority of those infected are working-class people, and the majority of them are women. The United Nations predict that the country's economic growth rate will decline by 0.3 per cent – 0.4 per cent a year, resulting in a 17 per cent drop in the gross domestic product. While the law prohibits discrimination against persons living with HIV/AIDS, these employees are often ostracised and bullied in the workplace, and there are brave attempts to break the silence.

9 Fraud, corruption, bribery and nepotism are a big problem in the government and private sector. Whistle-blowers' lives are in danger. A human resources manager at a prison in KwaZulu-Natal was recently murdered during an investigation of bribes for jobs at the prison. This is part of the 'moral decay' of the country that needs to be addressed, according to President Thabo Mbeki.

10 Poverty is a major problem. At COSATU's Seventh National Congress, held from 18–21 September 2000, the federation resolved that the state and business must contribute to the elimination of poverty through the establishment of a Basic Income Grant. It is envisaged that such a Basic Income Grant would help alleviate the poverty and oversupply in the labour market responsible for the exploitation of semi-skilled and unskilled workers.

11 Mergers and the regrouping of business concerns, retrenchments, and the privatisation of state-owned corporations remain highly emotional issues. During 2000 the minister of finance intervened and stopped the merger of two major banking groups, Nedbank and Standard Bank, as it would have led to 15,000 lay-offs with dire socio-economic consequences.

12 The working conditions and wages of particularly vulnerable groups of employees such as farm workers, domestic aids and those working in the 'informal sector' are poor. Recently incidents of violence against farm workers by their employers were highlighted, and this focused attention on the vulnerability and exploitation of this sector of the workforce.
13 While South Africans in general embrace the cultural diversity of the 'rainbow nation', the multiplicity of languages is not a significant factor in workplace bullying. English is, in general, the language of record. The racial discrimination laws in the Labour Relations Act are very explicit about any form of discrimination and harassment in this respect. Therefore bullying based on one's language proficiency or cultural disposition is very subtle, underhanded and difficult to prove. English is a second language to the majority of South Africans, resulting in accents and leading to ridicule in an underhanded manner. This happens behind closed doors and in politically 'safe' environments and is therefore difficult to expose.

These factors render South Africans extremely vulnerable to bullying in the workplace, and the high incidence of workplace bullying in the country is no surprise.

Activities against workplace bullying

Against this background the progress in research, raising awareness, designing interventions, challenging existing labour laws and changing the mindset of business towards the achievements in the battle against workplace bullying have been by no means modest.

Sensitising the public to the problem

Awareness of workplace bullying was reported in the regional newspaper The EastRander in July 1994. The term 'Corporate Hyena' or bully was coined, and a typology of office denizens described. The book Corporate hyenas at work (Marais and Herman, 1997) was published in 1997 and the then minister of labour, Tito Mboweni launched the publication. In a country with eleven official languages, the metaphor of the hyena cut through the cultural, racial and language divides with its narrative style. The limitations of the academic mould had been bypassed to reach workers at all levels. While the publication represented the first salvos in the battle against workplace bullying, the media played a major role in raising public awareness.

The years 1997–1999 represent a steady increase in public awareness although business was still to recognise the problem. It was therefore important that the second phase of awareness be introduced: that is, to obtain recognition from business.

Engaging business in South Africa

While the public was sensitised through the media and Internet, business clearly had to be reached in a different way. Interpersonal relationships and office politics formed part of presentations in communication. In the early stages of raising awareness, this method provided the opportunity to reach the business community with the message of workplace bullying. Once the door was opened, management listened and acknowledged that it was a problem. The specialised journals, communication skills workshops and the Internet initially opened corporate doors, and the free surveys conducted during the workshops provided the opportunity to intervene and influence business.

Business is realising more and more that there is a price to be paid for allowing bullying. Businesses are now readily accepting that through addressing workplace violence productivity would be boosted, and they are eager to receive training to prevent and combat bullying. It is expected that the ILO-based programme SOLVE would be extremely helpful in introducing these initiatives as part of the Employee Assistance Programme in the workplace.

Using surveys

Surveys are a powerful tool when dealing with business. To engage the business community, it is inevitable that it must be linked to their agenda: performance, productivity, profit, team building, conflict resolution, risk management, health and safety, corporate culture, communication and change management.

A joint research project initiated by the ILO, the WHO and the International Council for Nurses to investigate workplace violence in the health sector is currently under way in South Africa.

The work of the Foundation for the Study of Work Trauma (FOSWOT)

While the organised labour movement and state department are not yet taking any concrete action to combat workplace bullying, FOSWOT, a Section 21 non-profit company, is presently the only body in South Africa dedicated solely to the emotional well-being of employees, and in particular to the prevention of trauma caused by workplace bulling or violence (physical and emotional), by

- supporting postgraduates studying the phenomenon;

- giving the victims a voice. One of the aims of this organisation is to intervene, assist and empower victims;
- recommending appropriate interventions for business experiencing problems during change and restructuring;
- providing an expert witness service at labour tribunals;
- researching the phenomenon and providing the public with the necessary information;
- empowering business to deal with adverse behaviour in the workplace;
- networking with relevant organisations to benefit work trauma survivors;
- providing training in workplace bullying for trade unions;
- giving advocacy and actively participating in improving the labour laws and codes;
- exposing South Africans to debate with overseas colleagues and facilitating an international discussion group on work trauma;
- introducing ILO-based programme SOLVE to the workplace as a solution to emerging workplace problems; and
- guiding initiatives taking the following into account:
    - the socio-economic realities requiring a systemic approach to the problem;
    - the levels of education and/or what is at stake for the different audiences;
    - the fact that the restructuring of the South African workplace poses immense challenges and could affect interpersonal relationships;
    - the necessity of extensive consultation and networking with labour-related organisations and other stakeholders;
    - the fact that workplace bullying is not viewed as a personality issue, but a phenomenon, with its roots in the socio-economic dynamics of society and the restructuring and realignment of the world economy;
    - the knowledge that the importance of reconciliation in the process of normalising the workplace cannot be underestimated;
    - the multiplicity of languages and cultures in South Africa;
    - the need for a code of conduct to combat workplace violence and corporate aggression; and
    - the need to destigmatise retrenchees and play an advocacy role to create awareness that this group must have equal access to the labour market and, if necessary, be included in the Affirmative Action quota through an amendment to the Employment Equity Act (EEA)[8] to address the problem effectively.

Some major employers use the services of traumatologists to counsel employees after incidents like robberies, car hijacking and/or the death of loved ones through such incidents. The Foundation has built a network with traumatologists and psychologists to address the problem and provide employers with information.

The South African branch of the Commonwealth Open University is contemplating offering a course in workplace trauma, and has engaged the services of FOSWOT for this purpose. It is envisaged that the course would equip human resources specialists and therapists with adequate knowledge to recognise and to deal effectively with the phenomenon in the workplace, as well as the effects of emotional abuse on the individual(s).

The Council of Counsellors of South Africa is in the process of making the registration of work trauma counsellors compulsory under Act 110 of 1978, the South African Social Service Professions Act, as amended. FOSWOT is set to play an important role in this statutory body.

The role of forgiveness and reconciliation – the TRC perspective

Although South Africa may be a step behind northern hemisphere countries in taking action against workplace violence, this country can offer the world a very special wisdom that could be helpful to therapists in assisting the victims. The experiences of trauma survivors who witnessed before the Truth and Reconciliation Council (TRC), could very well become a blueprint for those dealing with the perpetrators and victims of workplace bullying. Debbie Kaminer of the Department of Psychiatry of the University of Cape Town has presented the research of the Medical Research Council's Unit on Anxiety and Stress Disorders on how useful the TRC is as a model for how to deal with victims and perpetrators in the workplace (Kaminer, 2000).

The aim of the study, which was conducted on the psychological effects of the TRC, was to assess whether testifying at the TRC led to psychological healing (that is, a reduction in psychiatric symptoms) and forgiveness. While trauma and specifically post-traumatic stress disorder (PTSD) is indeed the most worrying and devastating result of workplace bullying, this study is indicative of the treatment that could accelerate the healing process for victims.

A sample (n = 134) of survivors was stratified into three groups:

- those who gave public testimony at the TRC;
- those who gave a closed statement to a TRC official; and
- those who did not give any form of testimony to the TRC.

Each subject completed instruments measuring exposure to human rights abuses, exposure to other traumatic events, current psychiatric status and attitudes of forgiveness towards the perpetrator(s). Subjects who gave public testimony had significantly less PTSD than those who had given either a closed statement or no testimony, while those who had given a closed statement were as likely to have PTSD as those who had no contact with the TRC. There were no significant differences in the rates of depression and anxiety disorders other than PTSD across the three groups. There were no significant differences in mean forgiveness scores between the three groups. Admission of guilt, explanation of their actions, and an apology by the perpetrator did not lead to increased forgiveness on the part of the victim. The forgiveness of the perpetrator plays a significant role in the recovery of the victim.

The impact of this research with regard to the workplace context suggests that:

- when workplace bullying is discovered, it is important that the victim's experience be validated and that the victim be allowed to talk about the ordeal;
- organisations could explore the use of a restorative justice ritual, which may form part of the survivor–perpetrator mediation process. For example, there may be some behavioural ritual whereby the perpetrator expresses remorse and offers reparation (either material or symbolic), and the survivor enacts a ritual to show his/her forgiveness. Another alternative may be a ritual where the organisation as a whole offers material or symbolic compensation to victims, as the TRC is attempting to do on a national scale; and
- victims must be encouraged genuinely to forgive their perpetrators as this is very important for their recovery.

Code of conduct for South Africa

It can be expected that South Africa will, in the foreseeable future, accept a code of conduct to deal with workplace violence as part of labour legislation. The draft code of conduct developed by FOSWOT lists all the behaviours, physical and emotional, that would constitute workplace violence within the broad definitions accepted by international agencies of the United Nations, such as the ILO, the WHO and international trade union federations. The code distinguishes workplace violence from conflicts, personality clashes and office discipline. Provisions for grievance hearings as well as appropriate steps to rectify and prevent such behaviours are considered, with emphasis on therapy for both victim and perpetrator.

Once the process of consultation with all stakeholders has been completed, a consolidated report would see the light and be referred to the tripartite body NEDLAC, which represents government, organised labour and business, for comments and consideration that it be included in the Employment Equity Act.

Conclusion

Challenging workplace bullying is more complex in the developing world: there are more impacting factors, more urgent problems, and there is more resistance. Despite the shortage of resources and support for individuals and organisations, workplace bullying is receiving attention. South Africa has arrived and will play an increasingly important international role in the struggle against abuse in the workplace.

Workplace bullying is an infringement of human rights. Having dignity and respect adequately entrenched in labour laws and in corporate cultures require commitment, advocacy and the voice of the victims. In South Africa our beliefs and assumptions are challenged on a daily basis – this is a dynamic society and our salvation lies in our ability to change. This problem is being challenged with the same vigour as are racial discrimination and violence. The realisation that we can heal our society by starting in the workplace is gaining momentum. The work and the experience of the TRC, together with the new developments and the establishment of a statutory body, the code of conduct dealing with workplace violence currently being considered, are ample proof that South Africa is geared up to deal with this workplace challenge in an innovative way.

Notes

1 Information obtained from The Voice Clinic, Saxonwold, Johannesburg.
2 Act 75 of 1997.
3 Act 66 of 1995.
4 Extract from the paper by Dr Vittorio Di Martino, prepared for the conference described in 1 above.
5 Although the official figure is in the early 30s, organised business and labour estimate it to be 37 per cent while some would put the figure even higher.
6 COSATU's input to the SAHRC Public Hearings on Racism KwaZulu-Natal Region, 26 June 2000.
7 Congress of South African Trade Unions.
8 Act 55 of 1998.

Bibliography

Kaminer, D. (2000) Research into the psychological effects for trauma survivors who witnessed before the Truth and Reconciliation Council. Keynote address at the *International conference on work trauma,* Johannesburg, 9 November 2000.

Marais, S. and Herman, M. (1997) *Corporate hyenas at work*. Pretoria: Kagiso Publishers.

Part 5

Remedial actions

A critical outlook

19 Bullying from a risk management perspective

Anne Spurgeon

Introduction

During the last decade bullying has become the legitimate concern of a number of different professional groups involved in the welfare of people at work. One such profession is that of occupational health, comprising in particular occupational health nurses and physicians, but also a range of other specialist practitioners with expertise in health and safety management.

In the past, occupational health professionals have been concerned primarily with ensuring that workers are protected from harmful exposure to a range of physical, chemical and biological hazards which might be encountered in the workplace. With modern reductions in occupational disease, however, the focus has shifted from these more traditional concerns towards problems which may involve a range of psychosocial as well as physical determinants (Crawford and Bolas, 1996; Spurgeon *et al.*, 1997; Devereux *et al.*, 1999). Improvements in physical conditions at work, as well as in society as a whole, have meant that health expectations have risen beyond those which consist simply of an absence of disease. Instead there is now more frequent expression of the need for a more general sense of 'well-being', a concept which recognises both the psychological and physical components of health.

The role of the occupational health professional has thus become much wider than hitherto, and certainly now encompasses a remit to tackle problems which come under the broad heading of 'occupationally-related stress'. Included in this would be issues relating to various types of harassment.

Given their increasing involvement in this field, it is important to recognise the particular perspective which occupational health practitioners bring to psychosocial issues in general, and the problem of workplace bullying in particular. Modern health and safety practice, both in occupational settings and in the wider community environment, is carried out within a well-defined framework known as risk management, of which risk assessment is an essential component (Rampal and Sadhra, 1999). This approach has been applied successfully over many years to a wide range of physical and chemical hazards, and is one that is readily understood and endorsed by occupational health professionals throughout the world. Consciously or unconsciously, therefore, many such professionals are likely to approach psychosocial issues from a perspective which assumes the need for structured risk assessment. This raises some interesting questions about the applicability of such an approach to less traditional and psychosocially based health concerns. A number of elements of the risk management process need to be explored in this context, for example its aims and objectives, its underlying assumptions and its methodological approach.

This chapter therefore discusses these issues and attempts to describe how the principles and practices and also the language of risk management might be translated into a useful policy on workplace bullying. It considers the strengths and possible difficulties involved in such an approach, and the potential contribution it might make to prevention and remediation in this field.

The risk management approach

Before considering how the risk management approach might be applied to the problem of bullying at work it is important to be clear about the essential elements of the framework itself, its assumptions and its implications for employers and employees. A basic tenet of the approach is that risks in the workplace, as in society in general, cannot be entirely eliminated but can to a certain extent be controlled. Determination of the level to which they are controlled in a given situation requires an analysis of the perceived costs and benefits (economic, health and social) to those affected (HSE, 1993). The aim of risk management, therefore, is to minimise rather than remove the risk of harm, or more precisely to reduce the risk to a level which is acceptable to the people directly concerned. Thus an important distinction is made between a 'hazard', something which has the *potential* to cause harm, and 'risk', the *probability* that such harm will occur (HSE, 1993).

Effective health and safety management is normally considered to require a series of sequential steps, which operate in a continuous feedback loop and which contain the specific objectives of the process (Rampal and Sadhra, 1999). There is first a requirement to identify and define the problem, or in risk management terms, the hazard. Second, there is a need to assess the frequency and severity of the problem, the number of people affected and the factors present in the workplace which are likely to increase the risk of occurrence. This is essentially the process of risk assessment. On the basis of this assessment, some form of preventative or remedial action should be implemented, and finally there is a requirement to carry out an evaluation of the effectiveness of this action. Results of the evaluation thus inform further risk assessment. The primary emphasis, therefore, is on the prevention of unacceptable harm, with provision for continuous monitoring of the health of the workforce to ensure the effectiveness of the measures in place. In the UK this process is enshrined in health and safety law, to the extent that every employer is required 'to make an assessment of the risks to health and safety of employees of activities associated with potential hazards in the working environment' (HSE, 2000).

Clearly such an approach was initially developed with physical and chemical hazards in mind. To apply the process to a psychosocial issue therefore requires that the problem first be defined in risk management terms; for example, that workplace bullying is seen as an occupational hazard, in this case a *psychosocial* hazard. Many occupational health practitioners have in fact already chosen to view other aspects of occupational stress in these terms and have attempted to apply the process accordingly. Indeed, recent guidelines on tackling occupational stress published by the UK Health and Safety Executive approach the problem from this perspective (HSE, 2001).

An important underlying assumption of the risk management approach is that the hazard in question emanates from the workplace. Thus although the individual worker may in some circumstances be thought to have exacerbated the problem, for example by engaging in unsafe or inappropriate working practices, the ultimate responsibility for controlling exposure to the hazard lies with the employer. In this sense the approach is compatible with an organisational view of psychosocial issues, which emphasises the role of the workplace rather than the individual in generating particular problems (Cox, 1993). On the other hand, it lends itself rather less well to situations where the recipient of negative effects might be viewed as having themselves played an active or even passive role in the development of events or circumstances – one possible scenario where psychosocial issues are involved.

Consequent on this view of the workplace as the source of the hazard is the assumption that wherever possible hazards should also be removed at the source, rather than there be any requirement for the individual to make adjustments to cope with an unsatisfactory environment. In conventional health and safety terms this is called the 'hierarchy of control' whereby, for example, the use of personal protective clothing to reduce exposure to toxic vapours (gloves, masks, air-fed helmets, etc.) would be seen as the least acceptable option, to be resorted to only when all external engineering controls have proved to be technically impossible (Molyneux, 1999). Again, there are parallels with an organisational approach to stress management. In this context, training in individually based coping techniques (secondary prevention) would be regarded as an inadequate substitute for an approach which focuses on removing the organisational sources of stress (primary prevention). Similarly, the need for rehabilitation or treatment represents to some extent a failure of any preventative system. Viewed in these terms many of the approaches to workplace bullying which are primarily aimed at prevention (and notably organisationally-based prevention) would appear to fit very readily into a conventional risk management model. Conversely, however, rehabilitation and remediation, which in the real world would seem to be inevitable and necessary elements of any comprehensive policy on bullying, sit rather less easily within such a framework. Thus an important limitation of the risk management process as applied to psychosocial issues is its lack of emphasis on any form of therapeutic intervention. Essentially it is concerned with risk reduction and control, not with treatment. Within these limitations, however, how far can other elements of the process be usefully applied to the specific problem of bullying?

Problem (hazard) identification

Identifying the presence of a potential hazard constitutes the first step in any risk management process. It can be seen, therefore, that an immediate requirement is an agreed definition of the hazard in question, in this case a consensus within the workplace regarding unacceptable behaviours. Inevitably heavy reliance is placed here on workers' perceptions, which may vary not only between individuals, but also within the same individual over time. Further, notions of what constitutes bullying is likely to be culture-specific, both at a national level (creating particular difficulties for multinational companies) and at an organisational level, where, for example, there may be acceptance in certain 'tough' environments of behaviours regarded as intolerable in other settings (Pavett and Morris, 1995). Finally, the perceptions or 'mental models' of the professionals who encounter the problem are likely to have a significant influence on their interpretation of what a particular individual is telling them (Marteau and Johnston, 1990). These difficulties are likely to occupy the intellectual debate on bullying for some time to come and may never be entirely resolved. In the workplace precise definitions are clearly important within a legal or disciplinary framework. However, outside this framework the objective is essentially a practical one, that of developing a form of working definition which will assist in the day-to-day implementation of a preventative policy. As part of an overall risk management cycle, this definition would in any case be subject to continuous review and modification in terms of its practical utility.

The main aspects of bullying which tend to be considered in reaching a working definition are the characteristics of the behaviour and the time-scale in terms of frequency and duration (Björkqvist *et al.*, 1994). For example, some of the following characteristics have been suggested: threats to professional status or reputation, threats to personal standing, physical or social isolation and overwork and destabilisation. In terms of time-scale there appears to be general acceptance that a single incident of abusive behaviour does not constitute bullying, and most writers on this subject refer to behaviour which occurs persistently over a period of time. Obviously such characteristics vary considerably in terms of the extent to which they can be reliably identified, and experience in the longer established field of occupational stress suggests that a degree of flexibility of interpretation will be required if such descriptions are to prove useful in a risk management context. The basic objective would be to create a shared understanding in the organisation of what constitutes bullying and, equally importantly, what does not. Inevitably such an understanding will only approximate to the more precise definitions usually encountered in relation to traditional physical and chemical hazards. To use conventional health and safety language, there is no externally determined standard of what constitutes a safe or unsafe level of 'exposure' to a psychosocial hazard. However, many occupational health professionals have in recent years become entirely familiar with the notion that many hazards cannot be measured on a calibrated scale. Many ergonomic hazards, for example, are described qualitatively and thus subject to variations of interpretation (Pheasant, 1991). Questions of definition, therefore, while undoubtedly challenging, do not necessarily exclude psychosocial hazards from the risk management process.

Assessing the risk

The risk assessment phase of the risk management cycle is essentially about determining whether or not an organisation has a problem in relation to the hazard in question. To this end it seeks to establish the following types of information: the frequency of incidents of bullying in the organisation; the number of people affected; the characteristics of people affected (job type, level, age, gender, etc.); the frequency of incidents where someone's health has been affected; the amount of sickness absence or staff turnover which might be attributable to bullying; and finally the features of the organisation which constitute risk factors.

Answers to many of these questions will again contain large elements of subjectivity and, in most cases, individual perception probably represents the only valid measure of the event. This is something which is entirely familiar to those used to working in psychologically based fields, but may present difficulties for some occupational health professionals or more particularly for some of the managers and employers with whom they work.

Particular questions arise regarding the actual form of information gathering which is most appropriate in an organisational setting. While qualitative approaches are rapidly gaining in popularity in academic circles the sensitivity of this issue and the difficulties surrounding its open discussion in the workplace suggest that formal questionnaires which offer participants anonymity or at least confidentiality may be more productive (Rayner, 1997). Added to this, many employers and managers tend to prefer, and be more readily convinced by, numerical data as a basis for intervention.

Thus, while qualitative approaches may be appropriate in other contexts, notably remediation, the purposes of semi-formal risk assessment may best be served by a quantitative rather than an in-depth qualitative approach.

It is perhaps worth noting that where people express concerns about the validity of self-reported information this more often reflects unease about the possibility of over-reporting rather than under-reporting. Underlying such concerns may be an assumption that certain problems are individually-based and as such should be managed differently from those generated by factors originating in the workplace. Similarly, there may be a desire to separate specific problems of bullying from more general issues of occupational stress. These distinctions, although questionable (Vartia, 1996; Zapf *et al.*, 1996), are likely to represent a significant preoccupation of those operating within a risk reduction framework. Thus there may be a general belief that assessment based on self-report is likely to be overinclusive and hence overestimate the size of the problem. In practice, however, the possibility of under-reporting is probably a more important concern. Psychosocial problems in general and bullying in particular are often poorly understood at all levels of an organisation, which means that employers and employees are ill-equipped to recognise, still less to challenge or report, forms of bullying. This is equally true of those directly involved (bullies and victims) and those who might be termed 'onlookers' or witnesses (Lockhart, 1997). In terms of carrying out a satisfactory risk assessment, therefore, the problem for many organisations is more likely to be one of developing awareness and acceptance, in order that a more realistic estimate of the size and nature of the problem can be obtained.

Associated with the above is the question of staff expertise in what is a relatively new field. Those involved with the assessment of this type of risk need to have a good understanding of the possible health outcomes of continuous bullying and the organisational factors which might encourage certain individuals, groups or indeed the organisation itself to adopt a bullying style. For example, while many organisations will already routinely collect data on sickness absence and staff turnover, it may be necessary to re-examine and perhaps reinterpret that information in the light of emerging insights on the symptoms and sources of bullying. In common with many other workplace risks, therefore, the assessment of bullying requires a degree of competence, and is unlikely to be undertaken successfully by untrained staff (Crabb, 1995).

Training is also an important issue in relation to the immediate results of any risk assessment. Given that open acknowledgement of adult bullying is a relatively recent phenomenon, it is reasonable to assume that many organisations will uncover instances of the problem in the course of an assessment of which they were previously totally unaware. It has frequently been observed that monitoring and investigation of any problem tends to produce an apparent increase in its incidence (Watterson, 1999). Although this is generally accepted as more likely to indicate an increase in reporting than in occurrence, a failure to anticipate such an increase may leave organisations ill-prepared and ill-equipped for the scale of the intervention required. Thus they may find themselves in possession of what might be termed 'guilty knowledge' but lacking the resources to address the problem. All these issues require careful consideration at the outset of any risk assessment process, but more particularly where the territory is unfamiliar and controversial.

Prevention and control

The ultimate objective of the whole risk management process, as its name implies, is to *manage* workplace risks, and the next phase of the cycle is concerned with instituting measures which either prevent or control 'exposure' to the hazard in question. This involves not only carrying out specific actions to limit this exposure but also setting up systems to ensure the implementation and regular evaluation of the actions taken and their effects. Such systems underpin the effectiveness of any specific measures undertaken. Essentially they consist of mechanisms whereby remedial actions are formulated, instituted, maintained and continuously monitored. This allows for early identification of any inadequacies in intervention measures, and ensures that prevention or control is not a single event but an ongoing process subject to review and modification as new circumstances arise.

Primary prevention

This form of intervention is primarily concerned with changing factors in the working environment in order to reduce the possibility that a particular problem will occur. The process relies heavily on the thoroughness of the initial risk assessment in identifying negative features of the organisation, as well as in identifying vulnerable groups in the workforce. Where bullying is concerned, many such features may have been identified in the course of a more general psychosocial risk assessment, for example factors such as workload and content, quality of communication, leadership and opportunities for personal development (Cox *et al.*, 2000). Thus strains produced at both individual and organisational levels may be expressed not only in instances of bullying but also in other forms of conflict and maladaptive behaviour. Interventions are therefore likely to run alongside those organisational strategies formulated to reduce occupational stress at source, and, as such, are entirely compatible with a conventional risk management approach (Spurgeon and Barwell, 2000; Cox *et al.*, 2000).

Remediation and rehabilitation

As noted earlier, the development of comprehensive systems for remediation and rehabilitation represents a different focus from that encountered in the management of physical and chemical risks. There the system is primarily designed to be proactive in nature, with emphasis on prevention via workplace control. Within the general field of occupational stress individually based approaches are a somewhat contentious issue. Many would view them as part of more general management attitudes that require the individual to adjust to the environment rather than vice versa. Certainly it seems reasonable to argue that proactive policies which focus exclusively on the development of individual coping skills are questionable, and that emphasis should first and foremost be on designing a workplace which is conducive to comfort and health. However, to deny any requirement for treatment facilities is perhaps unrealistic. Just as psychosocial hazards can exist in any workplace, so individual distress can occur at any time, regardless of the sophistication of the stress management system in place. Strictly speaking, therefore, interventions based on treatment and rehabilitation, although part of the overall remit of occupational health, do not form part of a conventional risk management system. However, with the extension of such systems to a wider range of psychosocial as well as physical and chemical hazards, there may be grounds for introducing such elements under the general heading of prevention and control.

Evaluation

A final element in any comprehensive risk management system is the facility to evaluate the effectiveness of those measures which have been put in place (Cox *et al.*, 2000). Evaluation is usually considered to consist of three elements which address different aspects of the system:

1 *Content*: are there procedures in place which adequately address all aspects of the problem? In the case of bullying these would presumably consist of mechanisms to inform the workforce about the problem and what to do if they encounter it, a programme of primary preventative measures and perhaps mechanisms for appropriate remediation and rehabilitation
2 *Process*: do the mechanisms in place function as intended? For example, is there evidence that workers report bullying to the appropriate people in the appropriate way, and does their understanding of bullying coincide with that agreed by the organisation? Further, is there evidence that action is taken when bullying is reported. Where primary preventative measures have been recommended does their implementation appear to be under way?
3 *Outcome*: how successful is the system in reducing or eliminating bullying from the organisation? This tends to be the most problematic aspect of evaluation since for psychosocial problems the time between implementation of intervention and positive outcome may be months or even years. Thus there may be grounds for assessing more short-

term aspects in the first instance, for example the outcomes of specific situations.

Conventionally, evaluation processes involve quantitative data, and, as suggested earlier, these are often preferred by the managers who receive them. This is not necessarily a problem since most psychosocial issues are amenable to some form of quantification. For example, in the evaluation of workplace counselling services it is normal to report the number of people using the service, the type of problem presented and the degree of success in outcome, in addition perhaps to systematic surveys which canvass wider opinion (Cooper *et al.*, 1990). Regardless of the chosen methodology however, there is widespread agreement that systems for evaluation should be instituted at the inception of any policy, both to provide baseline data and to ensure that all elements of the policy are continuously monitored and reviewed. Evaluation is intended to inform future risk assessment and thus to close the loop of the risk management cycle.

The role of occupational health

It was observed at the outset that the process described here represents an approach which is familiar to most occupational health practitioners and represents the perspective they are most likely to adopt. It is appropriate therefore to discuss briefly the potential role of the occupational health specialist when the risk management approach is applied to psychosocial hazards.

To understand this role it is important to recognise that risk management is not an intervention strategy in itself but a structured framework within which specific preventive measures are contained. This framework provides the means by which particular strategies, whether these be reducing organisational stressors or providing counselling and rehabilitation, can be made to work effectively. Moreover the particular nature of the risk management cycle ensures the continuous evaluation, updating and improvement of those strategies. As in other fields of health and safety, therefore, the primary role of the occupational health professional is in developing and ensuring the implementation and monitoring of the overall risk management policy. Specific aspects of that policy, for example an initial survey or a particular intervention, may in some circumstances be carried out by the occupational health nurse or physician, but equally they may be carried out by other specialists (human resource managers, occupational hygienists, ergonomists, psychologists, counsellors, etc.) with relevant skills. In all but the largest organisations these functions tend to be sourced from outside the company.

In addition to policy development, however, there are two further ways in which an occupational health professional can become involved in specific elements of a remediation process. First, individuals may refer themselves to the occupational health department for assistance, either directly because of emotional distress or ostensibly with physical health complaints which are later found to be psychologically based. Alternatively, an occupational health professional may be asked by managers to see an employee because of an unacceptable sickness absence record. In the first case the physician or nurse is acting on behalf of the employee, and in the second on behalf of the company. These two circumstances involve quite different sets of obligations, with a number of wider ethical implications. Such situations can be complex but are not necessarily untenable and have usually been explored quite extensively in the ethical guidelines of the professional bodies concerned (Mason and McCall-Smith, 1999).

A specific consideration of relevance is that of personal confidentiality. This is in fact maintained in both circumstances since in the former, the normal rules of medical confidentiality apply, while in the latter a physician is only required to state whether or not absence is justified and not to provide details of the cause. However, although the medical practitioner is able to operate at one level under the protection of medical confidentiality there remains a duty of care to the workforce as a whole. Another important aspect of the role of any occupational health professional is to alert the company to health complaints that come to their attention which may be work-related. In both circumstances therefore, self or company referral, there may be a need to pursue the matter beyond the individual case. This can present particular difficulties where sensitive psychosocial issues are involved. It may, for example, be impossible to address the problem at an organisational level in cases of self-referral where an employee refuses to give consent for any breach of confidentiality. Managerial referral, despite its attendant company obligations scarcely makes this easier, since ethical considerations relating to disclosure of personal health information still apply. These difficulties are not confined exclusively to occupational health professionals however, and are likely to be encountered to varying degrees by all those assigned a role in a remediation process. Above all they point to the need for careful consideration of such factors at an early stage, in advance of policy implementation. All those involved, including employees, need to be clear about the ethical as well as the practical requirements and implications of any policy.

Conclusion

In describing the risk management approach reference has been made at various points to its feasibility when applied to the psychosocial as opposed to the physical or chemical environment. It has been noted that a number of difficulties may arise in this context, notably the need for reliance on self-report data and hence an acceptance of a degree of subjectivity, the lack of external standards of significant emotional harm and the need to incorporate elements of treatment as well as prevention. Within these limitations, however, and given a degree of flexibility, the process would appear to be broadly workable. Can it therefore make any useful contribution to tackling the problem of workplace bullying? In assessing this contribution it should perhaps be emphasised once more that risk management is not in itself a remedial action, but a framework within which to deliver such actions. In an occupational setting this framework has the distinct advantage of being both familiar and readily understood by a range of professionals concerned with health and welfare at work, and by the managers likely to be in receipt of their reports. Second, it places bullying and harassment firmly within the realm of potential hazards to health at work, thus creating a requirement to examine the contribution of workplace factors and to accept responsibility for these where appropriate. Finally, and crucially, it provides for continuous monitoring of the effectiveness of any measures undertaken. Thus it encourages a dynamic policy which is responsive to emerging situations and allows for the growth of a knowledge base. This would seem to be particularly important for psychosocial problems such as bullying where, in the workplace at least, there is currently a dearth of information about effective intervention.

Clearly this application of the risk management approach in its current form will involve a certain amount of adaptation. However, its current form is by no means immutable. Given the increasing prominence of psychosocial issues in today's workplaces, there is no reason why the process itself should not undergo change and development in the years ahead to accommodate and address these new and pressing occupational concerns.

Bibliography

Björkqvist, K., Österman, K. and Hjelt-Bäck, M. (1994) Aggression among university employees. *Aggressive Behaviour*, *20*, 173–184.

Cooper, C. L., Sadri, G., Allison, T. and Reynolds, P. (1990) Stress counselling in the Post Office. *Counselling Psychology Quarterly*, *3*, 3–11.

Cox, T. (1993) *Stress research and stress management: Putting theory to work*. HSE contract research report no. 61. London: HMSO.

Cox, T., Griffiths, A., Barlowe, C., Randall, R., Thomson, L. and Rial-Gonzalez, E. (2000) *Organisational interventions for work stress: A risk management approach*. HSE contract research report no. 286/2000. London: HMSO.

Crabb, S. (1995) Violence at work: The brutal truth. *People Management*, *1*, 15, 25–29.

Crawford, J. and Bolas, S. M. (1996) Sick building syndrome, work factors and occupational stress. *Scandinavian Journal of Work and Environmental Health*, *22*, 243–250.

Devereux, J. J., Buckle, P. W. and Vlachonikolis, I. G. (1999) Interactions between physical and psychosocial risk factors at work increase the risk of back disorders: An epidemiological approach. *Occupational and Environmental Medicine*, *56*, 343–353.

HSE (1993) *Quantified risk assessment: Its input to decision making*. London: HMSO.

—— *Generic terms and concepts in the assessment and regulation of industrial risks*. London: HMSO.

—— (2000) *Management of health and safety at work. Management of health and safety at work regulations 1999. Approved code of practice (revised)*. Norwich: HSE Books.

—— (2001) *Tackling work-related stress: A manager's guide to improving and maintaining employee health and well-being*. Norwich: HSE Books.

Lockhart, K. (1997) Experience from a staff support service. *Journal of Community and Applied Social Psychology*, 7, 193–198.

Marteau, T. M. and Johnston, M. (1990) Health professionals: A source of variance in health outcomes. *Psychology and Health*, *5*, 47–58.

Mason, J. K. and McCall-Smith, R. A. (1999) *Law and medical ethics*, 5th edn. Butterworth.

Molyneux, M. K. B. (1999) Organizing for risk assessment and management. In S. S. Sadhra and K. G. Rampal (eds), *Occupational health: Risk assessment and management* (pp. 22–37). Oxford: Blackwell Science.

Pavett, C. and Morris, M. (1995) Management styles within a multinational corporation: A five country comparison. *Human Relations*, *48*, 1171–1191.

Pheasant, S. (1991) *Ergonomics, work and health*. Hampshire: Macmillan Press.

Rampal, K. G. and Sadhra, S. S. (1999) Basic concepts and developments in health risk assessment and management. In S. S. Sadhra and K. G. Rampal (eds), *Occupational health: Risk assessment and management* (pp. 3–21). Oxford: Blackwell Science.

Rayner, C. (1997) The incidence of workplace bullying. *Journal of Community and Applied Social Psychology*, 7, 199–208.

Spurgeon, A., Gompertz, D. and Harrington, J. M. (1997) Non-specific symptoms in response to hazard exposure in the workplace. *Journal of Psychosomatic Research*, *43*, 1, 43–49.

Spurgeon, P. and Barwell, F. (2000) *Tackling the causes of workplace stress: A guide to using the organisational stress measure in the NHS*. London: Health Development Agency.

Vartia, M. (1996) The sources of bullying: Psychological work environment and organisational climate. *European Journal of Work and Organizational Psychology*, *5*, 203–214.

Watterson, A. (1999) Why we still have 'old' epidemics and 'endemics' in occupational health: Policy and practice failures and some possible solutions. In N. Daykin, and L. Doyal (eds), *Health and work: Critical perspectives* (pp. 107–126). Hampshire: Macmillan.

Zapf, D., Knorz, C. and Kulla, M. (1996) On the relationship between mobbing factors, and job content, social work environment, and health outcomes. *European Journal of Work and Organizational Psychology*, *5*, 2, 215–237.

20 Conflict, conflict resolution and bullying

Loraleigh Keashly and
Branda L. Nowell

Introduction

In their review of the workplace bullying literature, Hoel *et al.* (1999) argued for the importance of taking a conflict perspective on the problem of bullying. They suggested that the dyadic conflict literature was rich in insights on conflict development and escalation as well as the various procedures and processes for resolving conflicts. Their belief is premised on an implicit connection between conflict and bullying where severe bullying is likened to 'destructive conflicts going beyond the point of no return' (p. 221). Zapf and Gross (2001) concur, describing bullying situations as 'long-lasting and badly managed conflicts' (p. 499). Einarsen and Skogstad (1996) have also made a connection between bullying and conflict, but as distinctive constructs hinging on the ability of the involved parties to respond or defend against hostile actions. A key feature of bullying is the inability to defend oneself. If the parties involved are equally able to defend themselves, then the situation may well be a serious conflict, but it is not bullying. Einarsen (1999) further refines his earlier distinction by proposing that there are at least two types of bullying: predatory and dispute-related. Predatory bullying occurs when the victim has done nothing provocative that would reasonably invoke or justify the bully's behaviour. Dispute-related bullying, however, develops out of grievances between two or more parties and involves retaliatory reactions to some perceived harm or wrong-doing. If one of the parties becomes 'disadvantaged' during the dispute, he/she may become a victim of bullying. So a dispute may trigger bullying. In making this argument, Einarsen supports the idea that conflict and bullying are distinct yet related constructs.

Thus, while it is clear various authors consider conflict in their writings and research on bullying, the connection between these two constructs is unclear. In order to assess the value of a conflict perspective on bullying, we need to clarify what is the relationship between bullying and conflict. Thus, in the first section of this chapter, we will compare definitions of conflict and bullying in an effort to articulate their connection. We will then present and discuss several concepts from the conflict literature that may prove insightful for workplace bullying: individual conflict management strategies and the influence of power, escalation and de-escalation and co-ordinated intervention approaches. We will sum up with some cautions regarding the application of conflict concepts to the study and amelioration of workplace bullying.

Conflict and bullying: same or different?

Like so many concepts in the social sciences, 'conflict' has many definitions and there is no consensus on a common definition (Thomas, 1992). However, Putnam and Poole (1987) note that there are three general properties reflected in general definitions of conflict:

1 Interdependence between the parties, i.e. each has the potential to interfere with the other
2 Perception by at least one party that there is opposition or incompatibility among the parties' concerns
3 Some form of interaction between two or more parties.

There are also a number of definitions of bullying, with the term often encompassing a variety of situations. Like conflict definitions, these various definitions share certain key features (Einarsen, 1999; Keashly, 1998; Rayner *et al.*, 1999):

1 A pattern of repeated hostile behaviours over an extended period of time
2 Actual or perceived intent to harm on the part of the actor
3 One party is unable to defend him/herself
4 A power imbalance exists between the parties.

A cursory glance at the general properties of these two constructs reveals some broad commonalties and some differences. In terms of what is shared, it is clear that both conflict and bullying are referring to some form of interaction between two or more parties. Thus, it is not a conflict nor is it bullying if the scene is only played out in someone's mind. From a stress perspective, conflict and bullying as social interactions are similar in that they are examples of negative social stressors stemming from relationships between people (Zapf, 1999). Second, there is negative tone associated with both constructs, albeit of different strengths. In bullying, the reference to hostile behaviour and intent to harm is clearly and strongly negative. In conflict, the tone is subtler, captured to some extent in the perception of opposition or incompatibility. There is the potential for negativity but it is not a central defining feature for conflict. Thomas (1992) has indicated that general definitions of conflict permit the opportunity for a variety of non-adversarial types of strategies to be used by parties to deal with the conflict. Thus, conflict can be a constructive and positive process rather than a destructive process, unlike bullying.

In terms of differences, time is clearly central to bullying, but this is not so for conflict. Conflict can be quickly overcome, as when a misunderstanding over a work task is clarified as soon as it arises (i.e. a single episode). Conflict can also be very long-standing (a series of episodes), as indicated by the number of intractable conflicts that exist interpersonally (e.g. a prolonged marital dispute), organisationally (e.g. negative union–management relations) and internationally (e.g. Northern Ireland, the Middle East, Rwanda, and the former Yugoslavia). Bullying, by definition, is long-standing; it is the outcome of a series of episodes. In that sense, bullying may be similar to intractable conflicts that are noted for their long-standing and recurring nature (e.g. Coleman, 2000; Kriesberg *et al.*, 1989). The potential for a tie between (intractable) conflict and bullying may be strongest here as evidenced in the reference to bullying as 'long-standing and unresolved conflicts' (e.g. Hoel *et al.*, 1999; Zapf, 1999) or in Rayner's (1999) and Einarsen's (1999) reference regarding intervening before the conflict begets bullying. Another way to think of the element of time is that it provides a sense of a process of development from an initial episode to a series of events (Thomas, 1992). This is a place where the conflict literature can provide useful insight into the development of workplace bullying in terms of conflict stages, escalation, and intractability. We will discuss this more in a later section.

Intent plays an interesting role in both bullying and conflict. Einarsen (1999) specifically notes that bullying involves actual or perceived intent to harm. As noted in an earlier chapter (Keashly and Jagatic, this volume), intent figures prominently in workplace aggression and abuse research in North America. Intent is considered a necessary defining element, as it distinguishes these abusive and aggressive interactions from other forms of harmful behaviour such as incivility or accidental harm (Andersson and Pearson, 1999; Neuman and Baron, 1997). With regards to conflict more generally, intent is not a defining element; however, perceived intent may figure in the attributions of behaviour as the conflict proceeds, thus affecting response (e.g. Fisher, 1990). For example, attributions of intent to harm are associated with the choice of increasingly provocative behaviour that fuels an escalatory spiral (Fisher, 1990; Rubin *et al.*, 1994). Thus, intent is a key defining feature of workplace bullying and indeed of workplace aggression, but its significance for conflict is more as an important outcome of attribution processes that occur during the interaction of the parties.

The role of power differences is key to bullying but not essential to defining a situation as conflictual. Certainly, there are conflicts between parties of unequal power such as between child and parent, boss and subordinate, or majority and minority groups. While being unequal or equal in power does not define whether a conflict can be said to exist, it does define whether bullying can be said to exist. Even though power imbalance may not be a key defining feature of conflict, it is very influential in the progress of and response to conflict (e.g. Musser, 1982). We will discuss the influence of power on managing conflict and its implications for bullying in a later section.

Power imbalance is relevant to the ability to defend (Keashly, 2001; Keashly and Jagatic, this volume; Zapf, 1999). The inability to defend is not an inherent feature of conflict. In fact, the idea of interdependence, of being able to interfere with the activities of the other, speaks to the ability of parties to defend themselves, and to manifest some degree of influence and power. However, power imbalance does affect the types of responses one can or should make in the face of conflict with another. A feature of conflict escalation that is related to power and ability to defend is reciprocity or tit for tat (Andersson and Pearson, 1999; Fisher, 1990). Mutuality or reciprocity is perhaps the key distinction between bullying and conflict as it is generally defined. In conflict, parties mutually engage in exchange of behaviours and are simultaneously actors and targets (Andersson and Pearson, 1999; Glomb, 2002). In bullying, there is a clear actor who is the instigator and who is proactive, and a target who, in essence, cannot respond, or can respond but in a limited manner which does not protect him/her from harm or stop the actor's behaviours.

Bullying as defined is not a mutually engaged in, reciprocal process like conflict. However, some authors (e.g. Aquino, 2000; Einarsen, 1999; Zapf, 1999; Zapf and Gross, 2001) have argued that the targets or victims can be contributory to the bullying experience. While this argument has to be pursued cautiously, as it runs the risk of victim blaming, it is suggestive of the value of examining the extent to which reciprocity characterises the bullying process from its early stages of 'not-yet-bullied' (Rayner, 1999) to later stages of stigmatisation and traumatisation (Einarsen, 1999). In the conflict literature, reciprocity also encompasses a notion of mutual impact of these actions, i.e. both parties are affected, often negatively in conflicts, particularly escalated ones (e.g. Fisher and Keashly. 1990). Glomb (2002) has provided evidence that actors of aggressive behaviours show similar negative effects as the targets of these aggressive actions. Thus, something the conflict literature has to offer the study of workplace bullying is a fuller consideration of the contributory activities of, and effects on, both actors and targets, particularly in the development of escalated conflicts. For example, from a conflict escalation perspective, the inability of one party to defend themselves portends the figurative and literal disappearance and perhaps destruction of the other party (Fisher and Keashly, 1990). This description certainly maps on to the effects of severe workplace bullying such as psychological (reduced organisational commitment and job satisfaction) and actual withdrawal (leaving the job) as well as post-traumatic stress disorder (e.g. Barling, 1996; Hoel *et al.*, 1999; Keashly and Jagatic, this volume). This 'effects' connection between escalated and prolonged conflict and bullying highlights the value of discussing the characteristics of escalation processes and the role of various conflict management strategies, both individually and by third parties, that may alter these processes in more constructive ways.

While we have articulated several differences, they appear to be more matters of breadth rather than of qualitative differences. In looking across a variety of conflict situations that fall under the general definition of conflict, it would appear that bullying as defined is most like intractable, escalating violent conflicts between unequals. As noted above, this suggests that the concepts of individual conflict management strategies, escalation and intervention approaches would be the most useful to discuss in terms of their implications for the study and amelioration of workplace bullying.

Individual conflict management strategies

Research into conflict strategies has been conceptualised to differ across two dimensions: concern for self (satisfying own needs) and concern for others (satisfying the needs of others; Blake and Mouton, 1964; Rahim and Buntzman, 1990). When these two dimensions are crossed, five main strategies are identified:

1. *Problem-solving*: this strategy represents a high concern for self and the other. Through open exchange of information, common interests are identified to create integrative solutions meeting both parties' needs.
2. *Obliging* (accommodating, yielding): this strategy signifies a low concern for self but a high concern for the other by emphasising commonalties and downplaying differences.
3. *Dominating* (competing): representing a high concern for self with low concern for other, this style focuses on fulfilling one's own interests at the expense of the other.
4. *Avoiding* (withdrawing): represents a low concern for self and other. The objective is not to acknowledge or engage in the conflict situation.
5. *Compromising*: involves an intermediate concern for self and other. By developing solutions that meet somewhere in the middle, both parties get some, but not complete, satisfaction.

Although helpful in creating a framework to think about conflict behaviour, these conflict dimensions fail to predict actual conflict behaviour. This is because other salient elements of the conflict situation may limit strategy availability. Two elements of particular relevance to bullying are the type of conflict issue and relative status of the parties.

Type of conflict issue

It has been suggested that there are two broad types of issue over which conflicts arise: task/cognitive and socio-emotional/affective (de Dreu, 1997). Cognitive conflicts are conflicts over ideas and tasks. For example, a work group might have a conflict deciding between what strategy to pursue or how to allocate responsibilities. These types of conflicts are not only unavoidable, but they are also often highly fruitful and rejuvenating if managed correctly. Problem-solving approaches which allow participants to vigorously debate ideas through strategies that communicate high respect and concern for the other party are extremely productive in creating new solutions and enhancing relationships (Jehn, 1997).

Affective conflicts, on the other hand, involve issues that threaten one's identity and value system and are often characterised by intense negativity, friction, frustration and personality clashes. These types of issues are often perceived as non-negotiables and set the stage for a win/lose interaction. Research has found that affective conflicts are likely to reduce performance and satisfaction, as well as lead to aggressive behaviour on the part of one or both parties (Berkowitz, 1989; Jehn, 1997). Thus, these conflicts have a high risk for serious damage to both parties (de Dreu, 1997).

Because of the different nature of cognitive and affective conflict, it is not surprising that they differ in their response to management strategies as well. Indeed, de Dreu (1997) found that although problem-solving strategies were effective in the productive management of cognitive conflicts, they were ineffective and even harmful in managing affective conflicts. Bullying, like affective conflict, involves negative emotions, hostile actions and threats to identity; so much so that the target's identity is diminished as a result of prolonged exposure. Extending findings of managing affective conflict to bullying, problem-solving by the target will not be successful in managing bullying. Indeed, Rayner (1999) found that open discussion and information sharing with the bully increased the likelihood of the bully taking retaliatory action against the target.

Relative status

In addition to the type of conflict issue, the status of the parties to one another will also influence choice and effectiveness of management strategy. In the context of unequal status relationships, Musser (1982) proposes that the less powerful person (e.g. subordinate) will base their choice of strategy on three criteria: (1) his/her desire to remain with the organisation; (2) degree of perceived congruence between his/her attitudes and beliefs and those of the supervisor; and (3) perceived protection from arbitrary actions by the superior with whom the conflict exists.

Musser's (1982) model is useful when applied to bullying situations as it highlights some of the assumptions that are inherent in strategic choice of conflict behaviour in conflicts of unequal power. In addition, this model further dictates what strategies are available depending on the above conditions. *Problem-solving*, although generally perceived to be the most appropriate and effective strategy (Blake and Mouton, 1964), requires that both parties are able to participate openly in a non-hierarchical manner (Filley, 1975). If the subordinate values his/her position and perceives low congruence of beliefs as well as limited organisational protection, it is likely that problem-solving will not be a viable option due to potential risks. In support of this prediction, research looking at bullying and harassing behaviours notes that few targets directly confront the actor, often for fear of retaliation (Hoel *et al.*, 1999; Keashly and Jagatic, this volume).

Bargaining or compromising as a strategy option only exists to the extent that the subordinate has leverage in the situation. Since it is quite possible that the superior may control any available power resources, the subordinate may have no leverage effectively to utilise such an approach. *Competing* as a strategy option is inherently risky in any conflict situation as it defines the conflict as win/lose situation. Given that conflict behaviour is often reciprocated in a like manner (Deutsch, 1973), the competitive strategy will intensify rather than diminish the conflict. To the extent the conflict is occurring in an unequal power relationship, the risks are multiplied for the party of lesser power unless those risks are offset by a low desire to remain with the organisation and/or there is a significantly high degree of congruence between the two parties. Thus, *obliging or yielding*, due to lack of power and/or protection may be the only strategy available to a subordinate who has a strong desire to stay in their position. *Withdrawing*, either psychologically or physically from the relationship, may become an option as one's desire to remain with the organisation is diminished through the conflict. It may be characterised by increased apathy or actually terminating employment resulting from the perception that there is no chance of winning and costs incurred in staying in the relationship have begun to outweigh any benefits gained from employment. This management strategy is consistent with findings from bullying research that many victims leave their organisations as a result of the experience (e.g. Einarsen, this volume; Keashly and Jagatic, this volume; Zapf and Gross, 2001)

Thus, Musser's (1982) model offers some insight into explaining findings in the bullying literature. For example, Aquino (2000) found, contrary to expectation, that integrating was *positively* correlated with victimisation when the individual held a lower power position. Similarly, Richman *et al.* (2001) found that active coping strategies, such as problem-solving, were not only ineffective in stopping harassment but also increased negative personal outcomes for the targets. Zapf and Gross (2001) likewise found that bullied targets' initial attempts at active problem-solving were ineffective and were eventually abandoned for other strategies. The implication of these findings is highly significant for both the fields of bullying and conflict resolution. Regardless of the fact that problem-solving has long been heralded as the 'right' way to manage conflicts, research suggests that such an approach on an unequal playing field is not necessarily the most effective or most appropriate strategy, and may actually make things worse.

An interesting aside: the paradox of the bullying co-worker

In discussions of the centrality of power imbalance to bullying, it has been stated that this imbalance could be present a priori or develop during the course of the interaction (Einarsen, 1999; Zapf, 1999). Einarson (1999) suggests that bullying may evolve out of a conflict between equals (e.g. co-workers) if one party becomes 'disadvantaged' during the process. Supportive of this 'bullies as equals' phenomenon, several studies have found that co-workers are the most frequent source of hostile workplace behaviours (e.g. Cortina *et al.*, 2001; Neuman and Baron, 1997; Richman *et al.*, 1999). To this point, we have examined conflict literature in which unequal power status between disputants existed prior to the conflict. So what does the conflict literature have to offer in understanding how equals become unequal and hence increase the risk of bullying? We will look first from the perspective of the 'soon-to-be target' and then the perspective of the 'soon-to-be actor'.

Regarding the 'soon-to-be target', Deutsch (1985) suggests that when an individual experiences a frustration, he/she judges whether the actions involved are normatively or personally unjust. This assessment will influence their sense of capacity to respond. A frustration is assessed as normatively unjust if it is interpreted as a threat against a larger group to which the individual belongs (e.g. negative comments about workers generally). If the frustration cannot be attributed to group threat, it is experienced as directed at the self and is judged as personally unjust. In the case of normatively unjust, the person feels empowered and emboldened by the group reference and, hence, experiences an enhanced sense of their personal ability to respond. Personally unjust actions make the individual more aware of the limitations of his/her own personal power to resolve the issue of frustration. This makes the person feel weaker and less competent to deal directly with the source of the conflict. Behaviourally, when an individual is threatened at this level, his/her information processing is affected such that the person loses his/her repertoire of conflict management strategies and resorts to base levels of behaviour even if such actions are less appropriate or effective (Mack *et al.*, 1998). The resultant poor conflict performance puts the 'soon-to-be target' at an increasing disadvantage in the conflict situation, widening the power disparity and leading to possible escalation and/or eventual withdrawal.

In terms of the 'soon-to-be actor', Papa and Pood (1988; Papa and Canary, 1995) found that, in the context of a power imbalance, individuals are less likely to seek out others' views. Since actively seeking to understand the other's position is a prominent feature of productive resolution, lack of understanding of the other (i.e. empathy and perspective taking) on the part of the 'soon-to-be actor' may lead to increasingly aggressive behaviours on his/her part and, hence, escalation. In short, co-workers who become cognisant that they hold the upper hand are less likely to pursue mutually constructive resolutions due to their greater leverage in the situation. Thus, by examining the attribution processes and subsequent choices of management behaviour within a disputing dyad, it may be possible to track the development of inequality in resources to respond and, hence, make an assessment of the risk for bullying (Hoel *et al.*, 1999).

This discussion on individual conflict management styles provides a framework for considering some of the influences on both target's and actor's strategic and behavioural choices that are relevant to examinations of the dynamics of bullying. Unfortunately, this discussion has also revealed that recent evidence indicates none of the interpersonal management strategies available to targets are effective in stopping a bullying situation. For example, Zapf and Gross (2001), in examining the management strategies of bullying victims throughout the process of escalation, concluded that for some victims no strategy, whether active or passive, was successful, and that leaving the organisation was ultimately the only pragmatic option available. Richman *et al.* (2001) have argued that the problem may lie in expecting individual strategies to address a structural phenomenon. Given that the dynamic of bullying is largely hinged on someone of greater power acting on someone of lesser power, looking to the lower-power person to be empowered with the ability to remedy the situation is not likely to be fruitful. Thus, we turn our attention to ways in which others outside of the disputants (or bully and target) may influence the development of the interaction.

Conflict processes: escalation and intervention

Bullying has been characterised as a process moving from subtle, low-level aggression (not-yet-bullied), to bullying (direct and intense aggression over a period of time), to stigmatisation, and, finally, traumatisation (diminishment or destruction; e.g. Einarsen, 1999). The hope of many workplace bullying researchers is that there are ways to intervene early enough in this process either to prevent the interaction from reaching the stage of bullying or to stop the bullying, or at least to reduce its effects (Hoel *et al.*, 1999; Rayner, 1999; Zapf, 1999). Earlier, we likened the bullying phase and beyond to an intractable, escalated conflict characterised by violence. To the extent this comparison is appropriate given our earlier caveats, a discussion of conflict escalation and the work linking type of conflict intervention to stage of conflict development will provide some insight into possible ways of dealing with workplace bullying.

Escalation

A fundamental tenet of many conflict theories is that conflict is prone to escalate, that is to become more intense, hostile and competitive, to include more issues, to undermine trust, and to involve more powerful attempts at control such as engaging other parties in alliances (Fisher, 1990; Thomas, 1976). A variety of social psychological processes can fuel a conflict's intensity. They include elements such as negative and simplified stereotypes, selective perception, self-fulfilling prophecies, negative attributions, communication problems, zero-sum thinking, over-commitment and mistrust that build on each other; what Deutsch (1973) has characterised as a malignant social process.[1] In essence, conflicts can take on a life of their own, spiralling beyond even the parties' control to increasing levels of violence and destruction (Rubin *et al.*, 1994). Even when conflicts can be de-escalated to a more peaceful and constructive place, the experience of the conflict has produced fundamental structural changes in the parties (i.e. 'sticky' conflict residues). Unless these residues are specifically recognised and addressed, they will encourage further contentious and hostile responses, and inhibit efforts at resolution, generating the conflict anew. Conflict residues map on to the accumulative model of stress as articulated by Hoel *et al.* (1999) in which over time, the person is fundamentally and irrevocably changed. This is an important concept to consider in identifying what kinds of intervention efforts are needed to alter the conflict situation generally and workplace bullying specifically.

The perspective of conflict escalation has stimulated some interesting theorising and research into hostile workplace behaviours. Pearson and her colleagues (Andersson and Pearson, 1999) draw on the concept of conflict escalation processes to illustrate how even minor acts of workplace incivility can spiral down into increasingly hostile and violent behaviour on the part of co-workers. Glomb (2002), utilising an escalation framework, has found evidence for the linkage of seemingly low-level verbally aggressive behaviours with increasingly severe physical behaviours. It is important to note that the focus of these researchers was on understanding the development of increasing hostility between equal power co-workers. Thus, perceptions, orientations, attitudes and behaviours of both parties were characterised similarly: as mirror images. As noted earlier, given that bullying involves a power imbalance, it cannot be assumed that both parties will behave and perceive in a similar fashion. However, changes in these elements marking increasing escalation can be very helpful in assessing the current state of the parties to the bullying. For example, Zapf and Gross (2001) utilised Glasl's (1982) description of behaviours, attitudes and images at various stages of conflict escalation to determine where bullying in its most extreme form might fall. They suggest that severe bullying could be classified as a conflict at the boundary between the phase in which the relationship between the parties is severed and dominated by threats, and the phase in which destruction of the other becomes paramount.

Given the extreme detrimental effects of escalated conflicts, it is no wonder that many conflict researchers have focused on ways to alter the conflict situation in order to de-escalate to some more manageable and less damaging level. To facilitate their thinking regarding what to do and when, many theorists have chosen to 'package' conflict into discrete yet related stages in the escalation process. There are numerous models available regarding stages of conflict escalation (e.g. Fisher and Keashly, 1990; Glasl, 1982; Rubin *et al.*, 1994; Thomas, 1976). While these models tend to differ in the number of stages utilised, they essentially define each stage by changes in overt behaviour, patterns of interaction, perceptions and attitudes (Fisher, 1990; Glasl, 1982). The move to each stage heralds a new and more pervasive level of intensity.

These stages form the basis of the contingency approach to conflict intervention (e.g. Fisher and Keashly, 1990; Glasl, 1982; Prein, 1984). The basic premise of the contingency approach is that different management or intervention strategies would be appropriate and effective at different points in time (Fisher, 1990). Indeed, one of the reasons for the 'failure' of particular interventions in particular conflicts may be inappropriate application with respect to the stage of escalation. One of the earliest and most comprehensive models of this genre is Glasl's (1982) contingency model that consists of nine stages of escalation and six types of third-party intervention approaches. Zapf and Gross (2001) have drawn on this model in their discussions of how severe workplace bullying can be conceived as a particularly escalated form of conflict. Building on the formative work of Glasl (1982), Prein (1984) and Kriesberg (1989), Fisher and Keashly (1990) developed a four-stage model highlighting four main types of intervention strategies (see Table 20.1 and Figure 20.1). While this model was developed in the context of international disputes, it is unique in that it goes beyond the idea of matching to suggest that intervention approaches can and should be sequenced and co-ordinated in order to de-escalate and resolve the conflict. An additional reason for failure of some interventions may be the lack of co-ordinated follow-up interventions to deal with elements not addressed by the initial intervention. Because of their focus on co-ordinated action, we have chosen to describe the Fisher and Keashly (1990) model in more detail.

Table 20.1 details the significant negative changes that occur in parties' communication, perceptions and images of each other and their relationship, overt issues in dispute, perceived possible outcomes and, hence, the appropriate approach to managing the conflict as the conflict escalates. The stunning part is how images of the other devolve to something bad, evil, and other than human. The exclusion of the other from the human

Table 20.1 Stages of conflict escalation

	Dimensions of conflict			
Stage	Communication / interaction	Perceptions / relationship	Issues	Outcome / management
I. Discussion	Discussion / debate	Accurate /trust, respect, commitment	Interests	Joint gain / mutual decision
II. Polarisation	Less direct / deeds not words	Stereotypes / other still important	Relationship	Compromise / negotiation
III. Segregation	Little direct / threats	Good vs evil / distrust, lack of respect	Basic needs	Win–lose / defensive competition
IV. Destruction	Non-existent / direct attacks	Other non-human / hopeless	Survival	Lose–lose / destruction

Source: Fisher and Keashly (1990). Reprinted with permission.

condition opens up the possibility to treat them in inhumane ways (Opotow, 2000). Descriptions of severe bullying could be similarly characterised where the actor expresses contempt and disgust for the other and wants to be rid of him/her (either through exit or destruction; e.g. Leymann, 1996; Zapf and Gross, 2001).

Keashly and Fisher (1990) propose that effective efforts at de-escalation recognise the need to move conflict stage by stage rather than attempting to move directly from violence to rational discussion. Our presentation of the model will follow this de-escalatory sequence. At the *destructive* stage, the primary intent is to destroy or at least control the other by violence. Third parties function as peacekeepers by forcefully setting norms, defining unacceptable violence and isolating parties when necessary to keep the violence under control. In the workplace, these types of activities include zero-tolerance policies, moving parties to separate departments and behavioural contracts handled through Personnel. If the relationship can be stabilised and a commitment to joint effort is made, the way is cleared for using other strategies depending on the parties' receptivity and sense of critical issues (substantive or relational). At the *segregation* stage, competition and hostility predominate, and the conflict is perceived as threatening basic identity and security needs. Therefore, some immediate form of control is necessary to halt escalation and to demonstrate that agreement is still possible on substantive issues. Thus, either arbitration or power mediation can come into play. Many managers tend to use these latter styles quite readily. Sheppard (1984) describes the 'providing impetus' intervention style, where a manager tells employees either to

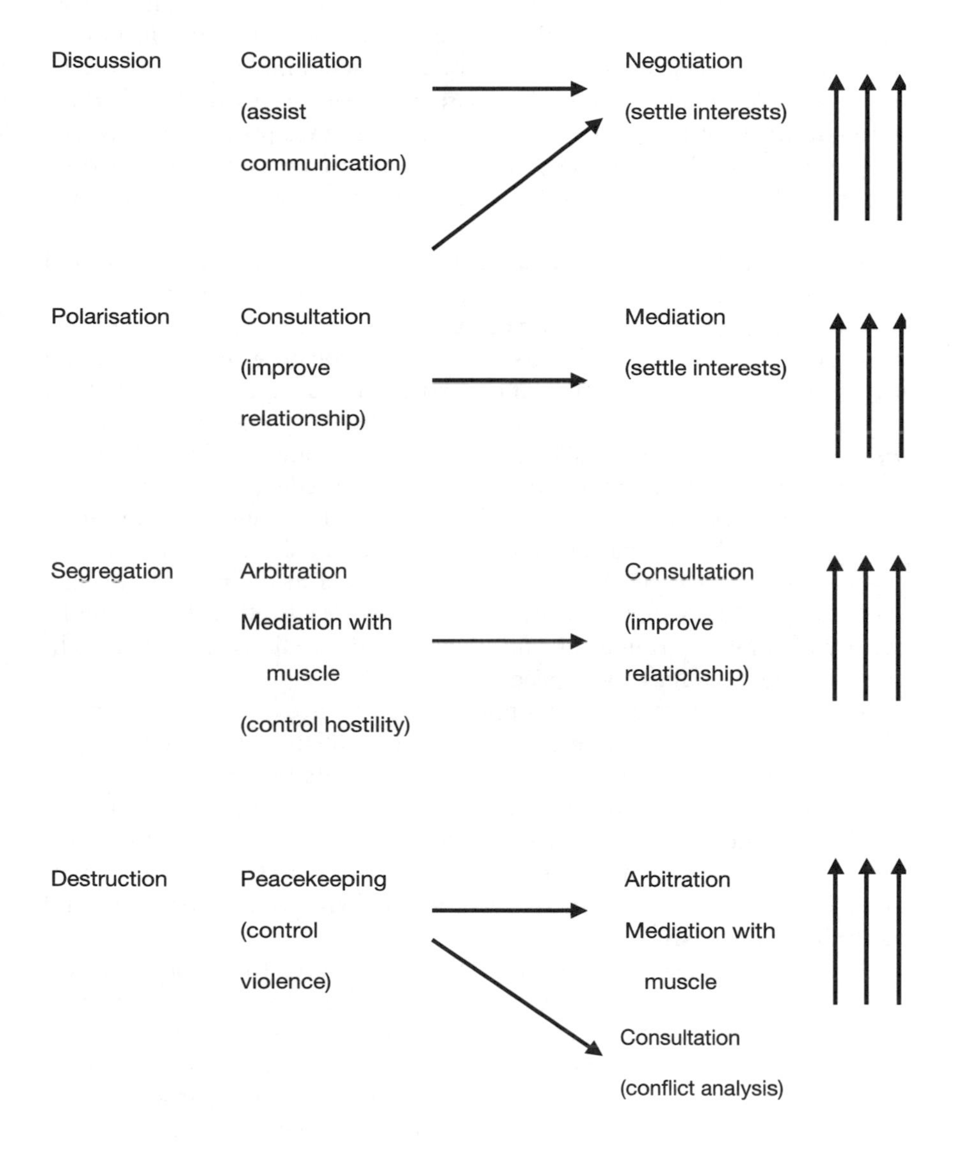

Figure 20.1 A contingency model of third-party intervention
Source: Fisher and Keashly (1990). Reprinted with permission.

resolve the situation themselves or the manager will resolve it for them. Once hostility is controlled, consultation would be provided to assist parties in examining the dynamics of their conflict and ground rules for improving the relationship back towards one of trust and mutual respect. This style of intervention is quite foreign in organisations because of its clear and direct focus on relational issues. However, the introduction of teams as a critical element of organisational functioning may be suggestive of the capacity among some organisational members to deal with these more sensitive yet critical issues. Employee assistance programmes as well as process-minded management may be able to provide this expertise, assuming their focus is at the level of the relationship as opposed to the person.

At the *polarisation* stage, relationship issues are central, as trust and respect are threatened and distorted perceptions and stereotypes emerge. At this point, consultation becomes most appropriate because it uniquely deals directly with relationship issues. Once a problem-solving orientation is established, work on the substantive issues can be handled by mediation, and hopefully the parties can move to negotiating on their own. At the *discussion* stage, the key challenge is to ensure communication is accurate and perceptions are grounded in reality. When needed, a third party can take a conciliation approach to facilitate clear and open communication on interests so that the parties can begin negotiating directly themselves. In organisations, this type of intervention is reflected in the myriad informal processes and skills that co-workers utilise with each other. For example, the use of summary statements and open-ended questions can often facilitate meetings becoming more productive.

So what does a contingency approach to conflict intervention have to offer the study and amelioration of workplace bullying? First, it highlights the critical role that other people outside the disputants can play in helping management of the conflict. We know from both the organisational conflict and bullying literatures that these interactions spill over and affect the people around them. So other people do have a stake in these situations being resolved constructively. The challenge is in mobilising and co-ordinating the efforts of these critical others.

Second, a contingency approach emphasises the need to thoroughly assess the history and current status of the bullying situation. Only by having a sense of what has occurred and where things are at present will it be possible to select methods of intervening that increase the chances of at least minimising the damage and at most, rebuilding the parties, particularly the target, and possibly the working relationship. While this is a seemingly simple suggestion, it is very difficult to implement as evidenced by the apparent blind eye that organisations often turn towards many conflicts in general and to bullying, specifically (Keashly, 2001; Zapf and Gross, 2001).

Third, the contingency approach may provide an explanation for why some intervention and management strategies may fail; that is, they were inappropriate for the circumstances. At this point, we would like specifically to address suggestions that have been made that mediation may be the appropriate intervention for bullying (e.g. Hoel *et al.*, 1999). Mediation has certainly gained privileged status among the dispute resolution processes, helping to facilitate solutions to seemingly unsolvable problems. Evidence of mediation effectiveness in contrast to litigation in a variety of settings (divorce, the environment, the corporate sector, etc.) has been strong (e.g. Kressel, 2000). However, mediation has been criticised for a number of failings. First, a basic assumption of mediation is that parties to the dispute are equally capable of negotiating with one another. This assumption of equal power is questionable in situations in which violence is involved such as domestic abuse, sexual and racial harassment, and violent crimes. The victims of these behaviours are often diminished and disempowered as a result of the experience, undermining their abilities to be assertive in dealing directly with the actor. As noted earlier in the discussion on individual conflict management strategies, problem-solving approaches fail to stop affective conflicts and in some cases serve to exacerbate them. Thus, the power imbalance, the inability to defend and the extensive undermining of the target's personal resources that characterise bullying relationships suggest that mediation may not only be inappropriate but that it may also be harmful.

A second criticism concerns mediation's focus on present and future relationships. Specifically, mediation does not address or punish past behaviours. In situations of mutually engaged-in conflicts, this is likely agreeable to both parties. However, in the situation of workplace bullying where one person is clearly the victim, this may do little to address their concerns for justice and recognition of the harm done. As damage to the targets is often cumulative in nature, the extent of the harm could easily be missed or delegitimised if the selected intervention focuses exclusively on a single episode without consideration of the history. In addition, the actor may well favour this approach as it does not require him/her to acknowledge responsibility for his/her actions.

The final criticism of mediation is that it keeps wrongdoings outside public scrutiny. One of the hallmarks of mediation is that it is a private, confidential process. Thus, information shared and decisions reached during the mediation are not available to anyone else. This confidentiality works against the identification of systematic patterns of conflict associated with a particular party, a particular unit within an organisation, or across the organisation (Rayner, 1999). Discussions of structural influences on conflict make clear that there are organisational ways of handling disputes that may not effectively and constructively result in necessary changes (e.g. Donnellon and Kolb, 1994). Thus, we disagree with Hoel *et al.*'s suggestion that mediation could help identify bullying patterns in an organisation. In fact, it may work to obscure them. This is particularly true if such case-by-case processing is used to cover up systemic or repeated violations.

The fourth advantage of taking a contingency approach is that it highlights the need to view dealing with bullying as a comprehensive and co-ordinated effort of a number of different activities and a number of different parties (Opotow, 2000). It requires the recognition of the need to co-ordinate short-term crisis management interventions such as separation with longer-term methods directed at fundamentally altering the parties' relationship specifically, and the system generally (Coleman, 2000). Such co-ordinated and comprehensive efforts require organisational awareness of bullying and a commitment to dealing with it directly (Hoel *et al.*, 1999). An instructive literature for these efforts regarding bullying would be dispute resolution system design (e.g. Costantino and Merchant, 1996; Ury *et al.*, 1988). This work speaks to how organisations can develop and implement multi-level and multifaceted systems for dealing with conflict in all its various forms and stages.

This notion of organisational involvement in dealing systematically with conflict and, by extension, bullying is worthy of further comment. A critical element of the development of such systems is that the organisation must examine the way in which its structures and methods of operations may cause, or contribute to, the proliferation of conflict among its members. This critical and reflective process clearly has relevance for workplace bullying. For example, several researchers (e.g. Rayner, 1999; Richman *et al.*, 2001) have argued that successfully dealing with bullying requires the recognition of structural culpability in permitting or protecting an actor (bully) through offering them broader resources of power. For example, it could be argued that flatter organisational structures where power is more widely shared would offer more protection against, and less support for, bullying than rigid, hierarchical structures where power is narrowly channelled. From a conflict perspective, flatter power structures do not reduce the amount of conflict and may in fact increase it (Kabanoff, 1991). This is a result of the equal power relationships encouraging individuals to engage in open communication, resulting in a greater exchange of information and hence the greater possibility for opposing ideas to be expressed. However, such conflicts are cognitive in nature and hence facilitate cohesion and solidarity among workers. Hierarchical structures with large power disparities tend to result in more hidden forms of conflict that manifest themselves in more covert and potentially counterproductive ways (Kolb *et al.*, 1992). Since bullying is initially characterised as consisting of highly covert and indirect behaviours, these findings from the conflict literature are supportive of the value of pursuing the connection between organisational structure and bullying, and the implications for managing bullying situations.

Finally, the contingency approach and our awareness of the sticky residues of conflict further highlight the limited means of handling severe bullying. This recognition underscores the importance of preventive measures and of addressing the harmful interaction early (the not-yet-bullied) before more damage occurs and when there is a chance of building a productive working relationship (Rayner, 1999; Zapf and Gross, 2001). Individual skill development (such as communication, anger management, perspective taking and negotiation) is considered one of the primary preventive dispute resolution efforts that may be relevant for the 'not-yet-bullied' situations (see Deutsch, 1993 for further information). Sheehan (1999) cogently argues for such skill development for managers who use bullying as a way of managing.

A final cautionary note

In an effort to identify places of connection between the conflict literature and workplace bullying, we feel that we have underplayed some critical differences that have less to do with defining or influential elements, and more to do with how conflict and bullying are generally perceived. Researchers and practitioners alike have expended considerable effort to promote and support the idea that conflict is inherent in social interaction and thus is common to organisational life. The corollary to this is that conflict in organisations can be a very good thing. When effectively stimulated and managed, conflict can result in improved relationships, greater creativity and innovation, and enhanced task performance. Thus, conflicts are normal and can lead to positive outcomes for all. To suggest that bullying is a conflict without attaching the numerous qualifiers we have noted above runs the risk of normalising this form of relationship and potentially providing justification that it makes people and organisations function better. While researchers and practitioners who engage people who have been bullied may not fall victim to this thinking, such thinking is functional for others who may wish to deny the extent and seriousness of bullying, as it minimises their need to take action or indeed their need to desist from such treatment.

Applying the label of conflict wholesale without qualification also creates the sense of shared responsibility for the bullying, and the victim may be expected to manage the situation on his/her own or, in some cases, be held accountable for the hostility exhibited by the other person. As discussed, this can only result in further damage to the target. The notion of shared responsibility is particularly troublesome if certain tenets of principled negotiation theory (e.g. Fisher *et al.*, 1991) are applied to 'bullying conflicts'. For example, one tenet is to 'separate the person from the problem'. As an approach to conflict, particularly between equals, this reflects an awareness that frustration over tasks (cognitive conflict) can easily generalise to more personalised and antagonistic frustrations with the other (affective conflict) which are difficult to resolve. Therefore, disputants are encouraged to focus on the task issues and not the ways in which the person is handling issues or dealing with them. But in the case of bullying, the person is the problem, and hence they cannot be separated. Trying to frame the situation otherwise can only be disastrous for the target as he/she will experience the inability to do so as failure on his/her part, contributing further to the diminishment of his/her sense of competence as a worker and as a person (Keashly, 2001).

For these reasons we caution researchers, practitioners and organisational members not to describe bullying as a conflict, but rather to consider what a conflict perspective might offer in understanding this extremely hostile and devastating phenomenon.

Note

1 A detailed discussion of the dynamics of escalation is beyond the scope of this chapter. Interested readers are strongly encouraged to read the reviews of Fisher (1990), Rubin *et al.* (1994) and Thomas (1976) for further detail.

Bibliography

Andersson, L. M. and Pearson, C. M. (1999) Tit for tat? The spiralling effect of incivility in the workplace. *Academy of Management Review*, *24*, 452–471.

Aquino, K. (2000) Structural and individual determinants of workplace victimization: The effects of hierarchical status and conflict management style. *Journal of Management*, *26*, 171–193

Barling, J. (1996) The prediction, psychological experience, and consequences of workplace violence. In G. VandenBos and E. Q. Bulatao (eds), *Violence on the job: Identifying risks and developing solutions*. Washington, DC: American Psychological Association.

Berkowitz, L. (1989) Frustration, appraisals, and aversively stimulated aggression. *Aggressive Behavior*, *14*, 3–11.

Blake, R. R. and Mouton, J. S. (1964) *Managerial grid*. Houston, TX: Gulf.

Coleman, P. T. (2000) Intractable conflict. In M. Deutsch and P. T. Coleman (eds), *The handbook of conflict resolution* (pp. 428–450). San Francisco: Jossey-Bass.

Cortina, L. M., Magley, V. J., Williams, J. H. and Langhout, R. D. (2001) Incivility in the workplace: Incidence and impact. *Journal of Occupational Health Psychology*, *6*, 64–80.

Costantino, C. A. and Merchant, C. S. (1996) *Designing conflict management systems*. San Francisco: Jossey-Bass.

de Dreu, C. K. W. (1997) Productive conflict: The importance of conflict management and conflict issue. In C. de Dreu and E. van De Vliert (eds), *Using conflict in organizations* (pp. 9–22). Thousand Oaks, CA: Sage.

Deutsch, M. (1973) *The resolution of conflict*. New Haven, CT: Yale University Press.

—— (1985) *Distributive justice: A social-psychological perspective*. New Haven, CT: Yale University Press.

—— (1993) Educating for a peaceful world. *American Psychologist*, *48*, 510–517.

Donnellon, A. and Kolb, D. M.(1994) Constructive for whom? The fate of diversity disputes in organizations. *Journal of Social Issues*, *50*, 139–155.

Einarsen, S. (1999) The nature and causes of bullying at work. *International Journal of Manpower*, *20*, 16–27.

Einarsen, S. and Skogstad, A. (1996) Bullying at work: Epidemiological findings in public and private organizations. *European Journal of Work and Organizational Psychology*, *5*, 185–201.

Filley, A.C. (1975) *Interpersonal conflict resolution*. Glenview, IL: Scott Foresman.

Fisher, R., Ury, W. and Patton, B. (1991) *Getting to yes: Negotiating agreement without giving in*. New York: Penguin.

Fisher, R. J. (1990) *The social psychology of intergroup and international conflict resolution*. New York: Springer-Verlag.

Fisher, R. J. and Keashly, L. (1990) Third party consultation as a method of intergroup and international conflict resolution. In R. J. Fisher, *The social psychology of intergroup and international conflict resolution* (pp. 211–238). New York: Springer-Verlag.

Glasl, F. (1982) The process of conflict escalation and roles of third parties. In G. B. J. Bomers and R. B. Peterson (eds), *Conflict management and industrial relations* (pp. 119–140). Boston: Kluwer-Nijhof Publishing.

Glomb, T. M. (2002) Workplace anger and aggression: Informing conceptual models with data from specific encounters. *Journal of Occupational Health Psychology*, 7, 1, 20–36.

Hoel, H., Rayner, C. and Cooper, C. L. (1999) Workplace bullying. *International Review of Industrial Organizational Psychology*, *14*, 195–229.

Jehn, K. A. (1997) Affective and cognitive conflict in work groups: Increasing performance through value-based intragroup conflict. In C. de Dreu and E. van De Vliert (eds), *Using conflict in organizations* (pp. 87–100). Thousand Oaks, CA: Sage.

Kabanoff, B. (1991) Equity, equality, power, and conflict. *Academy of Management Review*, *16*, 2, 416–441.

Keashly, L. (1998) Emotional abuse in the workplace: Conceptual and empirical issues. *Journal of Emotional Abuse*, *1*, 1, 85–117.

—— (2001) Interpersonal and systemic aspects of emotional abuse at work: The target's perspective. *Violence and Victims*, *16*, 3, 233–268.

Kolb, D. M., Bartunek, J. M. and Putnam, L. L. (1992) *Hidden conflict in organizations*. Thousand Oaks, CA: Sage.

Kressel, K. (2000) Mediation. In M. Deutsch and P. T. Coleman (eds), *The handbook of conflict resolution* (pp. 522–545). San Francisco: Jossey-Bass.

Kriesberg, L. (1989) *Varieties of mediation activities*. Paper presented at the annual meeting of the International Society of Political Psychology, Tel Aviv, June.

Kriesberg, L., Northrup, T. A. and Thorson, S. J. (1989) *Intractable conflicts and their transformation*. Syracuse, NY: Syracuse University Press.

Leymann, H. (1996) The content and development of mobbing. *European Journal of Work and Organizational Psychology*, *5*, 2, 165–184.

Mack, D. A., Shannon, C., Quick, J. D. and Quick, J. C. (1998) Stress and the preventive management of workplace violence. In R. W. Griffin, A. O'Leary-Kelly and J. M. Collins (eds), *Dysfunctional behaviors in organizations* (pp. 119–141). Greenwich, CT: JAI Press.

Musser, S. J. (1982) A model for predicting the choice of conflict management strategies by subordinates in high-stakes conflicts. *Organizational Behavior and Human Performance*, *29*, 257–269.

Neuman, J. H. and Baron, R. A. (1997) Aggression in the workplace. In R. A. Giacalone and J. Greenberg (eds), *Antisocial behavior in organizations* (pp. 37–67). Thousand Oaks, CA: Sage.

Opotow, S. (2000) Aggression and violence. In M. Deutsch and P. T. Coleman (eds), *The handbook of conflict resolution* (pp. 403–427). San Francisco: Jossey-Bass.

Papa, M., and Canary, D. J. (1995) Conflict in organizations: A competence-based approach. In A. M. Nicotera (ed.), *Conflict and organizations: Communication processes* (pp. 153–179). New York: State University of New York Press.

Papa, M. and Pood, E. A. (1988) Co-orientation accuracy and differentiation in the management of conflict. *Communication Research*, *15*, 400–425.

Prein, H. (1984) A contingency approach for conflict intervention. *Group and Organization Studies*, *9*, 1, 81–102.

Putnam, L. L. and Poole, M. S. (1987) Conflict and negotiation. In F. M. Jablin, L. L. Putnam, K. H. Roberts and L. W. Porter (eds), *Handbook of organizational communication* (pp. 549–599). Beverly Hills: Sage.

Rahim, M. A. and Buntzman, G. F. (1990) Supervisory power bases, styles of handling conflict with subordinates, and subordinate compliance and satisfaction. *Journal of Psychology*, *123*, 195–210.

Rayner, C. (1999) From research to implementation: Finding leverage for prevention. *International Journal of Manpower*, *20*, 28–38.

Rayner, C., Sheehan, M. and Barker, M. (1999) Theoretical approaches to the study of bullying at work. *International Journal of Manpower*, *20*, 11–15.

Richman, J. A., Rospenda, K. M., Flaherty, J. A. and Freels, S. (2001) Workplace harassment, active coping and alcohol-related outcomes. *Journal of Substance Abuse*, *13*, 3, 347–366.

Richman, J. A., Rospenda, K. M., Nawyn, S. J., Flaherty, J. A., Fendrich, M., Drum, M. L. and Johnson, T. P. (1999) Sexual harassment and generalized workplace abuse among university employees: Prevalence and mental health correlates. *American Journal of Public Health*, *89*, 3, 358–363.

Rubin, J. Z., Pruitt, D. G. and Kim, S. H. (1994) *Social conflict: Escalation, stalemate, and settlement*, 2nd edn. New York: McGraw-Hill.

Sheehan, M. (1999) Workplace bullying: Responding with some emotional intelligence. *International Journal of Manpower*, *20*, 57–69.

Sheppard, B. H. (1984) Third party conflict intervention: A procedural framework. *Research in Organizational Behavior*, 6, 141–190.

Thomas, K. W. (1976) Conflict and conflict management. In M. Dunnette (ed.), *Handbook of industrial psychology* (pp. 889–935). Chicago: Rand-McNally.

—— (1992) Conflict and negotiation processes in organizations. In M. D. Dunnette (ed.), *Handbook of industrial organizational psychology*, 2nd edn, vol. 3 (pp. 651–718). Palo Alto, CA: Consulting Psychologists Press.

Ury, W. L., Brett, J. M. and Goldberg, S. B. (1988) *Getting disputes resolved: Designing systems to cut the costs of conflict*. San Francisco: Jossey-Bass.

Zapf, D. (1999) Organizational, work group related and personal causes of mobbing/bullying at work. *International Journal of Manpower*, *20*, 70–85.

Zapf, D. and Gross, C. (2001) Conflict escalation and coping with workplace bullying: A replication and extension. *European Journal of Work and Organizational Psychology*, *10*, 497–522.

21 Bullying, emotions and the learning organisation

Michael J. Sheehan and Peter J. Jordan

Introduction

Existing research into workplace bullying has dealt with the legal implications of bullying (Yamada, this volume), the costs of bullying to the organisation (Hoel *et al.*, this volume) and to society (Sheehan *et al.*, 2001). There also has been a focus on the characteristics exhibited by bullies and victims (O'Moore *et al.*, 1998), the psychological consequences of bullying (Leymann, 1996), the physical and work-related consequences of bullying (McCarthy *et al.*, 1995) and the initial and long-term effects of being bullied (Einarsen and Mikkelsen, this volume). Whereas this research has mostly focused on the cognitive aspects of bullying and the consequences for individuals and organisations, we argue in this chapter that to develop workplace remedies to bullying, managers need to understand the emotional aspects of bullying.

A central tenet of this chapter is that bullying emerges from emotional as well as cognitive processes. From an individual perspective, Goleman (1998) argues that personal behaviour is more of a function of emotional regulation than of rational or cognitive processes. Ashforth and Humphrey (1995) and Fineman (2000) offer a similar view at the organisational level, arguing that work life is intrinsically emotional and value-based, and that ostensibly rational organisational behaviour reflects the extent to which organisational members are able to reconcile emotional issues in the workplace. We propose that understanding the emotional elements involved in bullying expands the number of potential remedies which may ameliorate the incidence of bullying in organisations. This approach is consistent with Ashforth and Humphrey's (1995) call for the incorporation of emotional variables in organisational behaviour research. We argue, first of all, that emotionally driven actions and reactions contribute to bullying. Second, we suggest that the principles that underpin an understanding of organisations as learning organisations may be used to address these emotional actions and reactions, and to ameliorate the prevalence and severity of bullying.

Einarsen *et al.* define bullying as

> all those aggressive actions and practices directed at one or more worker(s) which: are unwanted by the victim; may be done deliberately or unconsciously but do cause humiliation, offence and distress; and may interfere with job performance and/or cause an unpleasant working environment.
>
> (Einarsen *et al.*, 1998, p. 564)

Behaviour associated with bullying includes sarcasm, threats, verbal abuse, intimidation, bad-mouthing, manipulation, duplicity, exclusion, isolation and the assignment of staff to unpleasant jobs (McCarthy *et al.*, 1995). Both the definition of bullying provided by Einarsen and his colleagues and the types of behaviour outlined by McCarthy and his colleagues involve the generation and experience of emotions. From this point of view, exploring the link between emotions and bullying becomes imperative.

While there has been considerable research conducted into the role of emotions in organisations (e.g. Ashforth and Humphrey, 1995; Fineman, 2000; Weiss and Cropanzano, 1996), to our knowledge the issue of emotions and bullying has not been addressed. Research into emotions in the workplace has primarily focused on affect-driven behaviours such as impulsive acts, organisational citizenship behaviours and transient effort (Weiss and Cropanzano, 1996). As bullying impacts directly on workplace relationships, we sought a theoretical framework that would support the link between emotions and relationships. As a framework for considering remedial actions in relation to bullying in the workplace, we will draw on the theory of bounded emotionality (Mumby and Putnam, 1992).

Bounded emotionality

Putnam and Mumby (1993) introduce the notion of 'bounded emotionality' as a foil to Simon's (1976) notion of 'bounded rationality'. The concept of 'bounded emotionality' is predicated on the idea that emotional variables underpin organisational behaviour. Whereas bounded rationality identifies involuntary limitations in human information processing (Simon, 1976), bounded emotionality refers to voluntary behaviours implemented within the workplace to ensure relationships are maintained and enhanced (Mumby and Putnam, 1992). One of the main ways to enhance relationships within this framework is by both increasing and constraining emotional expression. The implementation of a regime that only relies on the constraint of emotion could be used to facilitate bullying in organisations. Bounded emotionality, however, also encourages the appropriate expression of emotion in organisations. In this context, inappropriate or bullying behaviour can be confronted with the appropriate expression of emotion, thus enabling bullying to be acknowledged and addressed. For instance, in addressing a bullying incident there would be a focus on describing how that incident made the victim feel, with the emotional expression being focused on maintaining the relationship between the protagonists rather than, for example, gaining revenge. The focus of this behaviour would be to identify the unacceptable behaviour and to reduce the possibility of escalating the confrontation.

The application of a bounded emotionality perspective to the phenomenon of bullying allows the emotional antecedents and consequences of bullying to be explored within a framework that acknowledges the importance of emotions for maintaining and enhancing relationships in the workplace. For example, there may be a need for the bully to constrain their emotional expression of feelings such as frustration, anxiety and anger that may result in a reduction of bullying behaviours such as sarcasm or verbal abuse. On the other hand, for victims of bullying, there may be a need to increase their emotional expression to become more assertive, rather than avoiding or withdrawing from the situation. In this case the victim, as a way of addressing the bullying, can express to the bully or to others the feelings of intimidation or threat that emerge during a bullying episode. Either way, we propose that the introduction of bounded emotionality will facilitate improved relationships in the workplace.

Martin *et al.* (1998) argue that organisations can change their existing behaviours and practices through the management of emotions. The key point for Martin and her colleagues is that employees can be encouraged to constrain their emotional expression to create effective relationships with their fellow workers. That is, if people are required to become more attuned to their emotions and the emotions of others and to consider the implications of their actions, then this may result in better working relationships. This, however, is not a common occurrence in workplaces today. In order to engender new ways of working, therefore, we need to examine methods for introducing such ideas into organisations. We propose that the processes involved in the learning organisation may facilitate the implementation of such ideas.

Bullying and the learning organisation

We propose that the utilisation of the framework of a learning organisation (Senge, 1992) may be an appropriate method for advancing remedial strategies to minimise bullying, emphasising the need for organisations and employees to learn and continuously improve their skills and abilities. In that way they move beyond simply adapting to new challenges and into the realm of generative learning. Senge (1992) suggests that generative learning can be developed through the framework of the learning organisation. He identifies the fundamental disciplines of the learning organisation as comprising personal mastery, mental models, systems thinking, team learning and building a shared vision for the organisation.

We propose that each of the five disciplines advanced by Senge (1992) have a role to play in ameliorating the incidence of workplace bullying. It is important to note, however, that in adopting the processes of the learning organisation to introduce an emotionally aware workplace, we are advocating that both the victims and those who bully need to take action. The philosophy that underpins the learning organisation suggests that all employees need to question and address their behaviour and reactions, not just specific groups. In the following section, we outline the contribution that each of the five disciplines can make to provide remedies for bullying in organisations, addressing each of these in some detail.

Implications and remedial actions

We consider that understanding the emotional elements that contribute to bullying can assist in identifying different ways of reducing the incidence of bullying in organisations. Although we acknowledge the importance of organisational level interventions, in this chapter we want to focus on remedial actions at the individual level. Importantly, the approach we adopt in this section follows a developmental, rather than a punitive method for dealing with those identified as workplace bullies. This ensures that all parties who have a role to play in addressing the problem are treated with a measure of respect and dignity. We also suggest that the victims need to play an active role in addressing bullying. A developmental approach avoids the risk of turning the perpetrators into victims, or the victims into perpetrators, an outcome against which McCarthy (2000) cautions.

Personal mastery

Senge (1992) describes personal mastery as involving the enhancement of the technical skills that many organisations focus on as they seek to improve efficiency and effectiveness, as well as enhancing individual personal skills. In enhancing personal skills, we argue that the understanding and development of emotional abilities is an area that is often overlooked in organisations. The adoption of a bounded emotionality framework within an organisation would require individuals to take responsibility for personal relationships and the expression and control of emotions within the workplace. This would apply to both those who bully and their victims. We contend that bounded emotionality involves people becoming more attuned to both their own emotions and others' emotions, but also regulating those emotions to enhance relationships between employees. Taking responsibility for one's own emotional displays becomes an important part of emotional regulation. For example, in the case of the bully, this may require the individual to acknowledge the impact of their emotional outbursts and behaviour on others, and to be able to apologise for such behaviour. It may also help them to develop new, more appropriate actions for displaying their emotions. In the case of the victim, it requires them to understand the impact of the bully's behaviour on themselves and to take steps to address the impact of those behaviours, such as reporting the incident to others and ensuring their well-being is maintained in the face of bullying.

Within a workplace practising bounded emotionality, workers may become more assertive and less aggressive with other employees. Such a style has positive implications for reducing bullying behaviours in the workplace, and for the establishment and maintenance of more appropriate relationships with fellow workers, and may extend to improved relationships with customers and suppliers. All of these factors have the potential to reduce the incidence of workplace bullying.

Mental models

Senge (1992) also identifies the importance of addressing existing organisational mental models for individuals who are seeking to adapt to a changing environment. This means challenging personal assumptions and conventional approaches to managing in organisations. Rayner (1998) suggests that the acceptance of bullying stems from two sources; the first is the direct observation of others being bullied, and the second is the assumption that the organisation tolerates bullying behaviour. If an individual has observed bullying to be successful in achieving goals or if they have had a successful experience of gaining an objective through bullying, this increases the possibility that this technique will be used again (Archer, 1999). If individuals can openly assess past successes, and not only the goals they attained but also the consequences of their actions, then they can address these mental models. The development of appropriate learning development strategies and programmes that embrace critical reflection as part of the learning and development process would enable this to occur. The concept of the learning organisation assumes that individuals in organisations are willing to engage in questioning their collective assumptions (Vince, 2001). Vince further argues that such questioning requires a re-examination of power relationships in organisations. Addressing dominant mental models in the organisation could attain this aim.

Systems thinking

Senge (1992) identifies the importance of considering organisations as systemic entities where each action produces an interrelated consequence. In the learning organisation, critical questioning of systems and processes that appear to support, or fail to address, bullying by organisational members should be encouraged. Conflict needs to be identified as a part of organisational life, with systems developed to ensure that conflict is resolved, instead of leading to the creation of new tensions, conflicts and differences, resulting in a lack of trust (Sheehan and Jordan, 2000). Organisations need to develop systems that support learning and change to replace those that inhibit learning (McCarthy *et al.*, 1995).

Sheehan and Jordan (2000) have suggested that to address workplace bullying, organisations need to move towards a co-operative rather than a confrontational workplace. This is not to suggest that conflict should be eliminated from the workplace, but rather that conflict should be functional rather than dysfunctional, with a focus on problems and issues rather than personalities. An environment needs to be encouraged in which information about the problem of bullying is shared rather than controlled. Progressive systems are required to address bullying, rather than rigid and inflexible systems that are unable to deal with the problem (Field and Ford, 1995). By 'progressive' we mean systems that are prepared to confront problems as they arise. By 'rigid and inflexible' we mean those systems that continue to abide by policies, procedures and rules even when, in the face of prevailing evidence, those systems are inappropriate for dealing with the problems of workplace bullying.

Team learning

In stressing the importance of team learning, Senge (1992) asserts that it involves changes to the ways team members communicate. In part, members need freely to be able to explore important issues by listening carefully to each other, which involves suspending one's own views to enable empathic listening to occur. In order to address the consequences of their actions, and in particular the emotions that have been engendered as a result of their actions, it is important that employees develop their ability to empathise with others. Some discussion of remedial actions to address bullying includes requiring the bullies to consider the consequences of their own actions (O'Moore *et al.*, 1998; Sheehan, 1999). Within a learning organisation framework such discussion is considered to be single loop learning. We propose that 'double loop' learning (Argyris and Schön, 1996), or seeking to understand the attitudes and emotions that lead to the action, is a more appropriate approach to ameliorate bullying. If double loop learning is applied to bullying, the foci become the emotional and cognitive antecedents of bullying, rather than the specific actions that are determined to constitute bullying. This type of approach may assist employees to become more empathetic in dealing with others in the organisation (Korth, 2000). An empathetic approach to dealing with others could then lead to the employee seeking to regulate the emotions that contribute to their instigation of actions that are considered bullying behaviours (Sheehan and Jordan, 2000), thus enabling team learning to occur.

Shared vision

Senge (1992) argues that the employees of a learning organisation hold a vision that is future-oriented in terms of its philosophy, values, mission and goals. By achieving a shared vision and an understanding of the core set of competencies required to achieve that vision, the organisation is able to differentiate itself from its competitors (Field and Ford, 1995; Senge, 1992). The vision requires top-level commitment to address aspects of management practices and culture that inhibit performance. One of the challenges that organisations currently face is dealing with the problem of workplace bullying, an acknowledged contributor to lower performance and poor workplace relations (Sheehan *et al.*, 2001).

To address this issue, the organisation needs to develop a shared vision of a workplace in which bullying is not a part of the workplace culture and in which even isolated incidences of bullying are not condoned. The shared vision, in this instance, relates to the values that each worker ascribes to in the workplace. In practice, this may involve the incorporation of value statements during organisational planning cycles, such as the importance of respecting others in the organisation and the rejection of overt and covert forms of coercion. The inclusion of such value statements, however, needs to be inculcated in the organisational culture to prevent them becoming platitudes or tokenistic gestures. The development of codes of conduct may also assist in this regard.

To draw these ideas together, we suggest that organisational learning programmes incorporate the aforementioned disciplines within the framework of bounded emotionality learning programmes.

Bounded emotionality learning programmes

An awareness of the social context in which bullying occurs has been identified as being important to the learning process in organisations (Einarsen, 1999; Rayner, 1999). The social context includes role modelling and mentoring by managers and supervisors. It also facilitates the provision of learning and development programmes that could include sessions devoted to understanding the impact of emotions on employees in organisations. A positive and supportive work climate aids the application of learning in the workplace. People who are given work roles or tasks relevant to their new learning are more likely to apply that learning in the workplace (Burns, 1995). Role models should be identified and employees encouraged further to develop their own abilities as a way of modelling preferred organisational behaviours, activities and actions.

A consideration of the importance of emotion in the learning and development process should also be recognised. We acknowledge and accept that there is a developmental progression in ways of dealing with the role of emotion in learning (Mayer and Salovey, 1997) and understanding the impact of learning (Sheehan, 2000). A programme of the type suggested here would seem to be a way of helping employees to learn to cope better with this aspect of learning and development. Within such a programme, however, careful attention needs to be paid to the development of a learning environment in which the participants feel that they are able to trust each other and the learning facilitator. Appropriate mechanisms for doing so ought to be structured into the learning programme.

Action is also a key part of a learning process (Sheehan, 2000). Clearly, therefore, programmes of the type suggested here need to include a range of specific activities incorporating the opportunity to gain understanding about workplace bullying, the actions of the perpetrator, the impact on the victims, and the use or misuse of emotion. It may be argued that learning does not end at the conclusion of a learning programme (Sheehan, 2000). Nor does the learning only embrace topics or experiences covered during such a formal programme. Rather, further development needs to be embraced. Such development should occur by way of an advanced learning programme, by attending other programmes of a similar nature, by senior managers mentoring learning participants, or by regular and constructive feedback about the use of their learning from senior managers and peers. As suggested by Milliman *et al.* (1994), the use of 360-degree feedback could be helpful for such a process.

Employees should also be given self-management strategies to help prevent them from lapsing into previously learned, and often less effective, actions. Self-management strategies enable people to discuss lapses openly, and to explore ways to overcome those lapses. Open discussion of emotional reactions and feelings provides a catalyst for further personal development and self-improvement, and enables the development of abilities or actions to overcome obstacles to the application of bounded emotionality in the workplace (following Senge, 1992).

More specifically, encouragement should be given to developing management and work group sensitivity to a range of issues. There are four avenues managers could explore. First, the importance of the critically reflective process could be embraced, with encouragement of employees to engage in such reflection after formal bounded emotionality learning programmes have been completed. Second, support for their staff in the development of their new learning ought to be forthcoming. This expectation could be made explicit in their performance criteria. Third, they need to be aware that they can expect a range of reactions to bounded emotionality change programmes from themselves and from their staff. Reactions may range from doing nothing to being fully motivated to implementing all that they have learned or resisting the change process. Fourth, both their own and their employees' reactions need to be addressed to ensure ongoing development. Clearly, therefore, managers need a range of intrapersonal and interpersonal skills to enable them to perform this part of their role appropriately (Sheehan, 2000).

Limitations

Our first concern is that calls for constraining emotional expression may lead to the manipulation of individuals in order to achieve the desired outcomes for the bully. While the idea of introducing a bounded emotionality framework within the process of the learning organisation is aimed at creating a co-operative workplace, the approach requires the collaboration of all parties. Attention needs to be paid by members of the organisation to ensuring that bounded emotionality is not being misused in workplace relationships. From another perspective, the constraint of emotions may have undesirable outcomes for bullies as well. This may prevent the individual from free expression of their emotion, resulting in stress and undesirable consequences for the health and well-being of the bully. From this point of view, the organisation has to facilitate other ways for the bully to express those emotions.

Our next concern is that the perpetrator may not always be aware of their actions or understand the impact of their actions on others. They may be unable or unwilling to accept feedback about their actions and the impact they have had. They may be unwilling to change. While we advocate the development of personal mastery in the workplace that involves a level of self-awareness, and challenging existing mental models, this may not always be self-evident. In this case, the introduction of a new corporate culture using a shared vision of the values of the organisation may assist in overcoming entrenched and unrecognised behaviours, and eliminating them from the perpetrator's repertoire.

A related issue is the impact of individual differences on workplace behaviour and, in particular, responses to bullying. We acknowledge that individual differences that arise from personality and attitudes will have a major bearing on the individual's propensity to engage in a bounded emotionality organisation. Some employees will feel uncomfortable with enacting some of the skills required to reduce bullying in the workplace. Our intention is not to make these employees uncomfortable in their work environments, but rather to enable the rest of the workforce to challenge the mental models that prescribe one set of behaviours as being appropriate within the organisation.

Finally, we need to acknowledge that power and authority have an influence on working relationships that remains unaccounted for in this chapter. While we have suggested that the victim confront the perpetrator, we acknowledge that those who perceive themselves as being bullied may find it difficult to carry out this action. The victim has a right to ask for what they want, namely for the bullying behaviours to stop. Those who are tasked with dealing with the problem, such as managers, supervisors or the perpetrators, need to realise that allowing the victim to give voice to their concerns is an empowering strategy that will help overcome the problem and lead to healthier and more productive workplaces.

Conclusion

The purpose of this chapter is to present remedial actions to address workplace bullying using a bounded emotionality framework and principles from Senge's concept of the learning organisation. Our intention has been to focus on a neglected area of understanding relating to emotional reactions to workplace bullying.

We argue that the phenomenon of workplace bullying should be addressed proactively. Focusing on the outcomes of those who have experienced bullying, while important and necessary, does not necessarily lead to addressing constructively the problem at its cause. Rather, we argue, a remedial approach to bullying could be determined using bounded emotionality, and within the framework of organisations as learning organisations. We contend that organisational-level intervention in culture, attitudes and beliefs will provide a supportive environment that can minimise the incidence of workplace bullying. But this tactic needs to be supported by acknowledging the potential emotional reactions of individual employees to change. We argue that a bounded emotionality response, by way of a learning and development programme, embracing improved emotional and social skill development, would contribute to alleviating bullying.

Bibliography

Archer, D. (1999) Exploring 'bullying' culture in the para-military organisation. *International Journal of Manpower*, *20*, 1/2, 94–105.

Argyris, C. and Schön, D. A. (1996) *Organizational learning II: theory, method and practice*. Reading, MA: Addison-Wesley.

Ashforth, B. E. and Humphrey, R. H. (1995) Emotion in the workplace: A reappraisal. *Human Relations*, *48*, 2, 97–125.

Burns, R. (1995) *The adult learner at work*. Sydney: Business and Professional Publishing.

Einarsen, S. (1999) The nature and causes of bullying at work. *International Journal of Manpower*, *20*, 1/2, 16–27.

Einarsen, S., Matthiesen, S. and Skogstad, A. (1998) Bullying, burnout and well-being among assistant nurses. *The Journal of Occupational Health and Safety – Australia and New Zealand*, *14*, 6, 563–568.

Field, L. and Ford, B. (1995) *Managing organisational learning: From rhetoric to reality*. Melbourne: Longman.

Fineman, S. (ed.) (2000) *Emotion in organizations*, 2nd edn. London: Sage.

Goleman, D. (1998) *Working with emotional intelligence*. New York: Bantam.

Korth, S. J. (2000) Single and double-loop learning: Exploring potential influence of cognitive style. *Organization Development Journal*, *18*, 3, 87–98.

Leymann, H. (1996) The content and development at work. *European Journal of Work and Organizational Psychology*, *5*, 2, 165–184.

Martin, J., Knopoff, K. and Beckman, C. (1998) An alternative to bureaucratic impersonality and emotional labor: Bounded emotionality at the Body Shop. *Administrative Science Quarterly*, *43*, 2, 429–470.

Mayer, J. D. and Salovey, P. (1997) What is emotional intelligence? In P. Salovey and D. J. Sluyter (eds), *Emotional development and emotional intelligence: Educational implications* (pp. 3–31). New York: Basic Books.

McCarthy, P. (2000) The bully-victim at work. In M. Sheehan, S. Ramsay and J. Patrick (eds), *Transcending boundaries: Integrating people, processes and systems* (pp. 251–256). Conference proceedings 6–8 September 2000, The School of Management, Griffith University, Brisbane.

McCarthy, P., Sheehan, M. and Kearns, D. (1995) *Managerial styles and their effects on employees health and well-being in organizations undergoing restructuring*. Report for Worksafe Australia. Brisbane: Griffith University.

Milliman, J. F., Zawacki, R. A., Norman, C., Powell, L. and Kirksey, J. (1994) Companies evaluate employees from all perspectives. *Personnel Journal*, *73*, 11, 99–103.

Mumby, D. K. and Putnam, L. A. (1992) The politics of emotion: A feminist reading of bounded rationality. *Academy of Management Review*, *17*, 3, 465–486.

O'Moore, M., Seigne, E., McGuire, L. and Smith, M. (1998) Victims of bullying at work in Ireland. *The Journal of Occupational Health and Safety-Australia and New Zealand*, *14*, 6, 569–574.

Putnam, L. L. and Mumby, D. K. (1993) Organizations, emotion and the myth of rationality. In S. Fineman, (ed.), *Emotion in organizations* (pp. 36–57). London: Sage.

Rayner, C. (1998) Workplace bullying: Do something! *The Journal of Occupational Health and Safety – Australia and New Zealand*, *14*, 6, 581–585.

—— (1999) From research to implementation: Finding leverage for prevention. *International Journal of Manpower*, *20*, 1/2, 28–38.

Senge, P. (1992) *The Fifth Discipline*. Milsons Point: Random House.

Sheehan, M. (1999) Workplace bullying: Responding with some emotional intelligence. *International Journal of Manpower*, *20*, 1/2, 57–69.

—— (2000) *Learning and implementing group process facilitation skills: Individual experiences*. Unpublished doctoral thesis, Griffith University, Brisbane.

Sheehan, M. and Jordan, P. (2000) *The antecedents and implications of workplace bullying: A bounded emotionality analysis*. Paper presented at the Association Francophone De Gestion Des Ressources Humaines (AGRH) Congress, Paris, 16–17 November.

Sheehan, M., McCarthy, P., Barker, M. and Henderson, M. (2001) *A model for assessing the impacts and costs of workplace bullying*. Paper presented at the Standing Conference on Organisational Symbolism, Dublin, July.

Simon, H. A. (1976) *Administrative behavior: A study of decision-making processes in administrative organization*, 3rd edn. New York: Free Press.

Vince, R. (2001) Organizational learning: A personal view. *International Journal of Management and Decision Making*, 2, 1, 2–7.

Weiss, H. M. and Cropanzano, R. (1996) Affective Events Theory: A theoretical discussion of the structure, causes and consequences of affective experiences at work. In Barry M. Staw and L. L. Cummings (eds), *Research into organizational behavior*, vol. 18 (pp. 1–74). Stanford, CT: Jai Press.

22 Bullying and human resource management

A wolf in sheep's clothing?

Duncan Lewis and Charlotte Rayner

Introduction

Of all the functional departments that deal with bullying at work, it is the personnel or human resource department that is likely to have the greatest involvement. Typically, when we think of workplace bullying and the human resources (HR) function, we think of policy, procedure and a mediating role. Our approach in this chapter is to examine critically the role of human resource management (hereafter HRM) and its relationship to bullying at work. Our perspective is to examine the very essence of HRM as a managerial ideology and not simply as a functional area of organisational structure.

Our discussion will not imply that the HRM function itself has ignored workplace bullying. On the contrary, many HR managers have embraced policy and procedure to deal actively with workplace bullying. Our thesis, derived from an analysis of the philosophy and components which underpin HRM, will be that 'the' philosophy and components of HRM may create an environment in which bullying can remain unchallenged, allowed to thrive or actually encouraged in an indirect way. Therefore the managerial paradigm of HRM may be a source of bullying (not its constituent membership), and our efforts to understand and tackle bullying at work must include it. In order to examine this hypothesis, we need briefly to describe the arrival of HRM and its raison d'être. Additionally, we would like briefly to consider some of the more generalist managerial ideologies within which HRM falls. For without an understanding of the managerial rhetoric, we can never fully grasp how HRM fits into the complex organisational environment. Finally, we would like to demonstrate how these elements can combine to mitigate against dealing with bullying *effectively*.

This chapter is written through a UK lens; the history of industrial relations and personnel management practice may be different from other countries. As Drumm (1994) noted in his evaluation of HRM in Germany, 'within any specific country, HRM as practised in companies is underpinned by nationally distinct legal frameworks of the constitutional structure of the firm and labour relations' (p. 35). This is particularly important when recognising the rights and responsibilities of both employers and employees, as well as the legal rights of worker representation by the workforce, where European Union policies are largely opposite to UK policies, which are restrictive and control-oriented (Richbell, 2001). For example, in the UK, workers have no legal right to board-level participation, which is different from the situation in Germany and other countries (see Mayrhofer, 1995, for example). We are not suggesting that the situation in the UK is unique, but that the political landscape is an example of international difference. Indicators of a changing UK industrial relations landscape are declining union membership, from 65 per cent in 1980 to 38 per cent in 1998, and a reduction in the number of working days lost to strike action from 330 days between 1980 and 1983 to a mere eleven days in 1997–1998. In addition, UK workers spend some of the longest periods at work in Europe, an average of forty-four hours per week (Richbell, 2001).

Welcome HRM

The establishment of HRM as a managerial ideology is largely attributed to the USA and the Michigan School (for example Beer *et al.* 1985). The impetus for HRM was much like any other managerial role, being about deployment and quantitative acquisition of assets. This in essence sums up the differences between HRM and its forebear, personnel management, which was established more as a paternalist, welfare function. A central tenet of HRM philosophy is that human resources are an integral component of an organisation and should therefore not be under the patronage of specialists. Similarly held is the widespread belief that human needs and dignified treatment will accrue if the workplace environment offers mutual respect within a context of efficiency. In a UK context, HRM was established in the 1980s against a backdrop of commitment and employee involvement, although a healthy debate continues as to some of HRM's philosophical underpinnings (see, for example, Storey, 1993; Hart, 1993).

Since those early days, numerous attempts have been made to define and categorise the constituent elements of HRM, and we intend using the four key elements as outlined by John Storey (1993). First, the 'beliefs and assumptions' of HRM are derived from the doctrine that the workforce themselves provides the organisation with a competitive edge and that thus the workforce should be 'nurtured'. Furthermore, individuals and the organisation employing them should aim for commitment rather than simple compliance. Second, the notion is that HRM deserves top-management attention and subsequently has a strategic role to play. Third, if people are critical towards organisational success, the execution of HRM should lie with line managers rather than simply with a personnel specialist. The line manager role is seen to encompass everything from individual performance appraisal through to discipline, quality and performance-related pay. Implementation of such initiatives is not the concern of personnel staff but of the line managers themselves. The fourth component of Storey's model concerns the nature of the implementation of HRM. The premise here is that culture and not rules would result in the achievement of commitment, consensus and flexibility. This would be achieved by redesigning jobs to allow and encourage empowerment and harmonisation of the major elements of employee development from selection through to training and reward systems.

Aside from these constituent components, debate has remained split on what Storey (1993) referred to as 'hard' and 'soft' approaches. The 'hard' model of HRM drives strategy and policy that are consistent with the organisation's objectives. Authors including Legge (1995) and Tyson and Fell (1986) suggest the 'hard' model sees human resources as factors of production at the 'right' price. By contrast, the 'soft' model of HRM, whilst not ignoring the objective-driving elements of the 'hard' model, treats employees as 'valued assets'. Authors such as Legge (1995) and Guest (1987) see the 'soft' approach as gaining competitive advantage through commitment where 'employees are proactive rather than passive inputs into productive processes'.

A final feature of HRM that requires discussion is the role of HRM with respect to industrial relations. The rise in HRM throughout the 1980s and 1990s and the demise of collective bargaining and trade unions, specifically in the UK, are more than simple coincidence. Guest (1998), however, suggests that acceptance of HRM as being anti trade unionist is too simplistic, and that too many parallels were drawn in the early HRM literature and to non-unionised American firms where HRM had originated. Instead, Guest (1998) suggests that HRM diminishes the need for trade union representation through the central HRM principle of commitment. By placing organisational and employee commitment at the centre of the employment relationship, there is an attempt to 'win the hearts and minds' of employees. Clearly, such a view, in which employers and employees work in 'unitary' harmony, is desirable, even if somewhat utopian. In countries where workers are protected by a raft of legislation covering every aspect of the employment relationship, from the right to withdraw labour through to worker representation at a corporate level, there is little to fear from such a philosophy.

The stark reality of a globalised, commercial world, dictates that organisations are increasingly favouring lower-cost strategies of survival where part-time, flexible employment ensures human resources can be managed in a rational, economic way. The concept of 'hearts and minds' may favour organisations that require a creative, innovative and knowledge-principled workforce, but this becomes much less of an issue in organisations facing economic challenges from producers in very low-waged economies. The bottom line will always win, even in those countries that currently enjoy good industrial relations protection.

If the 'hearts and minds' philosophy could be successful, there would be little need for a third party to represent the views of the workforce (in the traditional 'pluralist' perspective), and instead what would become important would be the relationship between employee and employer. If we correlate the demise of trade union power in the UK with the introduction of numerous pieces of labour legislation under successive right-wing UK Conservative governments, added to which the adoption of HRM as a new people-management ideology, the position of trade unions becomes superfluous. By focusing directly on staff, the 'hearts and minds' model effectively bypassed trade union representation (see Marchington and Wilkinson, 1996, p. 240).

Similarly, whereas personnel management had established itself as the buffer *between* management and trade unions operating under pluralist ideology, HRM, through the notion of commitment, favoured this unitary model of representation. Such an ideology tries to confer legitimacy on forms of managerial control, whereas the pluralist ideology favours a more discursive approach, traditionally derived from worker representation by trade unions. The belief that modern HRM derives from a unitary perspective legitimises the workforce as a resource or instrument to be used for organisational purposes. Payne and Wayland (1999) suggest such a view of employees as 'resource' can lead to first and second-class employees, which can be divisive within the 'hearts and minds' philosophy described above.

Recent UK legislation (1999 Employment Regulations Act) that allows for trade union recognition after an absence of two decades has yet to produce results, and we should expect to see further shifts in the industrial relations landscape. Suffice to say the survival of the trade union role should have alerted us that not all was well in the unitary HR landscape rooted in a 'hearts and minds' philosophy.

We do not need to look far for the reason for some of the breakdown in the ideology in practice. All UK surveys have shown that managers themselves are perceived as the 'bully' in large proportions of incidents, and that this is the case in the UK in particular (e.g. Rayner, 1998). Other data has shown that bullying is part of many people's daily lives and exchanges in communication and discussion (Lewis, 1999). Perhaps these reflect how the 'hard' side of HRM (reaching organisational targets at a human cost) can dominate over the 'soft' side for some managers and some organisations, and we will return to this later.

Changing managerial and organisational paradigms

These changes from personnel to HRM principles have not developed in isolation. Instead they feed into and take from contemporary managerial paradigms and organisational frameworks that have become recent features of organisational life.

An extremely broad and wide-ranging literature accompanies any potential 'cause' of bullying, whether the cause be organisationally, managerially or individually grounded. For example, changes to work and organisational life caused by electronic and communication media, drives to be more visionary, strategic and globally responsive, coupled with continuous change and demand for sustainability that satisfies numerous stakeholders, have altered modern organisations significantly (Mintzberg *et al.*, 1998). In parallel to the emergence of these organisational paradigms, a commensurate range of comment at the managerial level has occurred. The overall result has been a frenzy of management hype and hyperbole which self-perpetuates, feeding on the most recent management fad as a means of establishing newer ones. This 'Darwinian' evolution in management thinking has created an environment in which organisations, managers and employees are struggling to make sense of the complexities brought on by the latest paradigms. Authors such as Drucker (1988), Handy (1989) and Senge (1992), amongst others, have given us a truly heterogeneous description of the organisation of the future. Examples include 'The Learning Organisation'; 'The Shamrock Organisation'; 'The Networked Organisation'; 'The Virtual Organisation'; and 'The Crazy Organisation'. The key to these 'new' organisational concepts lies in the recognition that change is continuous, historical managerial competences are no longer valid, and the traditional paths of competition are no longer sustainable. Instead, organisations are encouraged to collaborate rather than compete via strategic alliances (Bleeke and Ernst, 1995), and to operate in a networked and boundaryless environment, as opposed to a rigid and inflexible hierarchy. Organisations are also encouraged to view the world as a global trading medium requiring an appreciation of cultures and impending paradoxes (Parker, 1998) and willingly to accept work as a series of discontinuous and unconnected, temporary and part-time jobs (Handy, 1989).

Within this changing environment we can appreciate the importance of the 'soft' aspects of HRM as glue that can hold us together as we hurtle through into the new millennium. A prime candidate to allow us to examine the 'glue' more effectively is the concept of organisational culture, reaching into the softer side of corporate life. Thus we need one further foray into the management literature to explore how organisational culture provides a platform for HRM to thrive.

Organisational culture

Organisational culture has received a great deal of attention during the last ten to twenty years, primarily from the recognition that it can be a source of competitive advantage. As Hoecklin (1995) stated, 'However, when understood and successfully managed, differences in culture can lead to innovative business practices and sustainable sources of competitive advantage' (p. ix). Although Andrew Pettigrew (1979), a UK academic, was one of the first to discuss culture in a business context, it has been the USA that has been at the forefront of organisational culture writings. Numerous American management writers and business gurus, such as Ouchi (1981) and Schein (1985), amongst others, have firmly established culture in the psyche of management. Much of these writings are anecdotal tales from American elite corporations, and, as Parker (2000) notes, 'Standard academic conventions are avoided in favour of shock-tactics, cultural and disciplinary eclecticism, flip chart subheadings and the seduction of a clever turn of phrase' (p. 15). Indeed, Hofstede (1994), himself a key researcher on culture and its international impact, suggests that it (culture) is a phase or fad that although it may eventually decline, will have left its mark.

Ostensibly, culture is historically established, has holistic tendencies, is formed by rituals and symbols and is difficult to change (Hofstede, 1994, pp. 179–180). Additionally Hofstede recognises that culture is socially constructed and difficult to comprehend. Parker (2000) states 'There is simply no compelling evidence here that organizational culture – whatever it might be – is related to profitability, efficiency, job satisfaction and so on' (p. 17). Yet despite such criticisms, it is apparent that organisations and managers have embraced culture with an evangelical zeal. This enthusiasm is easily explained as it provides a vehicle for organisations to establish at least a partial degree of commonality (winning hearts and minds) amongst its workforce. Clearly, such utilitarian tools as the vision or mission statement are as important as instrumental controls as the rule book or policy statements. Culture provides a platform from which unity can be attempted in a variety of ways.

The morality and justice considerations of HRM

This section does not discuss individual and personal morality or value systems in respect of workplace bullying, but rather examines the HRM role in dealing with workplace bullying in an ethical context. At the outset it is worth stating that for the purposes of this brief discussion, we see morality and ethics as directly interchangeable and as having similar meaning.

The latter half of the 1990s saw new challenges to HRM. Authors such as Legge (1998) amongst others have begun to question the morality of HRM. Evidence and questions are beginning to emerge, covering such diverse issues as whistle-blowing (Lewis, 2000), presenteeism (Simpson, 2000) and employee mental health under HRM (Doherty and Tyson, 2000). Payne and Wayland (1999) point to the ethical voice of stakeholder groups who may need to express their moral claims to organisational leaders. Our concern over the link to morality and HRM is primarily about the notion of 'voice' and claims of 'justice'. We have detailed above how unitarist HRM aims to capture the 'hearts and minds' of employees, and thus diminishes the need for trade union representation of the workforce. How then do employees, as one of the biggest stakeholder groups, give voice to non-ethical activities such as bullying? Who will speak for them if there is an absent staff representation or trade union conduit? To whom do they speak if their line manager is a bully, as well as their agent-representative to upper management, and also often a first 'port of call' in a bullying complaints process? By driving down the practices of HRM to line manager level, HRM as an ideology creates its own problems of morality and justice.

Writers such as Miller (1998) point to an absence of the consideration of ethics and morals in the existing HR literature, especially in terms of policies and procedures. Miller explicitly identifies three issues of 'justice' in the modern employment relationship, namely system justice, procedural justice and outcome justice.

In terms of system justice, Miller (1998) identifies a decline in morality of management in favour of 'market conditions' according to which the systems of management now favour employment conditions that meet the market needs. Whilst globalisation inevitably brings about market pressures, Miller argues that HRM decisions over working hours, pay and conditions for the majority of the workforce are largely immoral, especially when market pressures are the leverage tools. These link directly into some aspects of the workplace that are cited as contributory causes for bullying, such as low work control and the negative aspects of the job environment (Zapf *et al.*, 1996; Einarsen and Skogstad, 1996). It is in this area that the 'soft' side of HRM seems to struggle to achieve profile, perhaps because the links between productivity and organisational culture (etc.) are not quite proven (Parker, 2000).

Procedural justice is argued by Miller (1998) to be potentially immoral because there is an absence and/or erosion of employee representation by trade unions under HRM. The 'correctness' of managerial decisions remains largely unchecked under a non-unionised and unitarist HRM regime. If those managerial decisions include engaging in behaviours that are bullying or not dealing with cases of bullying, it is hard to see how this decentralised 'pure' form of HRM can function effectively. An organisation that does have policies and procedures may take account of this and utilise professionals (such as those employed in HR departments) to implement processes that deal with bullying at work. However, as these very professionals are also required to protect the organisation, this means they must tread carefully regarding the targets of bullying (such as not admitting fault) in case of action by the targets against the organisation at a later date. This is surely a conflict of interest. It is difficult to perceive how a unitary perspective can genuinely be taken by such players and highlights the need for proper balance in the justice procedure for dealing with bullying.

The third component proposed by Miller (1998) is outcome justice, where rewards, punishments and allocations are made. Under the philosophies of HRM, if line managers, empowered within the HRM rhetoric, make such decisions, they are judge and jury combined. If those line managers are also bullies, then it is hard to see how the outcomes will be just under true HR systems. Similarly, if procedures enforced by HR professionals are actually one-sided managerialism, then it is equally difficult to envisage the HR professional having the latitude to engage in true unitarism. Perhaps the most pervasive evidence for an imbalance of outcome justice is that often it is the more junior 'bullied' staff that are relocated, not the more senior 'bullies'. One can understand how targets of bullying who have successfully brought a complaint are confused when they have to change jobs internally and may therefore feel punished as an outcome (Adams, 1992).

Within the concepts of HRM and the absence of the justice elements referred to by Miller, there appears to be recognition of the potential problem of this approach to people management. Legge (1998) goes even further by adopting a Marxist critique of HRM, suggesting the potential morality and injustices of HRM as 'exploitation, alienation and the protection of the interests of the few at the expense of social justice for the many' (pp. 20–21). Legge examines HRM under three ethical perspectives:

1 Deontology: treat everyone with respect; moral rules must be universal and respect the dignity of all. Self-interest issues are not moral.
2 Utilitarianism: based on 'people are a means to an end' – if this premise is to the advantage of the majority. Actions are judged in terms of consequences.
3 Stakeholder theory: in principle no organisational stakeholders are to be complete losers while others are clear winners. However, management prerogative must be on the survival of the firm and on stakeholder interests (primarily shareholders).

Legge (1998) suggests the deontological thesis is highly unlikely except possibly in high value-added industries where intelligence is valued. Yet even here exploitation of people is inevitable, especially given the huge knowledge base in emerging economies such as India. This result, of course, fits the second ethical approach of utilitarianism. Unfortunately, a utilitarian view is unlikely to affect only the minority. As Legge (1998) suggests: 'When most of us are on non-standard, temporary, fixed term contracts, facing high employment insecurity and ever-increasing labour intensification, the fact that a minority have salaries like telephone numbers becomes unacceptable' (p. 28).

The final stakeholder view is cited by Legge as being clearly in tune with HR concepts of 'customer first', employee second. Therefore such an appraisal of HRM under three different ethical perspectives makes for grim reading. As Mabey *et al.* summed up in their edited text on the experiences of HRM:

> On the first question concerning the promise of HRM we have to conclude, on the evidence of this volume, that many of its prized goals (more satisfied customers, more empowered workers, more trusting employment relationships, more unified culture, greater workforce creativity and commitment) remain unproven at best, and unfulfilled at worst.
>
> (Mabey *et al.*, 1998, p. 238)

Discussion

If we accept the lexicons of the managerialists of the 1980s and 1990s, we can start to build up a picture of how an environment is created at work that can sustain a bullying culture. If we couple the environmental conditions of increasing competition with the cultural environment of organisations where utilitarian instruments driven by political and social ideologies are encouraged, then bullying may become a reinforced or at least unchallenged behaviour. Although we can at best make only an anecdotal link between political climate and organisational culture, there is a degree of irony in the timing of the first reports of bullying in the UK in 1992 (e.g. Adams, 1992) being coupled with the fall from grace of the then prime minister, Margaret Thatcher. Could it be that some of the managerialist rhetoric under the headings of 'culture', 'organisational structure' and 'HR management' were being questioned? Were the new 'humanistic' regimes of governance under attack?

The emphasis towards self-management through teams supported by an organisational culture of 'shared hearts and minds' and empowerment fits the HRM model well. The idea that bullying would best fit the 'hard' or rationalist model of HRM is obvious and plausible. The 'soft' model, whilst being less transparent with respect to bullying, could encompass an environment in which bullying exists, but within the subtleties of management rhetoric and corporate culture through 'shared' beliefs. Here the employee as a stakeholder is in possession of nothing more than their own individual voice. Collective spirit for the benefit of the organisation is one thing; an absence of collective voice as a result of a unitarist *managerial* ideology is something else.

In terms of industrial relations, the marginalisation of trade unions under HRM principles created its own set of conditions. Reporting on a major study of workplace industrial relations Milward *et al.* (1992) confirmed the existence of environments where few procedures existed, discipline was harsh and there were limited mechanisms for grievance. Even where procedures exist, it is likely under the HRM doctrine that line managers are the first point of contact. If those line managers are also bullies, then achieving true justice will be difficult. It highlights the need for line manager awareness raising and skills training if an HRM approach is being taken within the organisation. Although line managers may not know all the procedures and techniques, they are required to appreciate the broad parameters. However, evidence suggests that even this is in question. A major three-year study on employee relations reported by Cunningham and Hyman (1999) illustrated that subordinates working under HRM principles consistently ranked their managers as having insufficient competence in a range of responsibilities. These included such examples as appraisal, communication, motivation, teamworking, leading meetings, training and counselling. The research also revealed a respondent's comments on how a manager used 'Bullying tactics and if you didn't like it you know where the door is.' Other examples of comments included 'physical assaults and verbal abuse ... everyone is terrified of him' (Cunningham and Hyman, 1999, p. 198). The Cunningham and Hyman (1999) paper revealed instances of poor line manager training, continued bureaucracy and discomfort in dealing with issues such as absence, discipline and HR problems generally. Cunningham and Hyman (1999) and Thornhill and Saunders (1998) suggest that line managers are ill equipped for and do not desire the role which has been thrust upon them.

Conclusions

Given that the term 'bullying at work' emerged in the early 1990s, and the term 'HRM' over a decade before in the UK, could one contribute to the emerging awareness of the other? We have shown some links between the HRM doctrine and a lack of action regarding bullying at work. A central creed running through this chapter has been the role of HRM and the morality of its application in organisations. As a result of this consideration of the literature, it appears that there are clear signals from key UK academics and researchers that all may not be well in HRM. Yet readers of this chapter may have some difficulty with this because HRM may be conceived more positively where they live. Evidence to support these differences can be seen in this statement by Drumm (1994), who was speaking specifically of HRM in Germany – 'the paradigm of social responsibility for personnel has been of constant importance since the fifties' (p. 35). We hope we have explained how in a UK work context this may simply not be true.

By examining the evidence from existing empirical sources on bullying (such as the role of managers in the process (Rayner, 1998) and their negative behaviour at work (Milward *et al.*, 1992; Cunningham and Hyman, 1999)), we can see that many of the cited causes of bullying fit with HRM as a managerial tool. What is equally interesting is the critical appraisal of HRM by key writers at a time when the concept of bullying at work has emerged and established itself. The establishment of the central themes of HRM, such as commitment, unitarist negotiation, and the devolvement of power to line managers, seems to present an effective foil for dealing with workplace bullying.

It is clear that HRM as a functional discipline lies at the core of organisational design and practice, and is instrumental in shaping the way organisations operate. Commitment between all parties (unions, employers and employees) is central to our understanding of work, especially paid employment. If we accept the management theorists who indicate the diminution of this traditional model in favour of non-work and balanced portfolios, we must recognise that work no longer becomes a purely economic phenomenon. Additionally, it is a complex configuration of social construction requiring justice and fairness (Kouzmin *et al.*, 1999). HRM appears to be in need of a careful review in this regard.

Researchers and commentators on bullying at work need to exercise caution when citing 'the work environment' or 'leadership' as reasons why bullying behaviours are manifest. Those who comment critically on this subject must investigate fully why such factors are being labelled as causes of bullying, and take into account the political and managerial ethos operating in the workplace. Unless we all make clear our attempts to understand, in detail, the environment under which such behaviours emerge, there can never be a clear understanding of the rise of workplace bullying as an organisational phenomenon, or how to mitigate against it effectively.

Bibliography

Adams, A.(1992) *Bullying at work – how to confront and overcome it.* London: Virago.

Beer, M., Spector, B., Lawrence, P., Quinn Mills, D. and Walton, R. (1985) *Human resource management: A general manager's perspective.* Glencoe, IL: Free Press.

Bleeke, J. and Ernst, D. (1995) *Collaborating to compete.* New York: John Wiley and Sons.

Cunningham, I. and Hyman, J. (1999) The poverty of empowerment? A critical case study. *Personnel Review*, *28*, 3, 192–207.

Doherty, N. and Tyson, S. (2000) HRM and employee well-being: Raising the ethical stakes. In D. Winstanley and J. Woodall (eds), *Ethical issues in contemporary human resource management* (pp. 102–115). Basingstoke: Macmillan.

Drucker, P. (1988) The coming of the new organization. *Harvard Business Review*, *January/February*, 45–53.

Drumm, H. J. (1994) Theoretical and ethical foundations of human resource management: A German point of view. *Employee Relations*, *16*, 4, 35–47.

Einarsen, S. and Skogstad, A. (1996) Bullying at work: Epidemiological findings in public and private organizations. *European Journal of Work and Organizational Psychology*, *5*, 2, 185–201.

Guest, D. (1987) Human resource management and industrial relations. *Journal of Management Studies*, *24*, 5, 503–521.

—— (1998) Human resource management, trade unions and industrial relations. In C. Mabey, G. Salaman and J. Storey (eds), *Strategic Human Resource Management: A reader* (pp. 237–250). London: Sage Publications.

Handy, C. (1989) *The age of unreason*. London: Business Books.

Hart, T. J. (1993) Human resource management: Time to exorcise the militant tendency. *Employee Relations*, *15*, 3, 29–36.

Hoecklin, L. (1995) *Managing cultural differences: Strategies for competitive advantage*. Wokingham: Addison-Wesley and The Economist Intelligence Unit.

Hofstede, G. (1994)*Culture's consequences: International differences in work-related values*. Beverly Hills CA.: Sage.

Kouzmin, A., Korac-Kakabadse, N. and Korac-Kakabadse, A.(1999) Globalization and information technology. Vanishing social contracts, the 'pink-collar' work-force and public policy challenges. *Women in Management Review*, *14*, 6, 230–251.

Legge, K. (1995) *Human resource management: The rhetorics and realities*. Basingstoke: Macmillan.

—— (1998) The morality of HRM. In C. Mabey, G. Salaman and J. Storey (eds), *Strategic human resource management: A reader* (pp. 18–29). London: Sage.

Lewis, D. (1999) Workplace bullying – interim findings of a study in further and higher education in Wales. *International Journal of Manpower*, *20*, 1/2, 106–118.

Lewis, D. (2000) Whistleblowing. In D. Winstanley and J. Woodall (eds), *Ethical issues in contemporary human resource management*. Basingstoke: Macmillan.

Mabey, C., Skinner, D. and Clark, T. (1998) Getting the Story Straight. In C. Mabey, D. Skinner and T. Clark (eds), *Experiencing human resource management*. London: Sage.

Marchington, M. and Wilkinson, A. (1996) *Core personnel and development*. London: Institute of Personnel and Development.

Mayrhofer, W. (1995) Human resource management in Austria. *Employee Relations*, *17*, 7, 8–30.

Miller, P. (1998) Strategy and the ethical management of human resources. In C. Mabey, G. Salaman and J. Storey (eds), *Strategic human resource management: A reader*. London: Sage.

Milward, N., Stevens, M., Smart, D. and Hawes, W. (1992) *Workplace industrial relations in transition*. Aldershot: Dartmouth Publishing.

Mintzberg, H., Ahlstrand, B. and Lampel, J. (1998) *Strategy safari*. London: Prentice Hall.

Ouchi, W. G. (1981) *Theory Z: How American business can meet the Japanese challenge*. Reading, MA: Addison-Wesley.

Parker, B. (1998).*Globalization and business practice*. London: Sage.

Parker, D. (2000) *Organisational culture and identity*. London: Sage.

Payne, S. L. and Wayland, R. F. (1999) Ethical obligation and diverse values assumptions in HRM. *International Journal of Manpower*, *20*, 5, 297–308.

Pettigrew, A. M. (1979) On studying organisational cultures. *Administrative Science Quarterly*, *24*, 570–581.

Rayner, C. (1998) Workplace bullying: Do something! *The Journal of Occupational Health and Safety – Australia and New Zealand*, *14*, 6, 581–585.

Richbell, S.C. (2001) Trends and emerging values in human resource management: The UK scene. *International Journal of Manpower*, 22, 3, 261–268.

Schein, E. (1985) *Organisational culture and leadership*. San Francisco: Jossey-Bass.

Senge, P.M. (1992) *The fifth discipline*. London: Century Business.

Simpson, R. (2000) Presenteeism and the impact of long hours on managers. In D. Winstanley and J. Woodall (eds), *Ethical issues in contemporary human resource management*. Basingstoke: Macmillan.

Storey, J. (1993) The take-up of human resource management by mainstream companies: Key lessons from the research. *The International Journal of Human Resource Management*, 4, 3, 529–553.

Thornhill, A. and Saunders, M. N. K. (1998) What if line managers don't realize they're responsible for HR? *Personnel Review*, 27, 6, 460–476.

Tyson, S. and Fell, A. (1986) *Evaluating the personnel function*. London: Hutchinson.

Zapf, D., Knorz, C. and Kulla, M. (1996) On the relationship between mobbing factors, and job content, social work environment, and health outcomes. *European Journal of Work and Organizational Psychology*, 5, 2, 215–237.

23 Tackling bullying in the workplace

The collective dimension

Mike Ironside and Roger Seifert

Introduction

When peasants and cottage labourers were forced out of the fields and their homes to work in mines, factories and, later, offices the process was often brutal. Work in these new workplaces took on new forms, as the owners strove for predictability, order and control over both the quantity and quality of labour. Their need to secure the adaptation of workers from the rhythms of agricultural and domestic work to the discipline of factory production resulted in management regimes in which fines, beatings, sackings, and all forms of harassment and abuse were the daily experience of the majority. For most peoples this is a recent phenomenon: 'with the exception of Britain, peasants and farmers remained a massive part of the occupied population even in industrialized countries until well into the twentieth century' (Hobsbawm, 1994, p. 289).

Official reports and the experts' views provide us with vivid illustrations of how employers perceived the nature of workers and their attitudes to factory work – colliers were 'naturally turbulent, passionate, and rude' (Report of the Society for Bettering the Condition of the Poor, 1798, cited in Thompson, 1968, p. 393–394), and generally the problem of factory work was seen as 'in training human beings to renounce their desultory habits of work' (Ure, *The Philosophy of Manufactures*, 1835, cited in Thompson, 1968, p. 396). Strict discipline had to be enforced to counter these and other failings. For example, there was a one shilling fine for 'any spinner found with his window open' (Political Register for 30 August 1823, cited in Hammond and Hammond, 1966, p. 32).

Such disciplinary sanctions to enforce the rules of the workplace were not additional to the activities of overseers and progress chasers – they were, and still are, central to the employment relationship. Historians, as well as novelists, satirists, cartoonists and film-makers have illustrated in their own ways the widespread public understanding of how management-enforced compliance with employer-oriented norms of workplace behaviour is central to the shared experience of employment. In a capitalist labour market, employed work has the purpose of profit making, which can only be sustained through continuous exploitation. Bullying at work is best seen not as the careless and casual behaviour of individual bullies but rather as part of management's exercise of its collective will to enforce workplace discipline under the contract of employment. Hence we see all forms of bullying and harassment at work as part of the day-to-day routine of managing labour.

While this perspective does not explain all forms of bullying that may occur in the workplace, it does provide a focus on the workplace context. Workers may bully each other into adopting norms of behaviour aimed at maximising their earnings, either individually or collectively, at minimising the risk of managerial sanctions, and at securing personal advantage in the competitive labour markets. Even where bullying is carried out for no reason other than the enjoyment of the act of bullying itself, when it occurs in the workplace then it at least carries implications for labour management practice.

This chapter provides a brief overview of the nature of the employment relationship, and in particular of its changing nature in the public sector. It then provides some examples of how trade unions are involved in handling the issue as part of the mainstream of union representation, and we conclude that where union representation is strong at the workplace the best chance exists of reducing both the extent and the consequences of bullying.

The nature of the employment relationship

Our analysis starts with the labour market. For most of the time most workers are in a weak position when they offer their labour power for sale in the various segments and strata of the labour market. This weakness relative to the employers' strong position reflects two important characteristics. First, the labour market consists of many sellers but relatively few buyers. Second, employment has a different significance for individuals than it does for employers. Individual workers who cannot find an employer to employ them are personally affected, involving loss of status, hardship, deprivation and possibly long-term harm. An employer who cannot find an individual worker to employ does not suffer in the same way – banks, supermarket companies, car manufacturers and hospitals do not get hungry, depressed and isolated.

This inequality in the labour market is then reflected in the inequality of the employment relation embodied in the employment contract. The proponents of free markets and weak labour regulation argue that as the contract is a free and equal one we do not need regulation (and therefore we do not need unions) because if the employer allows managers to mismanage the workforce, for example by bullying them, then each and every individual worker is simply free to quit. If all the workers do this then the employer cannot function, and hence the employer will encourage managers to behave better in the interests of the enterprise. In addition, if the individual worker does not quit then, they argue, it must be because whatever they are experiencing is not that bad, otherwise they would quit.

The difficulty with this argument is that when there is unemployment amongst a given category of worker then quitting is not realistic. You are only free to quit in the same sense as you are free to leave a room when a person with a gun says that you are free to leave but that if you do they will shoot you. The misuse of 'free' should not disguise the material choices available. Furthermore, if all available employers behave equally badly then even if there is full employment in the sector, any move between employers fails to solve the problem, and there is no incentive for any employer to behave better. The legal fiction that the employment contract is freely entered into by parties of equal power only serves to hide the reality.

A variation on this argument suggests that bullying is against the interest of the employers, since a happy worker is a productive worker. Notions such as equity theory (Adams, 1963) and the psychological contract (Sims, 1994), the successors to the ideas of Taylor, Mayo, Maslow and others, are based on the assumption that managers can create conditions under which workers can achieve self-fulfilment through activity that meets managerial goals. Most textbooks on human resource management present this assumption as a basis for managerial action, but none of them cite any convincing hard evidence that being nice to workers makes them more productive.

The alternative view is that for most workers most of the time the only choice is to stay where they are for as long as is bearable, and then to quit, often for good. Three examples from schools give an idea of this. The first concerns a teacher who was shouted at in front of colleagues, ignored in meetings and had her competence questioned for six years before she finally went off sick and contacted the union. Eventually she took ill-health retirement. In the second case a head of department was asked to draw up a report for governors within a time-scale that she could not meet. She contacted her union official, whose request for an extension of time and a copy of the grievance procedure was ignored. She submitted an outline report, and the head reprimanded her and gave the job to her second. The union lodged a grievance on her behalf, and her second (also in the same union) lodged a grievance of bullying against her. She went off sick. A twelve-hour grievance hearing was inconclusive, and she subsequently retired on the grounds of ill health. Finally, a teacher with twelve years' experience at the school took a grievance against the head for speaking to her in an aggressive and intimidating manner. The grievance was not upheld, and she lodged an appeal. After an unsuccessful attempt to organise collective action by other union members, the teacher went off sick. She then left the school without waiting for the appeal hearing.

These examples, and the huge number of incidents reported weekly in union and professional journals, show that improvements in working conditions, including freedom from bullying, are unlikely to come from management initiatives, and that the best route is from pressures from within the workplace through the mobilisation of the countervailing power of workers, usually in the form of trade union organisation. For such action to survive, it is a necessary, but not sufficient, condition that such workplace trade unionism is supported by some national union organisation that exists independently of the employer, and again that such existence should have some state protection. Although some unions argue for specific legal protection against bullying (UNISON, 1997b), legislation is frequently poorly enforced and provides limited remedies. It only helps when there is strong union organisation already in place.

It is our proposition that bullying tends to become worse when the balance of inequality increasingly favours managers as agents of the employer, and when the employer comes under increased pressure to hit either profit targets in the private sector or performance targets in the public sector. We believe bullying to be a basic part of the management of labour, and that most of its forms are accepted as part of the daily experience of employed work. It comes to the notice of managers, trade unionists and commentators if its form changes in particular cases and is deemed to be outside the normal tolerance of bullying behaviour, and/or when the norm itself changes in relation to pressure from unions, government agencies such as the Advisory, Conciliation, and Arbitration Service (ACAS), the Commission for Racial Equality (CRE) and the Equal Opportunities Commission (EOC) or the workforce. One survey of public and private sector employers 'found some evidence that rising work pressures and tight staffing levels are contributing to the emergence of an intimidatory management style, which in turn is leading to a rise in bullying within some organizations' (IRS, 1999, p. 5). Other recent research suggests that bullying is more prevalent in the public than in the private sector (Hoel and Cooper, 2000), and the National Workplace Bullying Advice Line reported that 20 per cent of its callers were from education, 12 per cent from the health services and 10 per cent from social work.

We have written elsewhere about the Conservative counterrevolution in health (Seifert, 1992), education (Ironside and Seifert, 1995), and local government (Ironside and Seifert, 2000). Below we summarise how this resulted in more oppressive management of labour, leading to well documented increases in absence, turnover, quitting and poor performance. As management in the public services has become more like management in the private sector, this illustrates our argument that bullying is endemic in the labour management practices associated with making a profit.

Theoretical and empirical model of change in the UK public sector

Our main argument runs as follows. The creation of competing business units within the public sector replicates some of the private profit sector conditions under which public sector managers have to operate. Such units compete either with each other, or with private sector units, or even with virtual private sector units, while income streams are controlled within tight limits. Under these conditions, service managers must respond to competitive pressures. Private sector managers might close the business, alter its core function, dramatically alter price levels, product range or quality, or horizontally slash costs, especially wage costs. In the public sector, despite all the reforms, these options are not available. This means that for most purposes most of the time competition is based on reducing unit costs. In a labour-intensive sector this means cutting unit labour costs by changing wage rates, systems and structures, and by intensifying performance, usually depicted as 'improving' performance.

As middle layers of both management and organisation are stripped away, so the burden of intensifying performance at lower unit costs falls increasingly on site line managers. The nostrums of neo-liberal market policies are mobilised to argue that business units compete with each other in terms of product quality and production costs. This then ensures that the most *efficient* in process terms, the most *effective* in management terms, and the most *economic* in cost terms will survive. The rest 'go to the wall'. The discipline of the market therefore guarantees the survival of the businesses with the best scores in those three Es. Owners have incentives to invest since there are profits to be made, managers to manage better since they can receive both bonuses and shares, and workers to work since they get some security of job and pay. These Taylor-like incentives (carrots) are matched with the punishment for getting it wrong – losses and job losses (sticks).

In this 'looking glass world' logic is inverted and a model of market competition rules supreme in the mind but, of course, not on the ground. Once adopted as the dominant conventional wisdom, it can be applied, willy-nilly, to all other organisations. The obvious conclusion is that all sectors should be privately competitive, but as this is socially and politically impossible (at least for now), the asserted best practices of the private sector are to be installed within those services surviving in the public sector. So internal markets were invented, managers were paid more with larger bonuses, workers were subject to tighter controls and worse national pay and conditions of service. All were threatened with privatisation if they did not conform to best value, compulsory competitive tendering (CCT) and marketisation. Hence followed the changes in the management of the labour process, mainly with retrogressive outcomes, for UK public sector workers. Similar processes are taking place throughout Europe (Bach *et al.*, 1999).

Consequences of the new regime for managers in public sector

The management of labour is a central issue for the delivery of labour-intensive public services. The workforce has a high level of unionisation, and it includes large numbers of qualified and professionally qualified occupational groups. This is a context in which it is difficult for managers to implement changes in pay and performance with outcomes that are acceptable to both the community and the political leadership. Central control systems have been toughened through a variety of regulatory and inspection devices. The result is oppressive nationally set targets and performance indicators linked to locally driven incentives for senior management teams.

The results are predictable. If, for example, the government wishes to reduce the waiting times for patients referred by GPs to hospital consultants, then, assuming (and this is a big assumption) both the reporting systems and practices are honest, how can we proceed? We can increase the numbers of consultants – but to train new ones takes ten years, so shorter-term measures can be used, to recruit from overseas and/or to promote those on sub-consultant grades more rapidly. Or we could increase the hours of labour rather than the numbers of staff through changing their contracts and making them work longer hours. We could also achieve this by removing some tasks from them deemed to be suitable for completion by more junior and/or less skilled (trained) staff. This can happen but the impact on quality of outcome must be seen to be at worst neutral. The mechanisms open to managers to get consultants to change their working practices include some carrots in the form of more pay, and some sticks in the form of threats, including actual bullying by both managers and peers. A report in the *British Medical Journal* stated that a professor of medicine had been struck off 'for bullying and threatening junior colleagues' (*British Medical Journal*, 9 October 1999, p. 319). The rarity of such action compared with the widespread practice of bullying in hospitals illustrates the depth of the problem (Quine, 1999).

Teachers, including head teachers, are subject to closer scrutiny than before, through tougher audit from Ofsted, performance targets published in league tables, and now performance-related pay. In a union survey, one teacher expressed the consequences very concisely: 'The head bullies his Deputies, who then bully middle managers who in turn bully members of their Departments' (NASUWT, 1996, p. 10). A health service union officer says that 'in the new-style NHS staff with any kind of authority are under a lot more pressure and when people are over-worked they are more likely to take it out on colleagues further down the hierarchy' (Snell, 1997, p. 24). A similar pattern is true in further and higher education. Many workforce surveys reach the same conclusion as that of a local college union committee: 'the principal finding of this survey is that the vast majority of respondents feel that they are working under a regime which they characterise as dictatorial and bullying'. Senior managers become the arm of the state, pushing down on conditions of service while requiring more quantity of labour at lower unit costs. In local government, requirements such as Best Value are enshrined in statute law, with the need to improve built into the penalties on councils.

The change is implemented through appraisal, promotion, selection for performance-related pay and training, and the tyranny of the one best way. The pressures on such managers are intense and the main pressure is to deliver an improved service, narrowly defined by senior managers, with fewer resources. The public sector union UNISON argues as follows: 'Anyone connected with public services will know the pressures that staff and managers have been under, including: competitive tendering, privatisation, budget and staffing cuts, changes in management structures and reorganization. This puts pressure on managers and their staff' (UNISON, 1997a, p. 3).

This leads to contested definitions of what constitutes good practice and performance. Unilateral attempts to stamp management views of performance on the staff have met with sharp resistance in the form of quitting, non-compliance with administrative duties, failure to apply for senior posts and union action. In some cases employers and government have backed down, but in others they have pressed on with the changes. Conflicts already existing at work are intensified, resulting in increased stress, absence, turnover, and poor performance, leading an ever-more perplexed government to veer from one policy to another in an incoherent effort to patch up a failing set of services.

The consequence is an endless attempt to adopt and adapt private sector management practices, and to grasp at any tried and untried set of practices from TQM and quality circles through Japanisation to empowerment, employee involvement schemes, teamworking, appraisal, performance pay and of course an increased use of disciplinaries. For UNISON, 'bullying has become part of the management culture of many public service employers, and is often allowed to happen and carry on unchecked' (UNISON, 1997a, p. 3).

Bullying in practice: policies, actions and remedies

The power realities of everyday working life mean that some of the time in some circumstances the only way to achieve the ends required is through some form of increased bullying. This applies both to managers who bully their subordinate staff and to employees who bully their peers – in both cases the purpose of bullying is to alter the work-related behaviour of others in the workplace. Bullying can be stopped by formal grievance taking, itself sometimes leading to formal disciplinary action, when the union is well organised and aware, and sometimes it is stopped by collective action (within or without the disputes procedure). Grievance, disciplinary and disputes procedures are the formal procedural alternatives to collective mobilisation against management, and while some details of practice may vary between countries there are some clear underlying practical principles involved.

A major practical problem concerns workplace understanding of how policies, procedures and practices on bullying are linked with grievance and disciplinary procedures. For example, trade unions and employers have adopted similar definitions of bullying. They emphasise behaviour that is regarded as unacceptable, as an attack by one person on the dignity of another, and the adverse consequences that ensue. The definition used by the Andrea Adams Trust includes most of the possible permutations that appear in management and union policies:

> Workplace bullying can be defined as unwanted, offensive and deliberately humiliating behaviour towards an employee or group of employees. Bullying can entail unpredictable, unfair and often vindictive attacks on an individual's personal or professional performance. Bullying usually, but not exclusively, entails an abuse of management power or position. Bullying can lead to anxiety, loss of self-confidence, stress-related illness, physical ill health, mental distress and suicide.
>
> (IRS, 1999, p. 10)

These definitions are so general that the words 'workplace' and 'management' could be left out, and 'employee' could be substituted by 'person' without weakening the meaning. The specific aspects of workplace bullying that make it different from bullying under other circumstances are not covered by the definition. When bullying is located in the context of the employment relationship to become workplace bullying then it acquires certain particular characteristics, becoming enmeshed in the fabric of workers' and managers' rights. This is what makes a clash between a bully and a victim in the workplace a collective issue – a management issue and a trade union issue. The extent to which managers do or do not act to prevent bullying depends on the existence of both management commitment and trade union organisation, and on the willingness of managers and trade union representatives to treat the issue as one of workers' rights.

According to the International Labour Organisation, a grievance exists when a 'measure or situation' affecting conditions of employment

> appears contrary to provisions of an applicable collective agreement or of an individual contract of employment, to works rules, to laws or regulations or to the custom or usage of the occupation, branch of economic activity or country, regard being had to the principles of good faith.
>
> (ILO, 1967)

According to the governmental agency the Advisory, Conciliation and Arbitration Service, employers have a legal duty of care of all their workers: 'Employers are usually liable in law for the acts of their workers, and this includes bullying or harassing behaviour'. Furthermore, 'failure by the employer to deal with any complaint of bullying or harassment, or failure to protect their employees from bullying and harassing behaviour' might give employees the basis to claim unfair constructive dismissal (ACAS, 1999b, p. 4).

Thus, employees have workplace rights, not only under the law but also under collective agreements, company rules, and custom and practice. Employees assert their rights either by taking collective action against the employer, or by asking the employer to remedy their grievance. ACAS advises employers that 'complaints of bullying and harassment can usually be dealt with using clear grievance and disciplinary procedures' (ACAS, 1999a, p. 7).

Agreements on preventing bullying at work usually set out the responsibilities of management and the rights of trade union representatives, as well as a formal procedure for dealing with complaints. Strangely, most management and union policies make only passing mention of the term 'grievance'. This is a clear error on their part. One of the teachers' unions, for example, suggests that complaining about bullying

> is not the same as invoking the grievance procedure because you are putting the onus on the head to take disciplinary action. If, however, the school management fails to take appropriate action, using the grievance procedure may be your next option.
>
> (ATL, 1999, p. 8)

Thus it seems to be widely assumed that complaints about bullying are best resolved through a formal complaints procedure that is different from a grievance procedure. In practice, the only difference is in the provision of a facilitator through which the bullied employee can discuss the issue with a specified employee (variously titled, for example, 'contact officer', 'workplace counsellor', 'harassment officer', 'confidential advisor', 'impartial HR manager' or 'advocate'), who has received some kind of training and provides informal advice, and facilitates an agreement. In many cases this person is supposed to provide support to employees who complain about bullying and to employees who are the subject of formal complaints (IRS, 1999). Such counsellors have no line management authority, can impose no sanctions, and are accountable to the employer and not to the employee. It is unclear what status any such agreement has, it usually leaves the bully unpunished, and it is part of a management process which provides a façade behind which bullying continues. Little thought has gone into further problems concerned with the use of any discussion/agreement in subsequent hearings, the bypassing of union structures, and the enforcement of any such agreement.

In all other respects, most model complaints procedures are indistinguishable from the formal elements of grievance procedures, usually identifying the following stages. First, the employee raises the issue formally with management, who must investigate the issue and respond. If the employee is dissatisfied with the response, they may appeal to successively higher levels of management. (For more detailed discussion of grievance procedures and their operation see Clancy and Seifert, 2000.)

The Royal College of Nursing's model procedure provides a further example of procedural muddle, declaring that

> bullying and harassment are not determined by the intention of the person who has caused offence, but by the effect it has on the recipient. It is up to that person to decide if they are being bullied or harassed because they find the behaviour unacceptable.
>
> (RCN, 2001, p. 9)

This is of course quite wrong. Bullying is a resolvable grievance only if it can be shown that the bully broke the rules of the workplace, and that the employer's agents, managers, also broke the rules by failing to stop the bully. Under this procedure the bullied employee would report the matter to a personnel manager, who would designate a senior manager to conduct an investigation including hearings at which the bully and the bullied employee would have the right to union representation. The senior manager would then decide whether or not there is a case to answer at a formal disciplinary hearing. This is effectively the first stage of disciplinary proceedings. There is no right for the bullied employee to challenge either the senior manager's decision about whether or not there is a case to answer, or the eventual outcome of any disciplinary action. Such rights could be exercised through a grievance procedure. Again the muddle: the bullied employee should first seek advice from the trade union steward, then if the union backs the case, it goes through the grievance procedure, and then, if appropriate, through the disciplinary procedure. This process is clearly recognised by other unions. The largest civil service union, the PCS, advises members that 'if you are being bullied ... talk to your union rep as soon as possible'; and another health union, the CSP, advises members that 'bullying is a problem which can ideally be tackled jointly by CSP stewards and safety reps' (CSP, 1997, p. 8).

The nature of grievance

Through a grievance procedure the employee can address the employer's representatives (the management). Even if the procedure is called a complaints procedure, it is important to distinguish grievances from complaints. While all grievances are complaints, not all complaints are grievances. In the specific context of the workplace, for a complaint to be a grievance it must meet two conditions.

First, a grievance is always concerned with the application or breach of the rules at work. In the case of bullying, several rules might be involved: for example, common law protection from threat and abuse, statutory protection from discrimination, rights under health and safety legislation, the employer's own policy on bullying and/or harassment, and the general duty of the employer to ensure fair treatment under the contract.

Second, while a grievance may arise from an act or omission of a particular manager who represents the employer, a grievance is never against another employee. For example, if hospital employee A bullies hospital employee B, and both are on the same grade, then the grievance is against the hospital Trust, not against A, for the management of the Trust has a clear duty to enforce the relevant rules and stop the bullying. If hospital managers are the guilty party, either for being bullies themselves or for failing to prevent employee A from bullying employee B, then the grievance is about their action as part of management, the collective representatives of the Trust.

Since any manager involved represents the employer, a collective entity, and since both the workers involved are representative cases of the state of workplace practices affecting employees collectively, any grievance held by any individual employee is a collective matter. It is not union involvement that makes such matters collective, but the nature of the employment relationship. Where unions do become involved, workers' rights, such as the right to be protected from bullying, may be established and enforced through collective bargaining. Grievance handling by union representatives, informally with managers, or collectively through industrial action and negotiations, or through a formal grievance hearing, involves collective bargaining. The aim of the grievance handling process is to reach an outcome that is agreed by representatives of the employees (a union official) and the employer (a manager).

For an agreement to be reached there must be a meeting or a series of meetings. First, it is necessary to reach an agreement on the facts of the case or, if these are in doubt, to agree the areas of uncertainty. Second, the aim is to agree in the light of the facts whether or not a grievance actually exists – has an act or omission of the employer resulted in an infringement of the employee's rights? Third, if it is agreed that a grievance exists, it is then necessary to agree on the action that the employer will take in order to provide an appropriate remedy. Agreements reached through this process are always collective in nature since decisions made in a particular case represent both the negotiated outcome between collective organisations and the likely application of such an outcome to other workers. Some of this is recognised in the 1999 Employment Relations Act, which allows for workers to be accompanied in grievance hearings by union representatives (Clancy and Seifert, 2000).

Only a minor problem?

That workplace bullying is endemic in the workplace is revealed by the difficulties involved in convincing managers that it exists at all. While survey evidence suggests that over half the workforce has experienced workplace bullying, one survey of employers showed that 93 out of over 130 organisations were not dealing with a single case (IRS, 1999). While the use of disciplinary measures to control the workforce has generally increased throughout UK public services, its use by employers against bullies is still very rare – further evidence of the really existing management practice of accepting bullying as part of labour management.

For the scale and the scope of workplace bullying to be limited it has to be recognised for what it is, and that recognition has to be understood from the perspective of the workers affected by it. If bullying is so widespread, then why is there so little action to prevent it? There is some evidence that when employees being bullied attempt to have the bullying stopped, it is more likely that they will be threatened with dismissal, or labelled troublemakers, or that the bullying will continue (UNISON, 1997b). Relatively few employees, less than one in ten of those saying they have been bullied, raise the issue formally through grievance or similar procedures (Hoel and Cooper, 2000).

UNISON (1997a) reports that 96 per cent of currently bullied people who made a group complaint said they were threatened with dismissal. When asked what they would do if they were bullied, 73 per cent of those not being bullied said they would see their UNISON representative, but only 26 per cent of those being bullied said they would do that. When asked 'what do you think causes bullying?', 70 per cent rated 'poor management' as 'very important', with 67 per cent for 'bullies can get away with it', and 64 per cent for 'workers scared to report it'.

Perhaps this explains the reluctance of trade unions to rely on grievance procedures. Teachers' union ATL

> has dealt with many cases where a procedure has been invoked against the person who is making allegations of bullying. For example, a head who is bullying a teacher may also invoke the disciplinary or competence procedure, securing the teacher's dismissal after due hearings.
>
> (ATL, 1999, p. 10)

UNISON makes the more general point:

> The normal grievance procedure will not always be sufficient as the facts of the case need to be established in a sensitive way and the bullying may be by the member's line manager, who is normally the person a problem is raised with in the first instance in a grievance procedure.
>
> (UNISON, 1997b, p. 9)

But this is also true of most other management actions that give rise to grievances, such as an instruction to work overtime, a refusal of compassionate leave, or rejection of a request for heating in a cold office. All of these actions may or may not be 'reasonable', depending on the circumstances. 'Reasonableness' is contested, primarily between management and employees. For example, a leaflet issued by the Employment Service attempts to explain to employees what constitutes 'firm and fair' management, while recognising that 'abuse of these procedures may constitute bullying behaviour' (IDS, 1999, p. 16) (see Box 23.1).

There has been very little systematic research into how employees feel about arbitrary management decisions on such issues, or about the extent to which the resultant grievances are (or are not) handled through informal contact with managers, through collective action by employees in the workplace, or through formal procedures. Available evidence on related matters such as the fate of whistle-blowers in the NHS is not encouraging (Vickers, 1995).

How to proceed

We now consider how an employee facing workplace bullying, who has exhausted all the other possible alternatives, might take up the issue

Box 23.1: Examples of legitimate actions of managers that also may be abused and hence constitute bullying

- Issue reasonable instructions and expect them to be carried out.
- Set and monitor expected standards of performance in line with appraisal guidance.
- Set and monitor standards of conduct in line with the staff handbook.
- Discipline staff for proven misconduct, following a fair and reasonable investigation.
- Consider inefficiency action where unsatisfactory work or unsatisfactory levels of sickness absence have been identified.
- Give a low performance rating to unsatisfactory performers.

through a formal grievance procedure with the support of an employee representative (usually a trade union representative).

A union representative who is convinced that there is a case to answer can take up the case with line management. If this fails to resolve the issue, the case will be taken to a hearing where the evidence will be presented to a senior manager whose role as the employer's representative cannot be considered independent. The outcome of the negotiations of the hearing is dependent upon the strength of argument and evidence, and perhaps even of the strength of the union. For the union representative, it is also essential to minimise the chance of retaliation against the complainant. If agreeing that rights have been infringed, the parties would consider possible remedies such as training and counselling of perpetrators, disciplinary action against the bully, reassigning the bully or the victim, or changing supervision and working practices. Throughout this process disagreements between the parties may exist, and it may be necessary to invoke the appeal procedure.

Thus the whole process of resolving a grievance, formally or informally, through a grievance procedure, or any other procedure expressly designed to deal with bullying, depends on achieving a negotiated set of outcomes. In some cases the outcome may be completely in line with the aims of both management and union representatives. An agreement that a bully whose activities are causing problems for the employer should be dismissed would represent an integrative solution to a joint problem. However, in most cases some compromise will have been involved. In order to reach an agreement, the union may have to moderate some demands, or management may have to include a commitment to allocate extra resources.

Conclusion

We have argued that when bullying occurs in the workplace it arises out of the employment relationship and it has the purpose of reshaping employee behaviour. Bullying is far too widespread to be the work of a small number of pathologically disturbed individuals who can be removed from the workplace, monitored, or controlled so as to prevent them from bullying. Most bullying in the UK seems to be carried out by managers, who in all other respects are similar to the employees who are bullied. Whether or not bullying occurs or reoccurs depends on the particular mix of management techniques in use and on the resources that management provides to control the behaviour of managers, which may in turn depend on the strength of the union.

Snell recounts a case handled by a union official, who told of a health visitor who had been off sick because of a problem with her joints.

> She was bullied back to work, against doctors' orders, with threats that she would lose her job. When she did go back life was made as

> difficult as possible for her. She described sitting in her car in tears outside clients' homes because she knew she couldn't climb the stairs.
>
> (Snell, 1997, p. 26)

She was one of twelve health visitors and school nurses employed by South Cumbria Mental Health NHS Trust who successfully argued that they were being bullied. The Trust updated its harassment policy, reviewed its nursing management structure, and gave the manager concerned a final written warning (Snell, 1997).

However, only a few cases of bullying are pursued to a definitive solution in favour of the bullied employee:

> Most cases of bullying take between one and four months to resolve, with the most likely outcome being the dismissal of the bullying allegations, or the resolution of the case at an informal stage following discussions between all the parties.
>
> (IRS, 1999, p. 5)

All too often bullying is not stopped, causing damage to individuals, groups and the entire organisation before deeper and more extreme consequences emerge. The National Union of Teachers (NUT), for example, lumps together bullying with excessive workload and stress as part of a growing set of related problems. It is pervasive and condoned at all levels of management, not in its crudest manifestations, but in the subtle and indifferent ways in which management's right to manage is asserted and implemented in a work world of intense pressure and increasing rewards for those who meet targets. The extent to which management will agree to commit resources against bullying is one of the central bargaining issues. Negotiations over a case of bullying are not isolated from other negotiations that involve other aspects of the distribution of the employer's resources. Thus, the way that bullying issues are handled, and the outcome of grievance hearings, both reflect the wider patterns of bargaining relationships between union representatives and managers in the workplace. It is the proposition of this chapter that bullying is less likely to occur, and more likely to be tackled when it does occur, if there is a strong and well-organised trade union presence in the workplace.

Bibliography

ACAS (1999a) *Bullying and harassment at work: A guide for managers and employers*. London: ACAS.

—— (1999b) *Bullying and harassment at work: A guide for employees*. London: ACAS.

Adams, J. (1963) Towards an understanding of inequity. *Journal of Abnormal Social Psychology*, 67, 422–436.

ATL (1999) *Bullying at work: A guide for teachers*. London: ATL.

Bach, S., Winchester, D., Bordogna, L. and Della Rocca, G. (1999) *Public service employment relations in Europe*. London: Routledge.

Chartered Society of Physiotherapy (1997) *Health and safety briefing pack: Bullying at work*. London: CSP.

Clancy, M. and Seifert, R. (2000) *Fairness at work? The disciplinary and grievance provisions of the 1999 Employment Relations Act*. London: Institute of Employment Rights.

Hammond, J. and Hammond, B. (1966) *The town labourer*. London: Longmans.

Hoel, H. and Cooper, C. (2000) Workplace bullying in Britain. *Employee Health Bulletin, April*, 14, 6–9.

Hobsbawm, E. (1994) *Age of extremes*. London: Michael Joseph.

Incomes Data Services (1999) *Harassment policies*, Study 662. London: IDS.

Industrial Relations Services (1999) Bullying at work: A survey of 157 employers. *Employee Health Bulletin, April*, 8, 4–20.

International Labour Organization (1967) *Examination of grievances, recommendation no. 130*. Geneva: ILO.

Ironside, M. and Seifert, R. (1995) *Industrial relations in schools*. London: Routledge.

—— (2000) *Facing up to Thatcherism: The history of NALGO 1979–1993*. Oxford: Oxford University Press.

NASUWT (1996) *No place to hide: Confronting workplace bullies*. Birmingham: NASUWT.

Quine, L. (1999) Workplace bullying in NHS community trust: Staff questionnaire survey. *British Medical Journal, 23 January*, 228–232.

Royal College of Nursing (2001) *Bullying and harassment at work: A good practice guide for RCN negotiators and health care managers*. London: RCN.

Seifert, R. (1992) *Industrial relations in the NHS*. London: Chapman and Hall.

Sims, R. (1994) HRM's role in clarifying the new psychological contract. *Human Resource Management, 33*, 3, 373–382.

Snell, J. (1997) Danger: Bully at work. *Nursing Times, 93, May*, 21, 24–26.

Thompson, E. (1968) *Making of the English working class*. Harmondsworth: Penguin.

UNISON (1997a) *Bullying at work: UNISON bullying survey report*. London: UNISON.

—— (1997b) *Bullying at work: Guidelines for UNISON branches, stewards and safety representatives*. London: UNISON.

Vickers, L. (1995) *Protecting whistleblowers at work*. London: Institute of Employment Rights.

24 Workplace bullying and the law

Towards a transnational consensus?

David Yamada

Introduction

As the chapters in this volume have demonstrated, we are witnessing the steady emergence of a global understanding that workplace bullying poses a serious threat to workers and employers alike. The work being done to comprehend and address workplace bullying is cross-disciplinary in nature, and one of the areas of focus is the law. More specifically, researchers and advocates are considering ways in which the law can be used to prevent and punish serious bullying behaviour. This chapter will identify some of the central themes concerning bullying and the law.

Any analysis of the role of the law in addressing workplace bullying should start by identifying the policy objectives that the law should advance within the realm of abusive work environments. Although such an assessment is inherently subjective, many researchers and advocates who are concerned about workplace bullying would generally concur with the following: first and foremost, the law should encourage preventive measures to reduce occurrences of workplace bullying. If bullying is prevented, then workers and employers alike reap the benefits, and the legal system is spared additional litigation. Second, the law should protect workers who engage in self-help to address bullying and provide incentives to employers who respond promptly, fairly and effectively when informed about bullying behaviour. This encourages the prompt, internal resolution of bullying problems.

Third, the law should provide proper relief to targets of severe workplace bullying, including compensatory damages and, where applicable, reinstatement to his or her position with the assurance that the bully has been either transferred or reformed. The law must be careful not to provide a cause of action for every bruised feeling or bad day at work, but it should enter the fray when bullying is severe in intensity and duration, and causes harm to the target. Finally, the law should punish bullies and the employers who allow them to abuse their co-workers. This provides the law with necessary 'teeth', and serves an important deterrent function.

It must be said at the outset that, when viewed from a global perspective, the law does not adequately respond to workplace bullying. A growing body of comparative legal research and analysis points to the conclusion that, particularly when evaluated against the backdrop of the policy objectives proffered above, the law lags behind our recognition of the serious effects of workplace bullying. However, there are hopeful signs that the situation is changing.

National examples

Among Western nations, Sweden has come the closest to fashioning a direct legal response to bullying in the form of its Victimisation at Work ordinance, promulgated by the National Board of Occupational Safety and Health in 1993.[1] The ordinance characterises victimisation as social phenomena such as 'adult bullying, mental violence, social rejection and harassment – including sexual harassment'. Under the ordinance, employers are obliged to institute measures to prevent victimisation and to act responsively if 'signs of victimisation become apparent', including providing prompt assistance to targets of abusive behaviour.

Unfortunately for targets of workplace bullying, Sweden is an anomaly when it comes to enacting legal protections. It is instructive and representative that none of the three nations examined below has enacted legal protections specifically in response to workplace bullying. Rather, efforts to obtain legal relief must be based primarily on a patchwork of statutory and common law measures governing discrimination, personal injury, wrongful discharge and workplace safety.

Three nations whose legal systems are historically grounded in the common law, Australia, the United Kingdom, and the United States, are the focus of the following discussion. Although each nation's employment laws have their own unique characteristics, they have many features in common. These include a variety of wrongful discharge and emotional distress causes of action grounded in common law tort and contract law, as well as an assortment of statutes covering discrimination, worker safety and collective action. Although it is beyond the scope of this chapter to provide a comprehensive examination of these legal standards, what follows is a more impressionistic look at the most salient features of each nation's employment laws pertaining to bullying.

Australia

Workplace bullying is gaining increasing attention in Australia in ways that relate to the law. For example, the Queensland Division of Workplace Health and Safety has published separate booklets on workplace bullying for employers and employees, both of which cover potential legal claims that may be initiated by bullied workers.[2] In addition, various unions and a non-governmental organisation, the Beyond Bullying Association, are drawing attention to the role that the law may play in responding to bullying.

Common law and statutory obligations concerning occupational health and safety

Australian common law imposes upon an employer a 'contractual duty to take reasonable care for the safety of employees while in the course of their employment'.[3] Employers also owe a duty of care for the safety of their employees under the tort law, which governs negligence and other claims for personal injuries. In addition, federal and state occupational safety and health statutes establish general standards of care owed to employees and provide specific measures for workplace safety. Typically, legal disputes invoking an employer's obligations to safeguard its workers have involved physical injuries. However, there are encouraging signs that this combination of an employer's common law and statutory obligations may provide grounds of legal relief for targets of bullying. A significant example is a 1998 Queensland Supreme Court decision, *Arnold v. Midwest Radio Limited*.[4] The targeted employee had pre-existing psychiatric problems that were exacerbated by being subjected to repeated verbal abuse by her supervisor. In holding the employer liable for the harm under duties of care prescribed by both common law tort theories and the applicable workplace health and safety statute, the court recognised the viability of actions 'brought by employees claiming to have suffered psychological or psychiatric damage as a result of being unnecessarily exposed to stressful situations in the course of their employment'.

Workers' compensation also enters the picture in situations involving work-related injuries or illnesses. Australian law requires workers either to pursue a legal claim against an employer or to apply for workers' compensation benefits. A typical workers' compensation statute is the Work-Cover Queensland Act 1996, which provides benefits to any worker 'who suffers an injury or disease as a result of [the] workplace'.[5]

Employment discrimination and harassment statutes

Federal and state anti-discrimination statutes prohibit employment discrimination on the basis of race, sex, disability, age, marital or parental status, pregnancy, family responsibilities, and political or religious belief.[6] In addition, these statutes prohibit sexual harassment. According to research by organisational behaviour specialists Robin Kieseker and Teresa Marchant,[7] 'some attempts have been made to bring bullying cases against employers under existing harassment or discrimination legislation'. They add, however, that because bullying is not specifically covered under these statutes, it is difficult to bring bullying claims under this legal framework.

An example of this is sexual harassment law. The federal Sex Discrimination Act 1984 defines sexual harassment as follows:

(1) the person makes an unwelcomed sexual advance, or an unwelcome request for sexual favours, to the person harassed; or,
(2) engages in other unwelcome conduct of a sexual nature in relation to the person harassed;

in circumstances in which a reasonable person, having regard to all the circumstances, would have anticipated that the person harassed would be offended, humiliated or intimidated.[8]

The plain text of the statute yields the conclusion that only verbal and physical conduct of a sexual nature is legally actionable. However, a campaign of extreme and severe harassment of a non-sexual nature – in other words, workplace bullying – that is directed at someone because of her or his gender does not provide grounds for relief under the statute.

United Kingdom

In the United Kingdom, there appears to be an expanding receptivity to workplace bullying-related claims by employment tribunals and the courts. Furthermore, a growing number of labour unions, human resources officials and researchers are incorporating bullying into the lexicon of the British workplace. Overall, awareness of workplace bullying is greater in the UK than in the United States or Australia.

Unfair dismissal

The law governing unfair dismissal offers one of the more promising sources of legal protection for bullied employees. The Employment Rights Act 1996 provides that employees may not be unfairly dismissed.[9] Poor performance, improper conduct, and redundancy are among the chief reasons that may justify dismissal.[10]

Of particular relevance to bullying is the concept of 'constructive dismissal'.[11] An employee is constructively dismissed when she voluntarily leaves her employment because the employer has fundamentally breached an express or implied term of the employment contract. Subjecting an employee to severe mistreatment can be a form of breach of an implied contractual term. Accordingly, an individual who leaves his job because of bullying could claim that he was constructively dismissed, and seek relief under the statute.

In fact, there is a growing line of employment tribunal and court decisions where former employees are claiming that they were constructively dismissed at least in part because of bullying. Although their record of success has been decidedly mixed, certain decisions show that bullying is being taken seriously within some legal circles. For example, in *Abbey National PLC v. Robinson* (2000), an Employment Appeal Tribunal upheld a finding of constructive dismissal where the worker's manager 'had been bullying and harassing her in the workplace to a degree she found insufferable'.[12]

In another case, *Storer v. British Gas PLC* (2000), an appeals court reinstated a claim of constructive dismissal that had been dismissed because the complainant failed to file within the statutory time limit.[13] The complainant had been 'victimised and bullied' by his manager, resulting in stress and depression that culminated in symptoms of post-traumatic stress disorder. The court ordered an employment tribunal to consider whether the complainant's mental condition rendered him unable to file in a timely fashion. Equally encouraging is the decision of an Employment Appeal Tribunal in *Ezekiel v. The Court Service* (2000), which held that an employee was properly dismissed under the Employment Rights Act because he engaged in severe bullying and mistreatment of several co-workers.[14]

Common law tort and contract duties

English common law imposes upon employers a duty to provide a safe workplace to its employees. This duty led to a significant court decision, *Walker v. Northumberland County Council* (1994), in which a public employer was held liable for a second nervous breakdown suffered by one of its supervisors as a result of an unmanageable workload.[15] Even though the facts underlying *Walker* may reflect management negligence rather than intentional bullying, the holding of the case indicates that employers can be held liable for subjecting employees to work conditions that foreseeably cause psychiatric injuries. In *Walker*, the plaintiff had suffered a first nervous breakdown after requests for more staff and assistance were denied. Unfortunately, the plaintiff's work situation did not improve upon his return to work, and he suffered another breakdown. The court held that it 'ought to have been foreseen by his superior that if he was again exposed to the same workload as he had been handling at the time of the first breakdown, there was a risk that he would once again succumb to mental illness'.

Anti-discrimination and health and safety statutes

UK anti-discrimination statutes prohibit employment discrimination on the basis of race, sex and disability. These statutes potentially protect an individual who has been subjected to bullying because of membership in a legally protected group. However, as scholar Brenda Barrett has demonstrated, these remedies are limited.[16] She notes, for example, that in sexual harassment cases, Employment Appeal Tribunals have rejected claims based on conduct that 'might have been offensive to either sex' and conduct that constituted 'intimidation without sexual connotation'. Barrett also recognises the recently enacted Protection from Harassment Act, a criminal statute designed to address stalking, but doubts that it will have much effect on workplace relations.

Under the Health and Safety at Work Act an employer must 'ensure, so far as is reasonably practicable, the health, safety and welfare at work of all his employees'.[17] The applicability of the act to workplace bullying appears to be largely untested. However, Britain's Health and Safety Executive, the governmental agency responsible for enforcing the act, recognises bullying as a cause of work-related stress, and recommends that employers develop internal policies and procedures to respond to it.[18]

United States[19]

Of the three nations examined here, the United States has lagged behind the other two in the legal recognition of workplace bullying, despite an extensive array of potential common law and statutory protections for employees. The term 'workplace bullying' itself has only recently started to enter the vocabulary of American employment relations, thanks largely to the efforts of a newly created non-governmental organisation, the Campaign Against Workplace Bullying, by Drs Gary and Ruth Namie.[20] It remains to be seen whether American employment law eventually offers a meaningful response to workplace bullying.

Common law tort claims and workers' compensation

In the US, the favoured (albeit seldom successful) tort claim for emotionally abusive treatment at work has been intentional infliction of emotional distress (IIED). Typically, plaintiffs have sought to impose liability for IIED on both their employers and the specific workers, often supervisors, who engaged in the alleged conduct. The tort of IIED can be defined this way:

1. The wrongdoer's conduct must be intentional or reckless;
2. The conduct must be outrageous and intolerable in that it offends against the generally accepted standards of decency and morality;
3. There must be a causal connection between the wrongdoer's conduct and the emotional distress; and
4. The emotional distress must be severe.[21]

Although on the surface the tort of IIED appears to be an ideal legal protection against workplace bullying, this author's extensive analysis of judicial decisions deciding IIED claims reveals that typical workplace bullying, especially conduct unrelated to sexual harassment or other forms of status-based discrimination, seldom results in liability for IIED. The most frequent reason given by courts for rejecting workplace-related IIED claims is that the complained of behaviour was not sufficiently extreme and outrageous to meet the requirements of the tort.

Perhaps the most stunning example of this is *Hollomon v. Keadle*, a 1996 Arkansas Supreme Court case that involved a female employee, Hollomon, who worked for a male physician, Keadle, for two years before she voluntarily left the job.[22] Hollomon claimed that during this period of employment, 'Keadle repeatedly cursed her and referred to her with offensive terms, such as "white nigger", "slut", "whore", and "the ignorance of Glenwood, Arkansas" '. Keadle repeatedly used profanity in front of his employees and patients, and he frequently remarked that women working outside the home were 'whores and prostitutes'. According to Hollomon, Keadle 'told her that he had connections with the mob' and mentioned 'that he carried a gun', allegedly to 'intimidate her and to suggest that he would have her killed if she quit or caused trouble'. Hollomon claimed that as a result of this conduct, she suffered from 'stomach problems, loss of sleep, loss of self-esteem, anxiety attacks, and embarrassment'. On these allegations, the Arkansas Supreme Court ruled for the defendant Keadle, holding that Hollomon's failure to establish that Keadle was made aware of her peculiar vulnerability to emotional distress was fatal to her claim.

Workers' compensation laws are also relevant to the potential application of tort law to bullying. American courts are split on the question of whether workers' compensation precludes a tort claim against an employer for intentional, work-induced emotional injuries, a category under which bullying may fall.[23] In any event, workers' compensation is a poor legal response to workplace bullying, as targeted employees can recover benefits only if they establish that they have become partially or fully incapacitated because of the offending behaviour.

Employment discrimination and harassment statutes

An assortment of federal and state statutes prohibit discrimination and harassment on the basis of an employee's membership in a protected class, such as race, colour, sex, national origin, religion, age and disability. Thus, bullying that is grounded in a target's protected class membership may be legally actionable. Of particular relevance to bullying is sexual harassment, which is considered a form of sex discrimination under Title VII of the Civil Rights Act of 1964.[24] The US Supreme Court has held that sexual harassment in the form of the creation of a hostile work environment is a violation of Title VII.[25] The court has also held that an employer can be held legally responsible for sexual harassment committed by its supervisory employees.[26]

However, law professor Vicki Schultz analysed the evolution of sexual harassment law under US federal law and concluded that 'the most prominent feature of hostile work environment jurisprudence' is the 'disaggregation of sexual advances and other conduct that the courts consider "sexual" in nature from other gender-based mistreatment that judges consider non-sexual'.[27] In other words, in considering sexual harassment lawsuits that allege the creation of a hostile work environment, the courts often disregard any harassing conduct that is not of a sexual nature. Therefore, similar to the situation in the United Kingdom, non-sexual harassment that is directed at an individual because of her gender is not legally actionable.

Because some of the worst workplace bullying comes in the form of retaliation towards those who file discrimination complaints, the anti-retaliation provisions of federal and state anti-discrimination statutes may be a source of legal protection under these scenarios. For example, in *Zimmerman v. Direct Federal Credit Union* (2000), a US magistrate judge upheld a retaliation claim where the plaintiff, after filing a gender and pregnancy discrimination claim, was subjected to 'a deliberate, calculated, systematic campaign to humiliate and degrade [her] both professionally and personally'.[28]

Workplace safety and health statutes

The federal Occupational Safety and Health Act of 1970 (OSHA) was enacted 'to assure so far as possible every working man and woman in the Nation safe and healthful working conditions and to preserve our human resources'.[29] However, the main concern of OSHA was the prevention of physical injuries, especially those occurring in the industrial sector, and 'OSHA's original emphasis on manufacturing and construction sites' remains the primary focus of the federal agency charged with its enforcement, the Occupational Safety and Health Administration.[30] While workplace safety advocates and others lobby for OSHA and its state counterparts to pay greater attention to occupational stress and abusive work environments, there currently is little in the statutes and accompanying regulations that will help targets of bullying.

Labour unions and legal protections for collective employee action

Labour unions and the legal frameworks that protect collective employee action constitute tangible and potentially important avenues for addressing workplace bullying. The National Labor Relations Act in the US, the Workplace Relations Act in Australia, and the Trade Union and Labour Relations (Consolidation) Act in the UK all provide, with varying degrees of specificity, legal frameworks for the existence of unions and various forms of collective bargaining and employee actions. Organised labour remains one of the strongest and most vocal sources of advocacy on behalf of working people, and there are encouraging signs that some labour unions are responding to workplace bullying faced by their members. These unions are raising concerns about workplace bullying at the bargaining table and in grievances, and they are supporting efforts towards law reform.

There are, however, several impediments to utilising unions and collective bargaining protections to combat bullying. Overall, union density in industrialised nations showed a steady decline during the last third of the twentieth century; with notable exceptions such as Sweden, unionised employees constitute less than half the wage and salary earners in these countries.[31] Accordingly, the majority of workers do not enjoy union representation. Furthermore, the nature of workplace bullying is such that individual employees often target individual co-workers for abuse. This dynamic can make it difficult to develop a collective consciousness among workers with regard to workplace bullying. Finally, collective bargaining processes typically assume that major workplace conflicts are between employers and rank-and-file workers. Bullying scenarios between union members are not easily addressed within this structure.

Potential transnational legal approaches

The globalisation of markets has created complex moral and legal issues concerning working conditions around the world. Critical issues such as the use of sweatshops and child labour in so-called third world countries have dominated that discussion. As bullying moves closer to the forefront of our discussions about workplace problems, it is possible that international policies and organisations will address it more fully. There is some evidence that this is happening already in the work of the International Labour Organisation.

The International Labour Organisation

The International Labour Organisation (ILO) is a tripartite agency, with representatives from employers, trade unions and governments, that has been affiliated with the United Nations since 1946. It has three primary functions: (1) promulgating labour standards dealing with the health and welfare of workers; (2) providing technical assistance to member nations on employment and industrial relations matters; and (3) conducting research and publishing studies on labour and employment issues. Most UN members also belong to the ILO.

The ILO has recognised workplace bullying in the broader context of violence at work. In a 2000 monograph, it observed that workplace bullying is behaviour that 'by itself may be relatively minor but which cumulatively can become a very serious form of violence'.[32] However, although the ILO's Declaration on Fundamental Principles and Rights at Work affirms the right of all persons to be free from forced labour, child labour and discrimination,[33] it does not address a right to be free from workplace bullying.

Even in the event that the ILO does enact a general standard against workplace bullying, it would be without authority to impose it upon a member nation. Unlike the UN itself, the ILO 'is not a supranational entity, and therefore may not impose obligations upon member States, except in so far as they have voluntarily agreed to them, thus accepting certain restrictions on their sovereignty'.[34] Thus, unless the ILO gains the authority to impose and enforce labour standards (an unlikely development), its ability to address workplace bullying on an international scale will be limited to a research, advisory and consciousness-raising role rather than a legislative one.

International trade agreements

Potentially, multinational trade agreements could be the basis for imposing legal obligations on the signatories to prohibit workplace bullying within their own borders. For example, the European Union and the North American Free Trade Agreement have incorporated various labour standards, often in response to pressure from labour advocates in their respective nations. At this juncture, it is premature to predict whether concerns over workplace bullying will make their way into such trade agreements. Because the political motivation behind most of these multinational agreements is to open up markets for business, one should be appropriately sceptical as to the possibilities for using them as vehicles for raising labour standards in general. In fact, there is considerable concern among labour advocates that such agreements will lower labour standards across the globe in the name of free trade.

Of existing multinational trade schemes, the European Union is the most promising example of a regulatory structure that recognises the importance of both commercial exchange and social protections. The EU's primary legislative body, the Council of Ministers, may issue directives that mandate certain policy objectives but allow the member nations to determine the means of achieving them.[35] Although the council has not issued any directives specifically related to workplace bullying, it has done so for workplace safety and health generally, and for workplace discrimination. As workplace bullying becomes more of a mainstream employment issue, it is possible that the EU might address it through a directive.

Towards a transnational consensus on law reform?

Within Western industrialised nations, our current societal understanding of workplace bullying is very similar to how we viewed sexual harassment some thirty years ago. Back then, even the term 'sexual harassment' was not part of our vocabulary. Such behaviour was frequent and often very hurtful, but for some women it was seen as part of the cost of being employed. The notion that someone should have a legal right to be free of such treatment was rarely discussed, much less accepted.

Similarly, workplace bullying has been seen as part of the cost of being employed. Consequently, the idea that the law should enter the picture when someone is subjected to bullying has only recently started to gain momentum. It should come as no surprise, then, that neither of the three nations examined in this chapter has truly comprehensive legal protections against workplace bullying.

But there are signs of change. For example, in the United Kingdom, recommendations to address the relevant shortcomings of British employment law are winning some attention, thanks in part to the work of labour unions, academics, and an informal network of researchers and employee advocates who are addressing bullying and work abuse. At least two tangible legislative proposals to combat workplace bullying have been set forth in recent years. In 1997, a British labour union drafted the 'Dignity at Work Bill', which has been introduced in the House of Commons but has not yet become law. The bill imposes civil liability on an employer for bullying and similar acts, including 'behaviour on more than one occasion which is offensive, abusive, malicious, insulting or intimidating'. The proposed law is designed to provide protections to bullied employees on a par with protections extended to targets of sexual or racial harassment.

In 2000 a panel of experts at Cambridge University released a comprehensive, independent review of the UK's anti-discrimination legislation.[36] One of the panel's recommendations was the enactment of 'a statutory tort of harassment and bullying at work':

> The elements of this tort should be that: (1) the act of other conduct is unwelcome and offensive to the recipient; (2) it could reasonably be regarded as creating an intimidating, hostile, offensive or humiliating work environment; and (3) the recipient has suffered or is likely to suffer some harm whether physical, psychological or emotional (including anxiety and injury to feelings).

Considering that public education on workplace bullying is still in its infancy in the US, it may be premature to assess the prospects of law reform. However, efforts have been undertaken to garner support in state legislatures for the passage of an anti-bullying bill drafted by this author. The bill holds employers liable when a worker 'has been subjected to an abusive work environment', defined for this purpose as a situation where an employee, 'acting with malice, subjects the complainant to abusive conduct so severe that it causes tangible harm to the complainant'.

It is noteworthy that efforts towards law reform, particularly in the United Kingdom and the United States, are centring around the creation of new statutory remedies that combine doses of common law tort theory with inspiration from statutory law concerning harassment based on protected class status. These proposals go a long way towards advancing the policy goals set out at the beginning of this chapter, particularly in the way that they encourage employers to engage in preventive measures and provide relief to targets of severe mistreatment. Furthermore, although apparently no such statutory proposal has been put forth in Australia, researchers such as Kieseker and Marchant are acknowledging that Australian law 'does not fully recognise workplace bullying'.[37] It also should be noted that when the Queensland Supreme Court, in *Arnold v. Midwest Radio Limited*, imposed liability on an employer for subjecting an employee to an ongoing course of abusive treatment, it relied on a hybrid combination of common law and statutory duties relating to employee health and safety.

Legal and policy developments and initiatives in these three nations, along with Sweden's Victimisation at Work ordinance, suggest that we are in the early stages of an emerging transnational consensus that workplace bullying is best dealt with legally by the creation of new protections, aimed specifically at bullying and abuse, combining elements and approaches from existing common law and statutory provisions. We are so early in the history of understanding and responding to workplace bullying that it is impossible to predict whether such legal protections will be enacted on a widespread level. In the meantime, targets of workplace bullying and their lawyers will have to make do with an uneven assortment of existing statutory and common law remedies.

Notes

1 *Victimization at work*, adopted 21 September 1993.
2 Division of Workplace Health and Safety (1998) *Workplace bullying, employer's guide*. Queensland Government, Division of Workplace Health and Safety, Department of Employment Training and Industrial Relations; Division of Workplace Health and Safety (1998) *Workplace bullying, worker's guide*. Queensland Government, Division of Workplace Health and Safety, Department of Employment Training and Industrial Relations.
3 Brooks, A. (1996) Occupational health and safety. In J. Golden and D. Grozier (eds), *Labour law*, vol. 26 of *The Laws of Australia*. Sydney: The Law Book Company Limited, p. 13.
4 *Arnold v. Midwest Radio Limited* (1998) Aust Torts Reports, paras 81–472.
5 WorkCover Queensland Act 1996; *Workplace bullying, employer's guide*.
6 Chapman, A. and Sivarajah, S. (2000) Discrimination. In Golden and Grozier (eds), *Labour law*, p. 13.
7 Kieseker, R. and Marchant, T. (1999) Workplace bullying in Australia: A review of current conceptualisations and existing research. *Australian Journal*

of Management and Organisational Behaviour, *2*, *5*, http://www.usq.edu.au/faculty/business/departments/hrm/HRMJournal/AJMOB-archives.htm.

8 *Sex Discrimination Act 1984* (Commonwealth).
9 *Employment Rights Act 1996*, sect. 94.
10 *Employment Rights Act 1996*, sect. 98.
11 *Employment Rights Act 1996*, sect. 95.
12 *Abbey National PLC v. Janet Elizabeth Robinson* (2000) WL 1741415 (EAT).
13 *Roger Storer v. British Gas PLC* (2000) WL 191091 (CA).
14 *Ezekiel v. The Court Service* (2000) WL 1274032 (EAT).
15 *Walker v. Northumberland County Council* (1994) WL 1062924 (QBD).
16 Barrett, B. (2000) Harassment at work: A matter of health and safety. *Journal of Business Law*, *May 2000*, 214–231.
17 *Health and Safety at Work Act 1974*, sect. 3.
18 Health and Safety Executive (1999) *Help on work-related Stress*. Sudbury: HSE Books.
19 Much of the information in this section is drawn from Yamada, D. C. (2000). The phenomenon of 'workplace bullying' and the need for status-blind hostile work environment protection. *Georgetown Law Journal*, *88*, 475–536.
20 Namie, G. and Namie, R. (2000) Workplace bullying: The silent epidemic. *Employee Rights Quarterly*, *1*, 2, pp. 1–12.
21 *Kroger Co. v. Willgruber*, 920 S.W.2d 61 (Ky. 1996).
22 *Hollomon v. Keadle*, 931 S.W.2d 413 (Ark. 1994).
23 Yamada (2000), 506–507.
24 *Civil Rights Act of 1964*, 29 U.S.C. Sec. 2000e-2.
25 *Meritor v. Vinson*, 477 U.S. 57 (1986).
26 *Burlington Industries v. Ellerth*, 524 U.S. 742 (1998).
27 Schultz, V. (1998) Reconceptualizing sexual harassment, *Yale Law Journal*, *107*, 1683.
28 *Zimmerman v. Direct Federal Credit Union*, Civil Action No. 97–12610-RBC (D.Mass. 2000).
29 *Occupational Safety and Health Act of 1970*, 29 U.S.C. Sec. 651(b).
30 Rothstein, M.A. (1998) *Occupational safety and health law*. St Paul, MN: West Publishing Co., pp. 4–5; Work is changing but observers wonder if OSHA is changing with it. *BNA Occupational Safety and Health Daily*, 24 May 1999, available in Westlaw BNA-OSHD database, 5/24/1999 OSD d8.
31 Adapt or die, *The Economist*, 1 July 1995, p. 54.
32 Chappell, D. and di Martino, V. (2000) *Violence at work*, new edn. Geneva: International Labour Office, p. 12.
33 ILO Declaration on Fundamental Principles and Rights at Work, 86th Session, Geneva, 1998, from *http://www.ilo.org/public/english/standards/decl/declaration/text/index.htm*, accessed on 22 Jan 2001.
34 Bartolomei de la Cruz, H., von Potobsky, G. and Swepston, L. (1996) *The International Labor Organization: The International Standards System and basic human rights*. Boulder, CO: Westview Press, p. 6.
35 Bellace, J. R. (1997) *The European Union*. In W. Keller (ed.), *International labour and employment laws* (chap. 1, pp. 8, 9). Washington, DC: Bureau of National Affairs.
36 Centre for Public Law and the Judge Institute of Management Studies (2000) *Equality: A new framework*. The report can be accessed at http://www.law.cam.ac.uk/ccpr/antidisc.html.
37 Kieseker and Marchant (1999).

25 Bullying at work

The way forward

Helge Hoel, Ståle Einarsen, Loraleigh Keashly, Dieter Zapf and Cary L. Cooper

Over the preceding twenty-four chapters, the results of fewer than ten years of research and practical intervention into bullying and abuse in the workplace have been explored. In what, by most accounts, must be considered a remarkably short timeframe, these issues have emerged from relative obscurity, to move rapidly up the organisational agenda in many Western countries. Moreover, stimulated by the initiative of individual researchers and campaigners, the issue of bullying is currently widening its scope to encompass developing countries, demonstrating that the phenomenon has a resonance with working people across the world.

Whilst the problem has only come to the fore in the last few years, there is no reason to believe that the issue is new. In the same way as children have been bullied and abused by their fellow pupils for generations, there have always been workers who have suffered persecution and victimisation at the hands of abusive managers or fellow workers. However, whilst the present interest in the issue may reflect different factors and antecedents in different countries, the greater role of work in most people's lives with respect to personal status and self-image may have contributed to bullying becoming a serious issue for an increasing number of people. Similarly, the combination of greater emphasis on individualism and a general weakening of the trade union movement in Western countries may have led to behaviours and acts, previously understood and interpreted in an 'us and them' framework reflecting conflicts of interest between manager and workers, taking on new meanings with connotations of blame and victimisation.

One express objective of this book has been to contribute to bridging the gap between two research traditions; the Scandinavian, or European, concept of bullying and mobbing, on the one hand, and the US tradition of emotional abuse and related concepts, on the other. It is our belief that academics and practitioners alike would benefit from shared insight into key aspects of the issues, despite these being a complex and multifaceted phenomenon. The alternative would be to allow this area of research to drift into further divisions, with the development of numerous concepts which, to all intents and purposes, cover the same or, indeed, very similar ground. This should not be understood as an attempt to curb conceptual development and debate. On the contrary, we believe that by establishing some kind of common ground, it will be possible to engage with a greater number of academics and practitioners alike, stimulating, rather than preventing, interest across academic disciplines. In addition, the development of a common core of shared knowledge permits us to expend our creative energies on addressing questions that hitherto have not been addressed, such as mediators of the impact of bullying on targets or the role of organisations as an impediment or facilitator of bullying. Such an approach is also more likely to translate into tangible outcomes, benefiting people at work.

Nevertheless, it is unlikely that a single definition of 'bullying' will emerge which would win unanimous support and which would be applicable uniformly across all organisational and cultural contexts. However, there appears to be considerable agreement that a definition of bullying would incorporate the following elements or features: repeated or patterned negative acts, prolonged experience over time (duration) and the presence of a power imbalance between perpetrator and target. Similarly, a distinction can be made between bullying *behaviour*, on the one hand, and the victimisation or *bullying process*, on the other. As far as bullying processes are concerned, their dynamics and non-linearity must be further emphasised to prevent researchers and practitioners alike from treating the problem in a 'one fits all' manner, as if process development and outcomes were predetermined.

Whilst we believe that it is in the wider interest of academics, practitioners, as well as people at work, to obtain greater agreement with regard to definitional and conceptual issues, people will continue to hold their own views and shape their own lay definitions regardless of the views of researchers (Liefooghe and Mackenzie Davey, this volume). However, by participating in a continuous social dialogue, researchers and practitioners may contribute to influencing and shaping the view of this important social phenomenon.

The book has presented considerable empirical evidence for the presence and prevalence of the problem, taking into account demographic factors such as gender, age, race and organisational status. However, if it has proved difficult to agree on a shared definition, it would be harder, even if it were desirable, to establish one common way of operationalising the concept under investigation. The widespread interest in measurement and surveys reflects a need to convince the general public and policy makers about the existence and prevalence of the phenomenon. Having succeeded in establishing, beyond doubt, that bullying is a social problem of considerable magnitude, the role of measurement is likely to shift from large-scale surveys to become diagnostic psychometric measures used to expose intra-organisational problems. Whilst sound scientific methods should apply to such in-house measurements, the need for uniform operationalisation is perhaps less important.

The explanation of bullying will remain a key issue of theory and practice. Even if researchers continue to attach different weight to explanations emphasising personality characteristics as opposed to organisational factors, or the interaction of the two, it is worth highlighting that the different explanatory frameworks are not necessarily mutually exclusive. Thus, future research may try to integrate these various frameworks which complement rather than contradict each other. Moreover, where individuals decide to put their emphasis may also be influenced by their role within the organisation. Thus, a clinical psychologist or therapist would look for and apply different explanatory models to an HRM manager or a union official. The nature of the victimisation process itself would also tend to stimulate interest in models and theories focusing on personal characteristics (and pathologies). These are valuable contributions to the debate. However, as the outcome of such approaches tends to feed initiatives such as personality profiling and testing, which may be used to exclude people from the organisation or from employment in general, researchers need to take some responsibility for preventing their findings from being exploited to serve as organisational 'knee-jerk' reactions to the problem of bullying.

As a reflection of the relative immaturity of the field, research and practice have so far paid little attention to how the problem may be experienced differently by different groups of people. Thus, in widening the scope, future research must explore how, and to what extent, the experience of bullying may differ depending on gender, age or ethnicity. Similarly, the impact of the organisational context must be better explored and understood. Our current understanding of the complexity of the problem also suggests that it is likely that a series of factors may interact, with further challenges for future investigation.

Naturally, the focus of investigation has so far largely been on the 'perpetrator' and the 'target' and their mutual interaction. However, recent findings suggest that we need to broaden our scope to include the experience and role of 'witnesses' and 'observers' in scenarios of bullying. The fact that they may be party to process development suggests a greater need for an understanding of the dynamics which facilitate negative involvement, as well as those factors which may have to be present in order for bystanders to interfere and bring the escalating process to an end. Any discussion of the role of witnesses and observers in the workplace also demands a consideration of the role of the organisation per se in the development of bullying scenarios. Such consideration leads us to think about the systemic nature of bullying and the view of Brodsky (1976) emphasising that bullying can only exist within a culture in which it is permitted. Thus, it would be a challenge to researchers and practitioners alike to identify those factors and circumstances which may inhibit bullying as well as those which increases the likelihood of its presence. Consequently, Pryor and Fitzgerald (this volume) provide strong evidence of the role of organisational tolerance in providing an environment for sexual harassment to thrive.

Keashly and Jagatic (this volume) have proposed that we need to broaden the net of potential victimisation itself beyond the workplace, to include family and friends of the target. The occupational stress literature suggests that any work stressors (including bullying) have a profound impact on these people. The endless hours of discussion and pleas to change jobs or to retaliate, as well as the increasing disability of the target, can take their toll on these supporters (and their relationships), who often end up needing help themselves. When the cost of this ripple effect is added to the other individual and organisational costs already documented, the enormous impact of this phenomenon will become even more powerfully clear and unacceptable.

Confronted by the type of issues outlined above, more attention needs to be paid to research method and design. This is true for both quantitative and qualitative research. Future investigations may call for a broadening of methodological approaches applied with a possible greater emphasis on qualitative research methods whose strength lies in examining the factors related to how those affected construct the meaning of their experience or the processes by which someone becomes engaged in bullying.

The nature of the problem debated throughout this book has also given rise to more politically and ideologically anchored approaches to the understanding of bullying and has introduced new and alternative perspectives (e.g. McCarthy, this volume; Ironside and Seifert, this volume). By approaching the problems under consideration from different angles and drawing on alternative concepts and explanatory frameworks, these contributions challenge many of our preconceived beliefs. Moreover, by emphasising the role of the wider socio-economic context in bullying, they enrich the debate. This development is encouraging and is likely to attract further interest in the issue across the social sciences.

In a section dedicated to 'best practice', different types of intervention, some specific, others more holistic in their scope, have been outlined and discussed. These best-practice models should be treated as examples of successful intervention more than as 'off-the-shelf' programmes which may be applied successfully in any context without further adjustment. As a general rule we would agree that, as far as possible, intervention programmes should be tailored to the needs of the specific organisation. This section on interventions also alerts us to the potential danger of researchers and practitioners operating in isolation from one another. With the growing public interest in the issue of bullying, there is a danger that interventions without any proper basis in research and theory will be implemented with possible damage to organisation and individuals alike.

Thus, local interventions need to be based on a sound understanding of the phenomenon, its causes and consequences. Similarly, from the research side, the danger is also not being aware of or paying attention to interventions which, by virtue of their success, provide valuable insight into the nature and causes of bullying that may further refine and, indeed, challenge our theories. It is in joint awareness and effort that a more comprehensive understanding of bullying becomes possible.

Without doubt research into bullying will benefit from an ongoing evaluation of the effectiveness of various interventions and intervention programmes. This is true as particular methods may turn out to have unintended outcomes. For example, as demonstrated by Keashly and Nowell (this volume), under certain circumstances the mediation process may be exploited by a perpetrator to further their own interest at the expense of a less powerful target. As far as undertaking sound evaluation studies is concerned, we also acknowledge that to apply rigorous scientific design when undertaking research in rapidly changing environments is fraught with difficulties. In this respect, one needs to bear in mind the views of Kompier *et al.* (2000), who argue that 'organisational members are not "study objects" but active organisers of their own working organisation' (p. 373). It may, therefore, in future evaluation studies be necessary to sacrifice some scientific rigour as long as general guidelines for good practice are followed.

Recent development has seen a number of cases of bullying brought to court. In an increasingly litigious society this trend is likely to continue. Frequent calls are also made for further intervention in the legal arena, though progress is likely to be slow and uneven. Whilst we welcome such a development, and the opportunity for victims effectively to seek compensation and redress, we would warn against any unrealistic optimism that legislation would remove or even necessarily reduce the problem.

However, in the end it is on the ground, in the individual workplace, that the battle against bullying is ultimately fought. In this respect, we would like to highlight the moral side of the problem. Whilst contextual and situational factors will influence the behaviour of perpetrators (and targets), we must all take responsibility for our own behaviour. However, it would be unfair to place equal responsibility on all members of the organisation. It follows, therefore, that managers, who have greater control over their own actions and the actions of those for whom they are responsible, have a commensurately greater degree of responsibility than those without managerial authority.

Bibliography

Brodsky, C. M. (1976) *The harassed worker*. Lexington, MA: D. C. Heath and Co.

Kompier, M., Cooper, C. L. and Geurts, S. (2000) A multiple case study approach to work stress prevention in Europe. *European Journal of Work and Organizational Psychology*, 9, 371–400.

Subject index